P9-EAN-893

This book stimulates awareness of the critical role of land resources in sustainable development, and the need to improve land management. It provides an authoritative review of the resources of soils, water, climate, forests, and pastures on which agriculture and rural land use depend, and assesses prospects for feeding future populations. Addressing the environmental issues of erosion, loss of soil fertility, deforestation, and desertification, it is critical of present methods of assessing land degradation and placing an economic value on land. It emphasises the links between land resources and wider aspects of development, including population and poverty. It shows that land available for food production is less than previously estimated, and that, unless action is taken, the developing world will face recurrent problems of food security and conflict.

The book should be read by all involved in rural development, including natural scientists, economists, geographers, sociologists, and planners, together with students of development studies. It provides a summary and perspective of the field of land resources, gives some forcefully expressed criticisms of current methods, and suggests improvements needed to conserve resources for future generations.

Land resources

Land resources

Now and for the future

Anthony Young
Honorary Research Fellow in Environmental Sciences
University of East Anglia

YNDICATE OF THE UNIVERSITY OF CAMBRIDGE
Street, Cambridge CB2 1RP, United Kingdom

CAMBRIDGE UNIVERSITY PRESS
The Edinburgh Building, Cambridge CB2 2RP, United Kingdom
40 West 20th Street, New York, NY 10011-4211, USA
10 Stamford Road, Oakleigh, Melbourne 3166, Australia

First published 1998

Transferred to digital printing 2000

Printed in Great Britain by Biddles Short Run Books

Typeset in Quadraat 10/14 pt [VN]

A catalogue record for this book is available from the British Library

Library of Congress cataloguing in publication data

Young, Anthony.
Land resources: now and for the future / Anthony Young.
p. cm.
Includes bibliographical references and index.
ISBN 0 521 59003 5 (hardback).
1. Land use, Rural – Planning. 2. Agricultural conservation. 3. Renewable natural resources. 4. Sustainable development. 5. Land use, Rural – Developing countries – Planning. 6. Agricultural conservation – Developing countries. 7. Renewable natural resources – Developing countries. 8. Sustainable development – Developing countries. I. Title.
HD111.Y68 1998
333.73'16–dc21 97–36654 CIP

ISBN 0 521 59003 5 hardback

Contents

Preface

In 1958, I began a three-year period as Soil Surveyor to the Government of Malawi, then Nyasaland, carrying out a reconnaissance land resource survey of the country; and, in 1993–5, completed consultancies on land use policy in Jamaica, land degradation in South Asia, and the preparation of an international programme to monitor changes in land conditions. Between those times I have worked on soil survey methods, land evaluation, land use planning and policy, and carried out research into land management, particularly through agroforestry.

This book is a review of land resources: their evaluation, management, and conservation, and their role in human welfare. Land resources are the environmental resources of climate, water, soils, landforms, forests, pastures, and wildlife, on which agriculture, forestry, and other kinds of rural land use depend. Renewable natural resources is an alternative name. Whilst details of the methods used differ from one kind of resource to another, many principles are common to their survey, evaluation, planning, and management. I also set down the opinions I have formed, in places more forcefully than was possible when writing as a consultant.

The objectives of the book are:

- to improve awareness of the critical role of land resources as a major element in the development of agriculture and the rural sector;
- to review the progress that has been made in different aspects of land resources, and to point to priorities for the future;
- to draw attention to the urgent need for action to improve the management of land resources, if they are to be conserved for the benefit of future generations;
- to show how land resources interact with wider aspects of development, including food security, poverty, and population policy.

Much of what is said has world-wide applications, for the need to care for resources is universal. However, most countries in the developed world have responsible and effective policies already in place. In contrast, there are many needs for improvements in policies and land management practices in developing countries, and it is to these that the discussion and comments are primarily directed. There is also much need for action to improve land resource management in the countries of the former Soviet Union.

The text is written so that it can be read without the interruption of references.

For those who wish to know the origin of statements, or seek further information, the sources can be followed up through the notes and references.

The book is intended to reach four groups of readers: scientists, the professional development community, students, and, if possible, policy-makers. Fellow resource scientists will already be familiar with much of what is written, but this wide-ranging review will give a perspective to their specialist knowledge. It is the wider community of development planners, including economists, geographers, sociologists, and technical specialists in agriculture and forestry, that are intended as the main audience; for them, the book provides a summary of progress and future needs in the area of land resources. I hope that it will also persuade them of the critical role which these play. All of this applies equally to the students who will take their places in the future. If possible, I should like also to influence policy-makers, directly or through those who advise them. To this end, executive summaries are given after each chapter heading, and the arguments are brought together in the final chapter.

A theme which runs through the book is that lasting improvements will ultimately depend on awareness, concern, and action by governments, scientists, and people of the developing nations. The international community can provide technical guidance and assistance, but lasting progress depends on political will in the developing countries themselves, arising from community awareness, and supported by financial provision and action. This book will accomplish its purpose every time it is read by citizens of developing countries, scientists, planners, policy-makers, and the students who will succeed them.

Acknowledgements

Travel for the preparation of this book was assisted by an Emeritus Fellowship from the Leverhulme Trust. Writing was begun during tenure of a Residency at the Bellagio Study and Conference Center of the Rockefeller Foundation. I am most grateful to both these foundations for their support.

Five institutions kindly granted me periods of residence for study and writing: Cornell University, New York; Stanford University, California; the University of Guelph, Ontario; FAO, Rome; and the World Bank, Washington DC. I thank their directors, and the many staff who spared time for discussion.

Many institutes also gave valuable information during visits: in The Netherlands, the International Soil Reference and Information Centre (ISRIC) and the Winand Staring Centre for Integrated Land, Soil and Water Research, Wageningen, and the International Institute for Aerospace Survey and Earth Sciences (ITC), Enschede; in the USA, the World Resources Institute and the Worldwatch Institute, Washington DC; in Kenya, the United Nations Environment Programme (UNEP) and my former employers, the International Centre for Research in Agroforestry (ICRAF), Nairobi; and in the UK, the Natural Resources Institute, Chatham. I thank their staff, too numerous to list, for generously giving their time.

It would be impossible to list the many kind friends who gave individual advice and help, over years of working on land resources. However, I should particularly like to thank the following friends and erstwhile colleagues for their advice and comments during the period of writing the book: Hugh Brammer, Robert Brinkman, Marion Cheatle, Malcolm Douglas, Dennis Greenland, Norman Hudson, Sir Charles Pereira, Christian Pieri, Maurice Purnell, Francis Shaxson, and Wim Sombroek; I must make it clear that those who were employees of international organizations gave advice in a personal capacity. I am equally grateful to colleagues at the University of East Anglia, including Neil Adger, David Dent, Michael Stocking, Ian Thomas and Kerry Turner. I should like to thank my copy editor Gillian Maude, for seeing what I was trying to say and, in a number of critical places, improving the way it was expressed. The Figures were kindly drawn by Phillip Judge, of the University of East Anglia. Responsibility for errors of fact and statements of opinion is my own.

Lastly, after 40 years of continuous help and love in many parts of the world, this book can only be dedicated to my wife, Doreen.

Note on acronyms and currency

Those working in international development swim in a sea of acronyms, in which NARS work with IARCs, who live in the CGIAR. Excessive use of these has been avoided. However, some institutions are recognized better by acronyms than by their full titles, therefore the acronym is given on its first appearance, but only used immediately following this. Acronyms for the institutes which belong to the Consultative Group on International Agricultural Research (CGIAR) are given on p. 214. Only the following are employed throughout the text:

FAO	Food and Agriculture Organization of the United Nations
UN	United Nations
UNDP	United Nations Development Programme
UNEP	United Nations Environment Programme
UNESCO	United Nations Educational, Scientific and Cultural Organization.

Following usual practice, the US dollar is employed as a standard unit of currency, taking its value in the mid-1990s.

World Wide Web site

The principal conclusions from this book, together with policy implications, supporting information, extracts from reviews, and updating, can be found on the Internet at World Wide Web site: http://www.land-resources.com

This site is personal to the author, and does not form part of the publications of Cambridge University Press. The Press is therefore in no way responsible for any use made of the material on this Web site.

1 Concern for land

With the rise in population placing ever-increasing pressures on scarce land, governments of developing countries should give high priority to rational land use, improved land management, and avoidance of degradation. At international level, public concern has been more with pollution aspects of the environment, and nature conservation, than with land as a productive resource. In developing countries, awareness by governments of the critical role played by land resources is poor, and institutions inadequately funded. Much progress has been made over the past 50 years in approaches and methods for land resource survey, evaluation, and management. What is needed now is more widespread and effective application of these methods. Sustainability, the combination of production with conservation, is a central concept in land resource management.

Management of land, of its soils, water, forests, pastures, and wildlife, has been central to human society from its earliest times. Land resources provide the basis for more than 95% of human food supplies, the greater part of clothing, and all needs for wood, both for fuel and construction. The developments of the industrial age have substituted coal, oil and minerals for some of the fuel, construction, and fibre needs, but have in no way removed the basic dependency of society upon the renewable resources of the land.

There has always been competition for land, sometimes reaching the level of conflict. In prehistoric times, among communities dependent on hunting and gathering, it would have shown in the kind of territoriality found amongst animal populations. As soon as the record of history begins, it is clear that there were great inequalities in land availability. Famine has never been absent, and migrations in search of better resources have extended from biblical times to the great world expansion from the nineteenth to the first half of the twentieth century.

Formerly, there was a solution to local problems of shortage of food and other basic necessities: to take more land into cultivation. Usually this was through clearance of forest, for forested lands are also the most fertile. In Europe in medieval times, in North America in the nineteenth century, and until quite recently in the tropical lands, forests and woodlands have receded, arable land has expanded, and land for nature has been constantly reduced. Even into the post-1945 period of planned development, new land settlement schemes were still possible as a solution to problems of crowding, small farms, and landlessness.

Most parts of the world are by now moving into a new era. The early taxation surveys of India recorded 'culturable wasteland', which today would be called land that is cultivable but not cultivated. Almost all such land has now been taken up. In 1965, the Malaysian Government, with international agency support, embarked upon the Jengka Triangle project, the objective of which was to clear 2000 km^2 of primary rain forest and plant it with oil palm, as a means of settling people from the overcrowded rice-growing deltas. Today, world opinion and the policy of international agencies strongly oppose any such clearance. Malawi, in East Africa, was a crowded country at the time it gained independence in 1964, but cultivation still stopped short at the foot of the hills; ten years later it had become widespread on hillsides and on the steeply dissected slopes of the rift valley. In Jamaica, needs for food have led to cultivation of slopes on which one can hardly stand, leading to severe soil erosion.

The massive increase in population which has led to land shortage started with the Industrial Revolution and has been accelerated by improvements in health and the introduction, since the Second World War, of international action for famine relief. From 2500 million in 1950, world population doubled by 1987 and will pass 6000 million in 1998. The current rate of increase is 88 million a year or 240 000 a day. At present average levels of crop yield, food alone for these extra people calls for an additional 80 000 hectares of land, nearly 30 by 30 kilometres, every day. In much of Asia and the Middle East, and a growing number of countries elsewhere, such extra land is simply not there. The world is becoming a 'full house'.[1]

Barring catastrophe, a further population rise over the next 30 years of at least 2000 million, and most probably 2500 million, is inevitable. Nearly all of this increase will be in less-developed countries. Much of it will be in families who are already very poor, many of them chronically undernourished. Because there is little spare land remaining, most of the added production required to provide basic needs for these people will have to be achieved by higher productivity from existing land. Higher crop yields form the largest component, but increased productivity from livestock and forests are also required. The productive potential of the land must be conserved, preventing erosion and other forms of land degradation. Land must be retained for forestry, water supply, and conservation of nature, checking a threatened reduction in the diversity of plants and animals. Lastly, people need land for settlements, not only housing but transport, industry, and recreation.

In this crowded world, with growing populations, severely limited land, and strong competition for its use, it is clearly desirable that governments should place high among their priorities:

rational land use: using different types of land in ways best suited to their potential;

improved land management, in agriculture and forestry, so as to secure higher productivity;

avoidance of land degradation – soil erosion, forest clearance, pasture degradation, and the like – so as to conserve resources for the future;

good data to guide decisions on the above, and research to advance knowledge on which improvements can be based.

These needs and pressures should mean, therefore, that land resources and land use lie at the centre of national policies, particularly in the developing countries. Regrettably, this is not the case.

Awareness at the international level

There is no lack of environmental concern at the international level. The growth of awareness that took place from the late 1960s was brought into focus at the first United Nations Conference on the Environment, held in Stockholm in 1972, leading to the foundation of the United Nations Environment Programme (UNEP). Among developed countries, the environment became an active element of national policy, not only through 'green' political parties but through the pressure which people in democracies placed upon their governing institutions. A proper regard for environmental questions was conceived as being a matter of self-interest.

The self-interest, however, was predominantly that of the developed countries. This is still the case. The issues most often given attention are global warming, the 'ozone hole', the dangers of nuclear and chemical pollution, and loss of biodiversity. Forest clearance is also widely discussed, but with respect to the role of forests in assimilation of carbon dioxide and in preserving genetic resources, not for their functions of water catchment protection or timber production in developing countries themselves. Following the prolonged droughts in the Africa sahel in the 1970s, desertification came temporarily to the attention of the world, but impetus for this concern was lost, partly because of the poor basis of scientific information on which discussion was based.

In the 1980s, there were two international commissions, the outputs from which are generally referred to by the names of their leaders, the Brandt and Brundtland reports.[2] Both drew attention to the growing gap in wealth between rich and poor nations, with the message that reduction in poverty in the latter was in the interests of all. Whilst attracting attention, these reports did not succeed in increasing the level of foreign aid from the developed nations, which has continued to fall as a percentage of national income.

In 1992, governments of the world met in Rio de Janiero for the United Nations Conference on Environment and Development (UNCED). Besides a review of

INTERNATIONAL MILESTONES IN AWARENESS OF LAND, POPULATION, ENVIRONMENT, AND DEVELOPMENT

1972 UN Conference on the Human Environment, Stockholm, leading to establishment of the UN Environment Programme (UNEP).

1977 First UN Conference on Desertification, Nairobi.

1979 World Conference on Agrarian Reform and Rural Development (WCARRD), directing attention to land tenure reform and social issues.

1980 Report of the Brandt Commission, *North-South: a Programme for Survival*.

1985 Establishment of the International Board for Soils Research and Management (IBSRAM).

1987 Report of the Brundtland Commission, *Our Common Future*.

1990 Four research institutes in land resource management, for forestry, agroforestry, irrigation, and fisheries, are admitted to the international agricultural research system.

1992 UN Conference on Environment and Development (UNCED), Rio de Janiero; *Agenda 21*, programme of action for sustainable development.

1993 Population Summit of the World's Scientific Academies, New Delhi, *Joint statement*.

1994 Third UN Conference on Population and Development, Cairo, *Programme of Action*.

1996 Second World Food Summit, *Rome Declaration and Plan of Action*.

progress in the twenty years since the Stockholm meeting, this was for the purpose of drawing up a programme of action, *Agenda 21* (meaning an agenda for the twenty-first century).[3] As compared with the Stockholm meeting 20 years earlier, which focused upon the pollution aspects of environment, there was a partial change of emphasis. Following upon the Brandt and Brundtland reports, and a growing realization of the problems set by population and poverty, development was given equal place with environment. This led to a change of emphasis between the different aspects of the physical environment. Its role in providing natural resources for production was given more attention, relative to that of a sink for waste products and associated problems of pollution. In terms of space allotted, there are 6 chapters devoted to land resources as a basis for production, compared with 4 on pollution and wastes, 1 on biological diversity, and 2 on the resources which are 'global commons', the atmosphere and oceans.[4] This meeting also brought to wider attention the concept of sustainable development, the conservation of resources for use by future generations. *Agenda 21* marked a big step forward in bringing the role of land resources to wider attention.

Further progress, although not as much as had been hoped, was made at the Third UN Conference on Population and Development, held in Cairo in 1994. It had been preceded by a meeting of the world's scientific academies, who very firmly called for urgent action to be taken to limit population growth: 'Ultimate success in dealing with global social, economic, and environmental problems cannot be achieved without a stable world population.'[5] The UN meeting did not go as far as this. It main advance was the recognition that population questions cannot be treated in isolation, but are closely linked with environmental resources and economic development. It called for programmes to limit population growth, with emphasis on the education and status of women, but did not go as far as to suggest that natural resources might set limits to population-supporting capacities.

Compared with the UNCED conference, the outcome of the 1996 Second World Food Summit was disappointing, even retrograde, in its lack of emphasis on the land basis of food production. In the *Rome Declaration and Plan of Action*, the main points of emphasis are responsible government, poverty, the rural sector, participation and sustainability, disaster relief, and the role of women.[6] In this high-level document, the function of land resources appears only indirectly as sustainability, which is coupled with participation by land users. Land availability is considered in the technical background documents but, even in these, loss of production through land degradation occupies a very subsidiary place. The critical role of water in food production is given more prominence, as 1 out of 15 Technical Documents.[7]

In 1997, a further UN Earth Summit meeting was held, to review progress in the 5 years since the Rio conference. The primary concern was with the issue of halting global warming, through reduction of greenhouse gas emissions, linked with a further resolve to check loss of the world's forests. In recent years, media attention has become increasingly focused on global warming, perceived to affect adversely the interests of Western nations. It is no answer to say that it is the scientists' views, and discussions in committee rooms, that really matter; the media strongly influence (some would argue, reflect) public opinion, and politicians must respond to this in allocating aid for environmental and development purposes.

Environmental issues: the balance of concern

In summary, the major environmental issues fall into five groups:

1. Land resources: the role of environment as a resource for production.
2. The role of environment as a sink: waste disposal and pollution.
3. The 'global commons': atmosphere and oceans.
4. Conservation of nature: biological and genetic resources.
5. Non-renewable resources: energy and minerals.

There are substantial overlaps between these issues. Climate is a major land resource, hence global atmospheric change potentially affects production. The genetic resources of nature provide a basis for future advances in plant productivity. Pollution of soils and water is one cause of land degradation, whilst energy resources place constraints on agricultural technology.

Since the environmental movement first arose in the 1960s, Western countries have always directed their interests primarily at the pollution aspects of the environment, including the specific aspects of atmospheric change and its effects on climate. Land resources for production have never attracted as much attention. The existence of a world food problem is perceived, but developed countries rightly assume that they will not run short themselves. The argument of 'only one world', that in these days of modern communications and advanced armaments, the welfare of the Third World is inseparable from that of developed countries, has failed to be sufficiently convincing to bring about action in the form of aid. Since the 1950s, there has been a steady fall in development aid as a proportion of national wealth, and, since 1990, less of this has been directed towards agriculture and the rural sector.

Awareness at national level

If the need for research and investment into the conservation, management, and development of land resources is to be made, the initiative has to come from governments of the countries concerned. In industrialized countries, particularly those in which agriculture contributes only a few percent to the national wealth, it would be understandable if land resources did not play a major role in policy. In fact, many such countries have balanced and responsible policies in these fields. The United States took a lead in soil conservation programmes in the 1930s. Australia made surveys of the resources of its northern and central areas from the late 1940s onwards, and a comprehensive land inventory of Canada has been completed. Deforestation was reversed in Europe from the 1920s onward.

In developing countries, the needs are very much more urgent. Many have a high dependence on agriculture for both food needs and exports, and supplement their food production with substantial cereal imports. Some have low crop yields, and most encounter problems of land degradation. Many countries have an average farm size of less than one hectare, and populations will rise by at least another 50%. One might expect, therefore, that they would place land resources at the centre of national policies and planning.

The actual position is very different. Natural resource survey organizations and land use planning departments exist, but the information they collect does not weigh strongly in development planning. The information basis for decisions about land is extremely poor. Many countries lack reliable data on even such basic facts as the areas cultivated and under forest. None have yet made systematic efforts to monitor changes in the condition of their soils, nor the extent of land degradation. The adoption of improved methods of soil management is held back by the weakness of the agricultural extension services. National development plans give due attention to agriculture and forestry, but rarely make more than passing reference to the land resources on which production in these sectors depends. Most national institutions for agricultural research are poorly funded. The wealth of soils, water resources, forests, and grazing lands is not taken into account in national budgeting. A forest may be cleared or a valley-floor pasture gullied, but these losses of natural capital do not appear in national accounts.

Landmarks of progress

Set against the negative aspects of awareness, there have been many substantial advances over the past quarter-century in knowledge of land resources, and their potential, management, and development in the Third World. Subsequent chapters review these, with an emphasis on what still remains to be done. With the objective of demonstrating the progress that has been made, a preview may be given of some major advances in approaches, methods, and knowledge at the international level. All advances are based on research, but their objective is practical action in land planning, development, and management. In the following list, general progress in each field has been linked to key publications of results, as landmarks of progress:[8]

1970 — The *Soil map of the world* at 1:5 million scale provides a common classification and the first estimate of world soil resources.

1976 — *A Framework for land evaluation* supplies a means to convert the results from resource surveys into estimates of land potential.

1977 — *Guidelines for predicting crop water requirements* provides a basis for planning efficient water use in irrigation.

1978–81 — The *Agro-ecological zones project* establishes a framework of reference for agro-climatological assessment, linked to a climatic database.

1980, 1990 — The first and second *World forest resources assessments* show the extent and rate of deforestation. The first leads to establishment of the *Tropical forestry action plan.*

1981 — The approach of rapid rural appraisal is formulated, later to develop into participatory rural appraisal, and diagnosis and design.

1984 — *Land, food and people* summarizes results of the first comprehensive comparison between the food needs and the food-producing potential of developing countries.

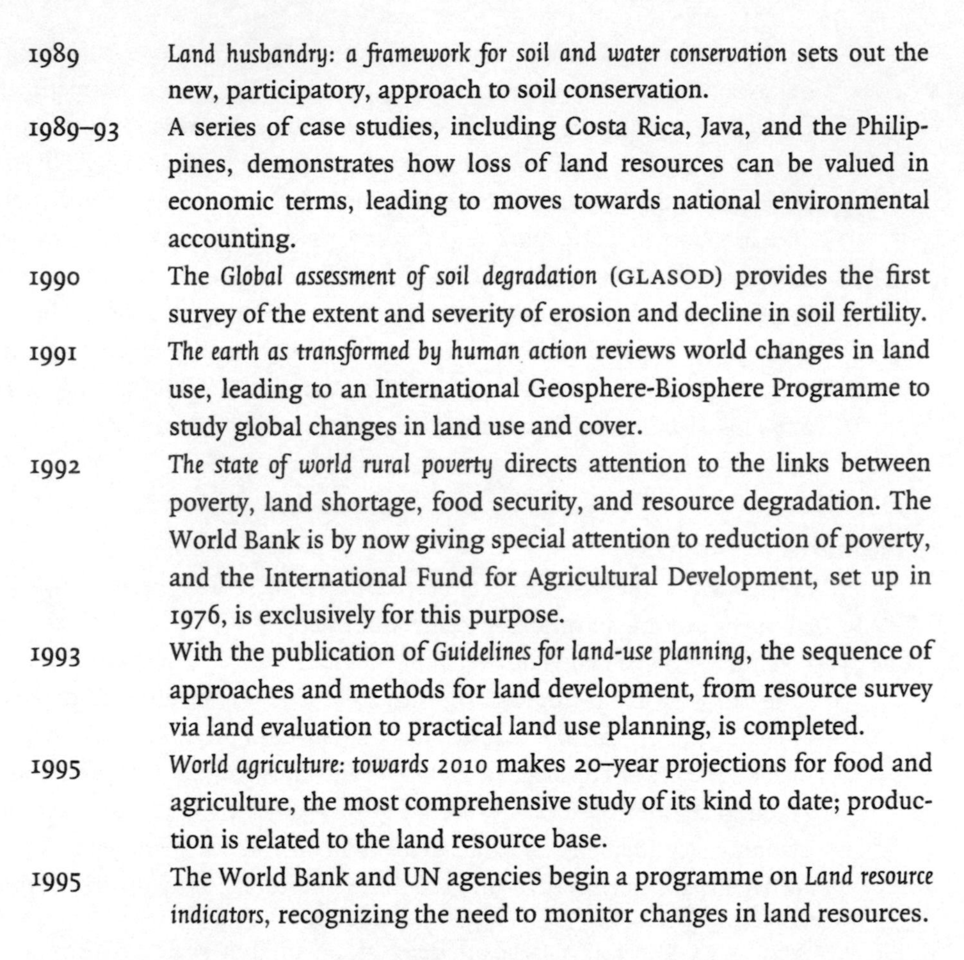

1989	*Land husbandry: a framework for soil and water conservation* sets out the new, participatory, approach to soil conservation.
1989–93	A series of case studies, including Costa Rica, Java, and the Philippines, demonstrates how loss of land resources can be valued in economic terms, leading to moves towards national environmental accounting.
1990	The *Global assessment of soil degradation* (GLASOD) provides the first survey of the extent and severity of erosion and decline in soil fertility.
1991	*The earth as transformed by human action* reviews world changes in land use, leading to an International Geosphere-Biosphere Programme to study global changes in land use and cover.
1992	*The state of world rural poverty* directs attention to the links between poverty, land shortage, food security, and resource degradation. The World Bank is by now giving special attention to reduction of poverty, and the International Fund for Agricultural Development, set up in 1976, is exclusively for this purpose.
1993	With the publication of *Guidelines for land-use planning*, the sequence of approaches and methods for land development, from resource survey via land evaluation to practical land use planning, is completed.
1995	*World agriculture: towards 2010* makes 20–year projections for food and agriculture, the most comprehensive study of its kind to date; production is related to the land resource base.
1995	The World Bank and UN agencies begin a programme on *Land resource indicators*, recognizing the need to monitor changes in land resources.

This list of landmarks is necessarily selective. It does not include the advances made in applications of remote sensing and computerized information systems to land resource development. The inclusion of social factors, people's participation, and poverty, in a summary of land resources is deliberate and important; land cannot be considered in isolation from the people who depend on it and manage it.

Comparable summaries could be made of advances in particular countries, both at national and district levels. The state of knowledge would be more fragmentary. Because national organizations are so poorly funded, progress has often been dependent on opportunities provided by development projects. For most countries, however, it would be possible to list land resource inventories at a national level and in district-level project studies. Much real progress has been made in land resource management through co-operation between scientists, extension staff, and farmers.

Sustainability

The term sustainability was introduced relatively recently, although it has long been a basic concept in land resource management. Farmers have always sought to pass on land to their children in at least as good a condition as they inherited it. The *Agenda 21* report of the UN Rio meeting is subtitled, *A programme of action for sustainable development*, and there are now few development projects which do not include reference to sustainability. Like many ideas which come into vogue, it has on occasion been misused, its meaning widened to almost anything that a writer considered to be 'a good thing'. Properly employed, however, sustainability is a valid concept of the highest significance, fundamental to questions of land resources.

Sustainability, or sustainable land use, has been variously defined, although the FAO definition has gained common acceptance. The essential feature is that sustainable land use achieves production combined with conservation of the natural resources on which production depends. This is expressed in the simplified definition. It can be compressed still further into a pseudo-equation, 'Sustainability = Production + Conservation'. For a land use system to be sustainable requires, first,

SUSTAINABILITY

Formal definition

Sustainable agriculture and rural development is the management and conservation of the natural resource base, and the orientation of technological and institutional change, in such a manner as to ensure the attainment and continued satisfaction of human needs for present and future generations. Such sustainable development conserves land, water, and plant and animal genetic resources, is environmentally non-degrading, technically appropriate, economically viable, and socially acceptable (based on FAO sources).

Simplified definitions

Sustainable land use is that which meets the needs for production of present land users, whilst conserving for future generations the basic resources on which that production depends.

Sustainability = Production + Conservation

that it should meet the needs of farmers and other land users; and, secondly, that it should achieve conservation of the whole range of natural resources, including climate, water, soils, landforms, forests, and pastures.

In this respect, there is a difference of emphasis from the environmental movement of the 1960s and after. In the latter, priority was given to conservation; if a human activity led to adverse changes to the environment, it was unacceptable. In the concept of sustainability, it is significant that production is listed first. The priority of farmers has to be to meet their present needs for food and cash income. Land use planning starts from the premise that production needs must be met; it then proceeds to consider how this can be made compatible with resource conservation. If farmers are gaining their livelihood from fragile environments, such as steep slopes or semi-arid climates, it is unacceptable to forbid such land use. Instead, ways must be found of making it environmentally acceptable.

Objectives and plan

The objectives of the following account are:

- to improve awareness of the critical role played by land resources in the welfare of developing countries;
- to summarize the present state of knowledge about land resources, methods for their assessment, and their present condition and productive potential, now and in the future;
- to draw attention to problems in land resources, including land degradation, and the need for improvements in management if further loss of productive potential is to be prevented;
- to consider the improvements in policy, institutions, education, and practice that are needed to bring about such changes.

The book falls into five general sections. Chapter 2 sets out the major concerns in sustainable land management, the land resource issues. Chapters 3 to 6 cover the spectrum of methods employed in rural development, commencing with survey and leading, via evaluation and participation, to practical land use planning. Chapters 7 to 10 cover different aspects of the central issue of land degradation, the lowering of the productive potential of resources. Chapters 11 and 12 are concerned with methods for the management of land, and research directed at improving these; the potential of research to increase land productivity is also discussed.

The final section places land resources into the wider context of development. Chapter 13 covers 'the great debate', on whether the developing world will be able to feed its future population. Chapter 14 widens the discussion into population increase and its consequences, not as a separate issue but as one so closely related as to be an integral part of land resource development. The final chapter summarizes the present and future problems of land resources, including the dangers if action is not taken to improve knowledge, planning, and management. Finally, it asks what needs to be done to bring about such action, both internationally and in the developing countries themselves.

The book is primarily about the land resources of the developing world, for it is there that the most urgent problems of the present day lie.[9] The developing countries lie mainly in the tropics and subtropics, but it is not climate that is the essential distinction. It is the greater need for improved conservation and management of land resources, and the low level of development of relevant institutions. Methods for the survey and assessment of resources are common to the developed and developing worlds. In general, however, developed countries have well-established institutions for resource assessment and land use planning, strong conservation movements, and sustainable land management practices. There could be improvements, but the needs are less by an order of magnitude than those of the developing world.

2 Land resource issues

Land resource issues are problems in land use planning and resource management, arising from the interactions between human society and the natural environment. This chapter identifies the key land resource issues for the 'three worlds of the tropics', the major agro-ecological zones: the humid tropics or rain forest zone, the subhumid tropics or savannas, and the dry lands, semi-arid and desert. Cutting across these climatic regions are two distinctive environments: steeplands and alluvial lowlands. Land resource issues are related to the essential nature of sustainable land use: combining the efficient use of resources to meet present needs with their conservation for the future. Many issues are concerned with the avoidance of land degradation.

The concept of land resource issues brings together two of the principal aspects of land use planning and management, making the best use of resources, and conserving them for the future – in short, sustainable use of the land resources. Land resource issues affect the state or condition of resources, but are not problems of the physical environment alone. They arise from the interactions between resources offered by the physical environment, the needs of land users, competition for land, and methods of land management.

Many other problems and policy issues have major impacts on land, but are not land resource issues as such. Examples are marketing facilities, supplies of agricultural inputs, veterinary services, and pricing policies. But all land resource issues are necessarily also 'people issues', arising from the relations between human society and the land.

Some land resource issues are found in most parts of the developing world. Food security in relation to population pressure is an example, or urban encroachment on agricultural land. Other issues are specific to particular types of environment, or are found more acutely in certain environments, such as water shortage in semi-arid and arid regions, or soil erosion on sloping land. This chapter is a summary of the major land resource issues in different regions of the developing world.

'Three worlds of the tropics', the major agro-ecological zones, are taken as a framework: the humid tropics or rain forest zone, the subhumid tropics or savanna zone, and the dry lands. These last cover the semi-arid and arid zones which extend from tropical to subtropical and warm temperate latitudes, including the summer-rainfall or Mediterranean zone.[1] In addition, there are two environmental regions based on landforms which cut across the climatic zones, steeplands and alluvial plains. For each region, brief accounts are given of:

the basic environmental conditions, and thus resources offered by the land;
the major systems of land use;
the land use issues, including problems of competition for land use, land management, and land degradation.

The humid tropics: the rain forest zone

To many people, the zone with a natural forest cover is the most 'typical' environment of the tropics: hot, humid, with dense vegetation and abundant biological potential. In fact it covers 23% of the region, a smaller area than either the subhumid zone or the dry lands. The main areas are the Amazon basin and humid parts of the Andes of Central and South America, the Zaire (Congo) basin and the West African coastal belt, much of Bangladesh, South-East Asia and Indonesia, and many equatorial island countries. The essential climatic feature is that there is no dry season or only a short one of 1–4 months, insufficient to dry out the soil profile. This leads to the natural vegetation of closed forest, wholly or partly evergreen. There are subzones determined by landforms: upland areas, dissected into a network of valleys with much sloping land, and alluvial basins. Soils are deep, highly weathered, and strongly leached, with fertility maintained under a forest cover by a high degree of nutrient recycling between plants and soils, often above 90%.[2]

Being hot and humid throughout the year, this zone has the highest productive potential of any part of the tropics. The problems lie in realizing this potential, since the potential for degradation is also high. The four major systems of production are:

shifting cultivation, of cereals and root crops;

swamp rice;

perennial crops (tree crops);

production forestry (hardwoods).

There are also major conservation functions: watershed protection, conservation of biodiversity, and the role of the zone as a regulator of atmospheric carbon dioxide.

Each system presents different land use issues. Indigenous systems of shifting cultivation require long fallows, as high as 20 years fallow for 1–2 years cultivation on the poorest soils.[3] This has become unrealistic under modern population densities, hence the search for alternatives. It is, for practical purposes, impossible

LAND RESOURCE ISSUES: THE HUMID TROPICS

Alternatives to shifting cultivation for annual crop production.

Soil erosion (water erosion).

Maintenance of soil fertility after forest clearance.

Maintenance of high productivity under swamp rice cultivation.

Efficient management for perennial crop production by smallholders.

Forest clearance and fragmentation.

Conservation of biodiversity.

to cultivate these soils continuously for annual crops, other than swamp rice; attempts to do so have led to serious soil degradation, for example in the 'transmigration' schemes for settlement of the outer islands of Indonesia. Cultivation of swamp rice (padi), rainfed or irrigated, has proved its sustainability by being continued for many hundreds of years in parts of Asia. This land use system receives less than its proportionate due in research and development attention, given that it provides food for nearly 40% of the developing world's population. Green revolution technology – improved crop varieties, fertilizers, and pest control – has raised yields above 5 t ha^{-1} (tonnes per hectare) in many countries. The major problems of this system stem from its success: how to maintain high production under increasingly intensive use, and the wider problem that productivity gains are constantly being nullified by population increase.

Perennial crops (tea, cocoa, rubber, oil palm, coconuts, etc.) are another solution to the problem of continuous land use in this zone, through the provision of soil cover by leaf litter and associated nutrient recycling. Dense multistorey tree and shrub systems have the potential to combine productivity with the protection of quite steeply sloping land from erosion, as for example on valley slopes in the Philippines and Sri Lanka. To achieve this requires good management: well-managed tea estates achieve excellent soil conservation, poorly managed ones are subject to severe erosion. Hence there is a technical and social need to transfer the better estate management methods to smallholders.

Two further issues are related to the forest cover: forest clearance, and conservation of biodiversity. Forest clearance is a complex issue linking production forestry, competition for land from agriculture, the indirect production function of head-water catchment protection, and biological resource conservation. This last forms an issue in its own right, the various forms of conservation – of plants, animals, ecosystems and genetic resources – being linked under the name biodiversity. National questions, such as use for forest production and water supply, interlink with the international concerns for conservation of biodiversity as a resource for humanity as a whole, and the perceived special role of rain forest in the atmospheric carbon dioxide cycle.

The subhumid tropics: the savanna zone

The subhumid tropics, also known as the wet-and-dry tropics or savanna zone, has a period of rainfall followed by a dry season of 5–8 months. There is a subzone with bimodal rainfall, two wet and two dry seasons. The subhumid zone occupies 30% of the tropics, including large areas of West, Central, and East Africa, areas north and south of the Amazon in South America, much of the Deccan or peninsular region of India, and part of tropical Australia. The plant cover is one of trees, predominantly broadleaf and deciduous, with an understorey of grasses, known in Africa as savanna or *miombo*, and in South America as *cerrado*; formations range from woodland to almost treeless grassland, with the percentage tree cover often reduced by cutting and burning. The characteristic landforms are gently undulating plains, ancient erosion surfaces which form a pattern of broad valleys often continuing for hundreds of kilometres. Soil fertility depends on rock type and relief, with fertile areas of limited extent, for example on volcanic rocks. The plains are dominated, however, by a highly weathered and poorly structured soil, difficult to manage; this is extremely distinctive although, strangely, not clearly identified on most major classification systems.[4]

Seasonality of climate is the basis for the productive potential of this zone. The dry season favours annual crops, whilst permitting crop production, livestock rearing, and some types of forestry. Hence there is flexibility of use, and hence competition for land. Contrary to what is sometimes supposed, it is nowadays rare to find shifting cultivation in this zone. Because of population pressure, most land is under nearly continuous cultivation, often maize monoculture. Some major land use systems are:

- annual cropping, especially for maize, sorghum and upland rice;
- nomadic and semi-nomadic pastoralism;
- certain perennial crops, notably coffee in the more humid subzone with two rainy seasons;
- plantation forestry, especially on highland areas and hills.

The most widespread land resource issue is soil management under annual cropping, both on the more fertile and poorer soils. This is a complex issue, combining achievement of acceptable levels of production with avoidance of soil degradation – in other words, sustainable use. It is very common to find a negative nutrient balance, removal in harvest exceeding inputs. It is possible to farm these lands

LAND RESOURCE ISSUES: THE SUBHUMID TROPICS

Sustainable soil management under annual cropping.

Erosion on sloping lands.

Fodder shortage in mixed arable-livestock farming systems.

Competition for resources between cultivators and pastoralists.

Competing uses for valley-floor grasslands.

Wood shortage, woodland degradation and clearance, finding land for forestry, and the role of agroforestry.

Agricultural impacts on biodiversity.

almost continuously for annual crops, but this requires good land husbandry and at least some fertilizers. Population pressure has led to cultivation being extended onto steeply sloping lands with consequent erosion. Many farming systems include livestock, and fodder shortage is widespread. A regional contrast is found: in India, even the poorest farmers devote some land to fodder crops for their draft animals, whereas in Africa these are expected to survive on common-land grazing. Where pastoral peoples live alongside cultivators there is potential conflict over land, as tragically illustrated in the Hutu–Tutsi conflict in Rwanda.

An issue related to a specific site in the savanna landscape is the use of the valley-floor grasslands, sometimes known by the vernacular term *dambos*. Traditionally these were for dry-season grazing, rested during the rains when cattle could graze on the hills. Vegetable gardens can be made at the margins. In places, overgrazing has led to gullying, with lowering of the water table and consequent loss of their distinctive role. Swamp rice or non-flooded rice can produce higher income than grazing, but with loss of the role of these grasslands in the wider farming system. This is an issue both of scientific interest and practical importance.

Forestry in the subhumid zone offers both problems and potential. With lower population densities in the past, cutting from natural woodlands could supply both fuelwood and domestic timber. Frequently nowadays the threshold has been reached where removal exceeds natural regrowth, leading to rapid and accelerating degradation. Forestry plantations, often of exotic or coniferous species, can give very much higher rates of growth increment. To the chagrin of foresters, good agricultural land cannot be spared for these; the role of afforestation is on the hillslopes and high-altitude plateaux. There are complex issues of forest control and management, whether by government, communities, or individuals. Agroforestry, growing trees on farms, has a role to play.

Conservation issues attract less attention in the savannas than in the forest and desert zones, but are no less pressing. The point of overall water shortage is being reached in a growing number of regions. The fact that so much of the land can be placed under managed ecosystems – cultivation, grazing, or forestry – leads to severe impacts on biodiversity. Conservation is focused upon national parks, but the exclusive use of these leads to a 'people versus parks' problem.

The dry lands: the semi-arid and arid zones

The third and largest 'world of the tropics' is the dry lands, extending beyond tropical latitudes and occupying over one third of the area of developing countries. This highly generalized description covers three zones: semi-arid lands with summer rainfall, desert, and the Mediterranean zone.

The semi-arid zone lies on the tropical margins of the desert. Besides the vast belt from West to East Africa, often called the sahel, it is found in West Pakistan and the adjacent Rajasthan zone of India, north-east Brazil, and equivalent areas around the tropic of the southern hemisphere. This zone has a rainy season of 2–4 months in the 'summer' (high-sun) period, followed by 8 or more totally dry months. Vegetation is adapted to drought and is often thorny, as in the typical acacia thorn scrub of Africa and its equivalents in the semi-arid zones of Brazil and India; by the middle of the dry season, herbaceous vegetation dies above ground level and only deep-rooting trees remain green. To the unaccustomed eye, it is remarkable that anything, plant, animal, or human, can survive the later part of the dry season. Soils range from almost pure sands, relict from former periods when deserts were of greater extent, to relatively fertile soils. The distinction between semi-arid and desert zones is that grazing and some non-irrigated cultivation is possible in the former.

As average rainfall becomes lower, year to year variability tends to increase. Hence drought years are a normal and expected feature of the dry zone environment. Not infrequently these occur several years in a row, as in the African sahel in the 1970s. Whether human-induced climatic change is increasing this hazard is not yet known.

Areas of winter rainfall and summer drought, the Mediterranean type of climate, range from subhumid to semi-arid. For convenience, we have included them among the 'dry lands', although the environment and resource potential differ in many ways. Many parts of this climatic zone lie in the developed world. The largest area in developing countries is the belt extending from North Africa to the countries known variously as the Near East, Middle East, or South-West Asia.

The scarcity and seasonality of water dominates the resource potential of this zone, and are directly or indirectly responsible for many of its land use issues. There are three major systems of land use:

livestock grazing, often with seasonal migration, nomadic pastoralism;
cultivation of annual crops by dry farming techniques;
irrigated agriculture.

A special function of grazing in this zone is to permit production from land that would otherwise be unproductive. If all the world's livestock were raised on feedstuffs there would be a large extra demand on cereals. Similarly, irrigation allows use to be made of the climatic and soil resources by supplying the missing element of water.

LAND RESOURCE ISSUES: THE SEMI-ARID AND ARID ZONES

Water shortage and competition between uses.
Lowering of the groundwater table.
Water management: rainwater harvesting, dryland farming, efficiency of water use.
Making provision for drought years.
Problems of irrigation management: water use efficiency, salinization.
Overgrazing, water and wind erosion (desertification).
Livestock nutrition and watering during the late dry season.
Encroachment of arable cultivation onto grazing lands.
Fuelwood shortage.
Food self-sufficiency.

First among the land use issues is water shortage and competition for its use. Many countries have reached the point of absolute scarcity. The problem is rendered more acute because irrigation is most needed in this zone, yet has high water requirements. There are many potential dangers of international conflict over water. A related and widespread form of land degradation is lowering of the groundwater table, in effect 'mining' water as a non-renewable resource. Planning measures are needed to take account of the inevitable drought years, since the former occurrence of starvation and warfare is no longer acceptable. Finding means of water conservation and its efficient use is a major issue of resource management in dryland (non-irrigated) farming. It is still more important, from the viewpoint of quantity of production, under irrigated agriculture, where water use efficiencies are often low, and salinization the major problem of soil degradation.

A further set of issues concern the rangelands and livestock production. Overgrazing, leading to pasture degradation and soil erosion, commonly called 'desertification', is certainly widespread, although the reliability of information about it is very low. A key problem is to sustain livestock through the long dry season.

Management of the livestock is not well integrated with conservation of the pastures on which they depend. Population pressure leads to the encroachment of arable cultivation onto grazing land, often the moister and most valuable pastures. Where different peoples are involved, this can be a source of conflict. Lastly, the semi-arid zone almost everywhere has a severe and chronic fuelwood shortage, linked with the slow growth rates of trees; as in the savanna zone, there is a consequent danger of passing the threshold where removal of wood exceeds replacement, leading to accelerating degradation.

Over and above these specific issues, estimates of food production potential in relation to population markedly identify this zone as one of food deficit or overpopulation. There is always a danger of famine, as recently in West Africa and Somalia. This long list of problems truly reflects the nature of this zone, identified by the United Nations as a 'fragile ecosystem'.[5]

Steeplands and alluvial lands

The steeplands

Two sets of environmental conditions determined by landforms cut across the climatic zones: steeplands and alluvial lowlands. Steeplands are the mountain, hill, and dissected valley areas, having in common the dominance of moderate to steep slopes. Examples are the major mountain ranges of the Andes, Himalayas, and South-East Asia, dissected zones associated with the African Rift Valley, and countless hill areas. They include many island environments, such as the West Indies, the former East Indies (Indonesia, The Phillipines, etc.) and volcanic islands.

Sloping lands in the dry zone are usually barren and of little resource potential, although those of the Mediterranean zone may not always have been thus.[6] In more humid areas, however, the soils may be fertile in their natural state, and have the potential for all of the major productive ecosystems: cropping, grazing, and forestry. Many such areas have become densely settled and show clear signs of overpopulation; examples are the endemic food shortage and recurrent starvation in the highlands of Ethiopia, and persistent out-migration of labour as in Nepal. At the same time, steeplands play a key role as water catchment areas and in biodiversity conservation.

The steeplands are the second 'fragile ecosystem' singled out by the United Nations.[7] Most of the land issues of the humid and subhumid zones are found on them, but three are of special importance: soil erosion, watershed management, and biodiversity. Soil erosion is an obvious hazard, yet it is socially and politically impossible to forbid cultivation. Ways have to be found of making arable use environmentally acceptable. In Asia, one solution is terracing. Banks of terraces

LAND RESOURCE ISSUES: STEEP AND LEVEL LANDS

Steeplands

Soil erosion, and finding methods for sustainable use of fertile soils on sloping land.

Watershed management.

Forest clearance and biodiversity conservation.

Alluvial lands

Maintenance of soil fertility and high productivity under continuous cultivation; both for swamp rice and dryland cropping.

Problems of irrigation management: water table lowering, water use efficiency, salinization.

Shortages of fodder and fuelwood in predominantly cultivated landscapes.

extending up mountainsides, as in Indonesia, the Himalayas, Yemen, and parts of China, are a remarkable achievement in land management, achieved by labour and organization extending over centuries. Attempts to introduce terracing in areas where it is not traditional have usually failed, as have many soil conservation projects based on conventional methods. Agroforestry techniques may offer an alternative. Watershed management combines different aspects of land resources through the links between forest cover, runoff and erosion, river flow, and sediment loads. In the Himalayas, forest clearance has led to the need to walk longer and longer distances to fetch fuelwood. Agroforestry techniques have a particular potential in this zone, in soil conservation and provision of wood. Humid mountain regions are often of the highest value for biodiversity conservation, yet forest clearance for cultivation and fuelwood are often severe. In Vietnam, 50 years of intermittent warfare coupled with a population of nearly 100 million have led to total fragmentation of the remaining forest areas, and there are now many endangered species. The forests of Madagascar are another critical area.

The alluvial lands

The alluvial lands, or river floodplains, terraces, and deltas, are a further major distinctive environment. Foremost among them are the major valleys of Asia, the alluvial floors and deltas of the Indus, Ganges, and Mekong, the rivers of China, the Nile, and the Tigris-Euphrates valley of South-West Asia. They are found in all climatic zones, the Indo-Gangetic plains extending from the arid deserts of the Sind in Pakistan to the humid delta zone of Bangladesh. The more humid areas were no doubt once forested but have been almost entirely cleared. Soils are geologically young, derived from recent sediments, and relatively fertile.

The alluvial lands are largely devoted to food production, most commonly dominated by rice in the tropics, and wheat and barley in subtropical and temperate areas. The two major land use systems are rainfed cultivation in the humid regions and irrigated agriculture in the dry zone, the latter divided into large-scale and small-scale irrigation. Double and triple cropping are common and rotations can be highly complex; a farmer in Bangladesh may be managing more than 20 controlled patterns of crop rotation.[8]

Through the combination of level land and major river systems, the alluvial lowlands have the sites of the big dam-and-canal irrigation projects. The sheer scale of the Upper and Lower Indus is amazing to behold; on leaving the reservoir or barrage, water passes through a bifurcating network of canal, major distributary, minor distributary, watercourse, before passing via a *nakkar* (flume) into the field distribution canals. It is small wonder that farmers on the lower end of these systems find that water does not reach them when it is most needed. Interspersed with these are areas served by the two kinds of small-scale irrigation, traditional and modern. Traditional methods are diverse and ingenious, ranging from the various systems of lifting water by human and animal power from rivers and shallow wells, to direct diversion from rivers by very gently sloping channels (anicuts) and even, in Iran, underground tunnels. The principal modern small-scale method is the tubewell, drilled and with power-driven pumping. The Punjab zone, both in India and Pakistan, is highly dependent on tubewell irrigation. In 12 countries, five of them in Central Asia, more than half the cropland is irrigated, and this proportion is 52% for China, 80% for Pakistan, and 100% for Egypt.

The major land use issues of the alluvial lands stem from their potential and success. Continuous cultivation, and the rising level of crop yields needed to support growing populations, are bound to lead to difficulties in maintenance of soil fertility. Fertilizers no longer bring the same yield responses as formerly, indicative of fertility degradation, such as loss of physical structure or micronutrient deficiencies. There is an endemic danger of build-up of pests. Of countries with large alluvial areas, only Thailand remains in food surplus. On irrigated lands, the big reservoir and canal irrigation schemes are notorious for management problems: siltation of reservoirs, loss of water through seepage from unlined canals, low water use efficiency at the farm level, and rising levels of saline groundwater leading to salinization. Tubewell irrigation appears highly successful when first practised, until overpumping leads to lowering of the water table. A further general issue, both on irrigated and dryland areas, is the shortage of fodder for draught animals and fuelwood, in regions where nearly all land is needed for food production. Over and above these issues, the sheer number of people dependent on these lands is a problem in itself: to take a step forward, say in agricultural extension, requires planning and inputs on a large scale, and drought or flood bring disaster on a massive scale.

A question which must sometimes arise in the minds of people of independent thought is: 'Do the forested basins found in South America and Africa have the potential, from a physical resource aspect, for transformation into highly productive rice lands such as in Asia?' If the present huge contrast in productivity were caused only by cultural differences, then truly there would be a vast untapped resource potential. There is, however, a basic difference of environmental conditions. The Ganges, Mekong, and other alluvial plains of Asia are formed from recent sediments, derived from geologically young fold mountains. The Amazon and Zaire occupy downwarped parts of ancient 'shield' areas, many of their soils being extremely highly weathered. These include deep sands, with no productive potential if the forest is removed. Whilst the basins of the Amazon and Zaire undoubtedly have some potential for swamp rice cultivation, neither better systems of water control nor prolonged inputs of labour can transform them into another Thailand or Bangladesh.

The land use issues outlined above are common to the major tropical environments in all continental areas. Many other issues are distinctive to particular environments. Thus, the black cracking clay soils (vertisols) offer a high potential fertility if their difficult structural conditions can be controlled by management of the soil water regime. Island environments present special problems, such as extreme land scarcity, urban encroachment, and water shortage. Other issues are linked to the socio-economic conditions of regions or countries. The contrast in land-population ratios between South America and Asia provides a different context for land resource issues. Nomadic pastoralism gives rise to resource problems particularly, although not exclusively, in Africa. Differing land tenure systems have impacts upon resource management. China, with 27% of the population of the developing world and extending from subtropical to temperate zones, has its own specific set of land resource issues.

The land resources of developed countries

This book is mainly about the land resources of less developed countries, which lie mainly although not entirely in the tropical and subtropical zones.[9] This is where the major problems lie, today and in the future. Contrasts in the socio-economic conditions of the developed world and most less-developed countries are so great that their problems of land resource planning and management are highly different. Managing farms of 200 hectares and upward offers opportunity not open on 1–2 hectares, although doubtless the respective farmers would find many problems in common. National systems of land use planning are much more developed in Western countries, and conservation interests far stronger. Of course, the environment is no respecter of national living standards, and land degradation is not uncommon in the developed world. It was the 'dust bowl' of the United States in the 1930s which first drew attention to the need for soil conservation. In Canada, erosion became serious for a second time in the 1970s. In Britain, continuous cereal production made possible by fertilizer inputs led to degradation of soil physical structure.[10]

Questions of land use and management in the developed world will not be treated specifically, but cannot be excluded for two reasons. First, the methods of resource assessment and the land use planning systems of developed countries have much to offer the developing world. Many methods of soil survey were first developed in the United States; the advanced land use planning systems of Canada and Australia could form a starting-point for developing countries. Secondly, the problems of developing countries cannot be considered in isolation from the rest of the world. This is most clearly seen in questions of world food supply and demand, and the potential and limits to the role of trade.

The present condition of resources and land use in developed countries will be treated as an 'external variable', and some broad working assumptions made about their future. These can be summarized for two groups of countries: the developed economies and the former Soviet Union.

In the developed countries, population is relatively stable and likely to remain so, as a result of which they will soon form less than 20% of the world's population. These countries have many and difficult land use issues, arising particularly from the intense competition between different types of use. Land use planning is more tightly bound by legal constraints, land management employs high inputs, and conservation interests are strong. For this group of countries, it will be assumed that:

- the extent of different types of land use – agriculture, forestry, conservation and settlement – will not greatly change, being firmly fixed by legal, institutional and economic forces;
- agricultural production could be appreciably raised, but this will only happen in response to economic forces, i.e. ability to pay;
- further land degradation will be limited in its extent and severity, and will not substantially lower productive potential;
- aid to developing countries will continue at approximately its present level – only a radical change in attitudes can alter this situation.

For the countries of the former Soviet Union, forecasting is problematic owing to the current state of political uncertainty. Under communist rule, land resource management was often not efficient or sustainable, conservation was sacrificed to the achievement of production targets, and serious degradation occurred. In some respects, many of the smaller republics in Central Asia now belong to the developing world. Aid funds have been diverted to this region. For this group of countries, it will be assumed that:

political stability and efficient government, and consequent economic progress, will be maintained;

there will be some further population increase, and a substantial rise in living standards leading to greatly increased consumer demand, including for food;

arising from better land resource and economic management, food production will increase, although not sufficiently to make this bloc a net food exporter.

3 Resource survey and land evaluation

Surveys of land resources – climate, water, soils, landforms, forests, and rangelands – are needed to avoid costly mistakes and to improve efficiency of investment. Valid techniques have been developed for all types of resource survey, and the method of land evaluation has helped in translating environmental data into terms of land use potential. Soil surveys are best carried out in two stages: a reconnaissance survey on a land systems or landscape basis, followed by special-purpose surveys as required for development. Soil monitoring should become an additional basic task for soil surveys. Information on water resources, forests, and rangelands is deficient for many developing countries, and institutional capacities are weak. The extent to which surveys have been applied is far from satisfactory. Natural scientists have failed to communicate the implications of their findings, and, conversely, planners and decision-makers are not sufficiently aware of the significance of natural resource information. The proper management of land resources is so fundamental to sustainability that it should permeate the whole fabric of development, from planning through implementation to monitoring of change. The lack of communication between scientists and planners needs to be improved by more broadly based education, and strengthening of institutions.

Most of the time, farmers do not need soil surveys. From time immemorial, when taking new land into cultivation they learnt to find good soils and avoid bad, often using what would now be called indicator plants. Shifting cultivators could recognize when land had recovered sufficiently to repeat the cycle of clearance and cultivation. Settled farmers have an intimate knowledge of variations in soils and water availability field by field, in far more detail and depth than an itinerant soil surveyor is likely to acquire. In recent years there has been a resurgence of interest in making use of such indigenous knowledge. Nomadic pastoralists know where good grazing resources and water are available at different seasons, and adapt movements of livestock to the vagaries of the weather. European settlers, finding sites for such crops as tea, coffee, rubber, sugar, or sisal, often established areas of cultivation which closely follow what are now known to be climatic limits; this was presumably achieved by trial and error, although for the most part the process has not been recorded.[1]

It was the post-1945 surge of foreign aid, in the form of rural development projects, which led to recognition of the need for land resource surveys.[2] In the early stages projects were often based on transfer of Western technologies to the tropics, not always successfully. A classic example was the 1947 Tanganyika (Tanzania) groundnuts scheme, which had the objective of meeting a shortage of vegetable oils. Three areas were selected for bush clearance and large-scale mechanized cultivation. It is now known that one of these areas was too dry for groundnuts, one too flat and marshy, whilst the third was affected by rosette, a disease which attacks groundnuts. Costly machinery was damaged by rock-hard and abrasive soils. There were many other planning faults on this scheme, not least the discovery that heavy machinery needs maintenance.[3] However, it served to draw attention to the need for a proper assessment of climate and soil resources before embarking on development.

Natural resource surveys are needed in situations where farming experience is not available: when new areas of land are brought into cultivation, and when new kinds of use are introduced to existing farms. The first of these was at the forefront of attention from the 1950s to 1970s, the latter is more often the case in recent years as unused land becomes scarce. In these circumstances, the knowledge and judgement of resource scientists, preferably those with previous experience of the environments and kinds of land use concerned, is an essential basis for planning.

Surveys have two main functions: to avoid mistakes and to improve efficiency. Avoiding serious mistakes, such as growing water-demanding crops on sites prone to drought, can make immense savings in development costs. Large-scale irrigation schemes, having the highest investment of all forms of rural development, have made the most expensive mistakes. Having selected broadly suitable land for a proposed development, the next stage is improving efficiency, the fine-tuning of land use and management. For development of arable land, a basic need is the selection of appropriate crop varieties and fertilizer treatments. There might be a local need for rice varieties tolerant of saline water, or deficiencies of sulphur or micronutrients on certain soils. Mapping such variations in resources, and matching these to appropriate kinds of land use and management, give agricultural extension services an initial basis, to which experience can be added. Forest development calls for selection of tree species and provenances suited to the local soils and climate, and choice of suitable planting techniques. The often difficult task of developing rangelands for livestock production calls for a knowledge of seasonal fodder resources and water availability.

The attitude that resource surveys are a luxury which cannot be afforded when development funds are scarce is still sometimes encountered. This could not be more mistaken. Economic analyses of soil surveys have shown benefit: cost ratios of the order of 40:1.[4] A fundamental reason is the rise in costs by a factor of ten or more between the pre-investment, planning, stage of development projects and their implementation. A range of natural resource surveys, properly selected as being appropriate to the kinds of development being planned, is unlikely to cost more than a quarter of the planning stage budget. If the total outcome of these surveys is an improvement in land use efficiency of as little as 5% during the first ten years of implementation, or if they lead to avoidance of just one single expensive error, the cost of surveys is justified many times over.

Over the past 50 years, a range of methods has been developed for the survey and evaluation of all kinds of land resources. These methods are scientifically sound, and have the potential to fulfil their intended purpose, of assisting rural development and land management. There are many technical questions of the best ways to conduct surveys, only a few of which can be considered here. More fundamental problems are whether appropriate surveys are in fact done, leading to an adequate base of knowledge; and whether the results are adequately applied to development planning.

There are two basic stages to the assessment of land resources: natural resource survey and land evaluation. Natural resource survey refers to the description, classification, and mapping of the physical environment: climate, water, geology, landforms, soils, vegetation, and fauna. In practice, geological survey is a separate task, and soil surveyors make use of whatever data are available. Landforms are generally surveyed jointly with soils. An alternative to assessing these factors individually is to map the physical environment as a whole, as in the land systems approach. In the second stage, land evaluation, the potential of the mapped areas for different kinds of land use is assessed. Historically, methods for land evaluation were developed after those for resource survey; today they form successive phases of land resource assessment.

Soil survey

'And we can save another 500 Lire by not doing a soil survey.' So runs the caption of a cartoon widely circulated among soil scientists, showing a medieval architect holding up to his client a drawing of a very vertical Tower of Pisa. That it refers to soil in the engineering sense does not detract from the message. In a land development project, the first and foremost reason for carrying out a soil survey is to avoid serious mistakes which it will be highly expensive, or impossible, to rectify.

The basic stages in soil survey are description and identification of soil types, classification, and mapping.[5] The standardization of methods for soil description was achieved largely by adoption of those employed in the United States; two elements were added, clay skins and weatherable minerals, important in tropical environments in showing whether the soil was still evolving or largely 'dead'. The FAO's *Guidelines for soil description*[6] was to be the first of a series of guidelines for developing countries to adopt, or adapt, as national systems.

Soil classification

To obtain international agreement on soil classification took longer and is still not fully achieved. For 25 years, 1950–75, classification was a focal issue among soil scientists. The root of the problem was that soils have more than 20 properties (texture, reaction, organic matter, etc.) which can be significant to land use. A debate of much intellectual interest took place between advocates of 'natural' and 'artificial' classification systems.[7] The former held that a number of natural, or ideal, soils existed, each of which was the outcome of specific sets of soil-forming processes (weathering, leaching, etc.). The task of classification was to group existing soils around these natural types, allowing that intergrades existed. The latter school held that such attempts were doomed to failure. Soils were a continuum, so artificial boundaries must be set up; the example often given was defining a bald man by the number of hairs on his head.

Partly owing to the arrival of computerized systems, which before the advent of 'fuzzy sets' required that a soil should be placed in one unit or the other according to exact values (mainly from chemical analysis), artificial systems won the day. There are currently two co-existing systems employed internationally, the American and international classifications, together with a third system common to francophone countries. In addition, many countries have their own national systems.

The American *Soil taxonomy*[8] was devised as a national system, extended internationally through a series of working groups. It is nothing if not definitive, a book of 754 pages weighing 2.4 kilogrammes. Besides extreme complexity, a feature is its rejection of established terms in favour of newly coined ones, the only definitions of which are those given in the *Taxonomy*. Unfortunately these give rise to extremes of jargon, such as 'an isohyperthermic epiaquic orthoxic tropohumult'. Another objection is the use of what are basically features of climate, in the guise of soil temperature and moisture regimes, to classify soils, thereby pre-empting independent assessments of climate and soil.

The international system originated in 1974 as a legend for the FAO-UNESCO *Soil map of the world*; it has twice been revised, most recently as the *World reference base for soil resources*.[9] It is of more manageable length and complexity whilst retaining quantitative definitions, intentionally modified to be compatible with the American system as far as possible. In the current version, there are 30 major soil groups (e.g. acrisols) divided into 153 soil units (e.g. ferric acrisols).[10] This system is gaining ground as the recognized international standard, to which national classification systems should be related.

There is a further system, fertility-capability classification, intended not as a mapping basis but for interpretation.[11] Soils are assigned letters showing where fertility limitations exceed defined levels, for example 'e' = low cation exchange capacity, 's' = salinity. This provides a system for characterizing the fertility limitations of soils, and thus for assessing, on the one hand, regional requirements for inputs, and, on the other, research priorities.

Soil mapping

The immediate output of a soil survey is a soil map, where 'map' includes the extended legend and text which accompany it. A question often asked is what is the purpose of a soil map? The first answer is, to enable statements to be made about soils in different areas, the soil mapping units, which are more accurate than those which could be made about the region as a whole: two soil types, A and B, are identified and mapped in order to be able to say that the area mapped as A has deep, sandy soils whilst those in area B are shallower and clayey. Whilst this may be true for much of the areas, soils are so highly variable over short distances that it is by no means always so. Ideally, a completed map should be tested by an independent 'soils auditor', who samples soils at randomly chosen sites and reports on the accuracy of the predictions. This has rarely if ever been done.[12]

But the second answer, and basic objective of soil surveys, is to provide guidance on land use and soil management. Typical purposes are to recommend which crops can be grown successfully, the kinds of fertilizer they will need, and over which parts of the region will drainage or soil conservation works be required.

Once this fact is recognized, there are two approaches, general-purpose and special-purpose soil surveys. General-purpose surveys are conducted as a pure mapping operation, seeking to map soil units that are as distinct from each other as possible. The properties employed to define these units need not be those which affect plant growth, notably the usefulness for field survey of soil colour. At a later stage the properties relevant to specific crops or other land uses can be investigated. General-purpose survey has the advantage that uses of land that were not foreseen at the time of the survey, waste disposal, perhaps, or suitability for a new crop hybrid, can be investigated without the need for a complete resurvey. In special-purpose soil survey, the intended land use is known and the surveyor collects only information which is relevant to it; for a project to develop smallholder tea production, for example, mapping is based on soil properties relevant to this. A little-known debate on these two approaches took place in the journal *Tropical Agriculture* in the 1920s, with Dutch scientists involved in sugar-cane development in the East Indies (Indonesia) arguing the case for special-purpose surveys.

The advocates of general-purpose surveys won the day, clinching the argument with the (supposed) fact that stable soil properties, those relatively unchanged by management such as texture, were employed as the basis for map units. The map could then be a document of lasting value, for 50 years or more. However, this basically sound principle led to difficulties when the results of surveys were

FIGURE 1 What the soil surveyor is conventionally looking at, the soil profile.

expressed almost entirely in pedological terms, with a meaning only to other soil scientists. Another limitation was that soils were treated as if they were static, whereas, in fact, many important properties are altered by management.

These approaches can be reconciled. Where there has been no previous survey, a general-purpose map is certainly needed. In preparing it, however, the surveyors should go to sites where performance data for various kinds of land use are available. This can be called the Hardy method, being pioneered by G. H. Hardy in the West Indies in the 1930s. On finding, for example, that an island's economy depended on coconuts, he would say, 'Take me to where your best coconuts grow, and also to some places where they are doing badly', and then he would get soil pits dug on these sites. There are many opportunities to examine soils on sites where quantitative data on crop performance or other land use is available: fertilizer trials,

FIGURE 2 What the soil surveyor should also be doing – crop performance data are essential for land evaluation.

farms with marketing-board records of yields (e.g. tobacco), forestry plots for site index measurement. By this means, soil surveys can provide immediate information relevant to land use, rather than requiring a second stage of interpretation.[13]

Integrated survey: the land systems method

Conventional soil survey, in which the surveyor proceeds from field to field, putting down auger borings, is thorough but extremely slow. There was no way that it could meet the needs for opening up new lands for settlement in the post-1945 era. A solution was found in the land systems method (also called soil landscapes or integrated surveys). This originated in Australia, where the government wanted to assess the resources of the largely unsettled northern and interior regions. A classic survey of the Katherine-Darwin region in 1946 established the basis of the land systems method, which has been widely adopted for rapid reconnaissance surveys.[14]

The basis of land systems mapping is that boundaries are drawn from air photographs (now supplemented by satellite imagery) in advance of field survey. In nearly all landscapes as seen on air photographs there is a clear distinction into spatial units at two scales: regularly repeating patterns which occupy tens or hundreds of square kilometres, the land systems, and the detailed elements which make up these patterns, called land facets. The climate is, for practical purposes, uniform across a land system. Landforms and vegetation, both directly seen on air photographs, are the main basis of mapping. Having delineated the provisional map units, the survey team makes field transects of the land systems, describing the landforms, soils, hydrology, vegetation, and land use. Detailed maps of land facets may be made for sample areas. If and when development planning is started, further mapping of facets can be carried out where needed.

The land systems method met the need of developing countries for rapid inventories of resources at a reconnaissance scale, typically 1:250 000. Conventional soil survey could never have accomplished this in so short a time.[15] A number of countries, Malawi, Uganda and Papua New Guinea for example, achieved complete coverage.[16] The colossal task of a reconnaissance survey of Ethiopia was accomplished by a land systems survey based on satellite imagery.[17]

There is still a need for many countries to complete land resource surveys at reconnaissance scale, and the land systems method is the best way to accomplish this. Strategic land use planning at national level can be done based on surveys at this scale, whilst it provides a framework for reference in more detailed surveys for project planning.

The present status of soil survey

The great era of soil survey in the tropics was the 1950s and 1960s. So little was known that at least a reconnaissance survey was essential to any form of land development. Currently there is less systematic soil mapping being done, as neither national governments nor funding agencies see it as a priority. The attitude is to make use wherever possible of such information as is already available. This would be more acceptable if survey coverage were completed, but this is not often the case. More than half of developing countries have 50–100% cover at reconnaissance scale, 1:250 000 or smaller, but only one in four has even a 25% cover at medium scales such as 1:50 000. There are some exceptions; Jamaica, South Korea, and Gambia, for example, have complete survey cover at scales needed for project-level land use planning, about 1:25 000.[18]

The present status of soil survey in many developing countries is unsatisfactory. It is not necessary to carry out detailed surveys on a routine basis. Complete national coverage at a reconnaissance scale, however, is essential to land use planning at the national level, and as a factor in selection of areas for development. In district-level development projects, a soil survey remains an essential basis, to avoid costly mistakes and to improve efficiency of investment in land resources.

Soil monitoring: a new task for national soil surveys

Arising out of recognition of the danger of land degradation, a new task for national soil survey organizations has arisen. This is to measure changes in the condition of the soils, to monitor the health of the nation's soil resources. The basic situation, set out in chapter 7, is that soil degradation, principally erosion and fertility decline, has reduced the soil's productive potential. It is known that these effects are widespread and serious, but quantitative information on the degree and extent of degradation is lacking. Indirect data, such as trends in crop yields, can only provide a partial indication. There is no way in which the severity of land degradation can be properly determined without measuring it in the field, and national soil survey organizations are the right institutions to do this.

A criticism of soil surveys in the past was that, probably by analogy with geological survey, they regarded soil resources as fixed and unchanging. It is now clear that this is not the case. The soils of intensively farmed areas, most of the Indian subcontinent for example, are in a very different condition from 50 years ago. In the 1960s, there was an attempt to inject a dynamic element into surveys, called the ecological approach. In place of the landform basis of land system surveys, this approach took changing patterns of vegetation and land use as its starting-point; an example used to illustrate the approach was the replacement of cacao by kola cultivation in south-west Nigeria, in response to a combination of crop disease and changing market prices.[19] Whilst undoubtedly valid in the hands of a skilled ecologist, this approach was never taken up widely.

The present unsatisfactory position in knowledge of soil degradation can only be overcome by measurement of changes in soil properties over time. There are established and replicable methods available to do this. The first need is to establish a sampling design, based on combining mapped soil series with defined patterns of land use, for example the Lilongwe soil series under continuous maize cultivation, unfertilized. Because of the high variability of soil properties, quite large sample sizes are needed, about 50 for each combination of soil and land use. A short-list of significant soil properties to be measured is then drawn up; this would typically include organic matter, nitrogen, phosphorus, some measure of soil physical properties such as pore space, together with properties known to be significant locally, for example salinity or micronutrients. These sites would be sampled at intervals of about 5 years, recording the previous land use and inputs. At a smaller number of benchmark sites, crop yields under known management and inputs are recorded.

An additional task is the measurement of erosion in the field, including loss of organic matter and nutrients in eroded sediment. The reluctance of soil conservationists to leave the scientific safety of erosion plot studies and get to grips with measuring rates on farmland must be overcome. Monitoring of small-catchment sediment loads may prove to be the most practicable method, although efforts should be made to link this with sample field measurement.

Measurement of soil changes over time has rarely been done other than on a research basis.[20] An exception is Japan, where 20 000 sites were resampled and analysed twice over a 20–year period. Being a country of intensive soil management with high inputs, the trends shown are the maintenance or increase in fertility, not its decline, but this example demonstrates the practicability of the technique. Switzerland has a repeated sampling programme to monitor soil pollution by heavy metals. In the USA, the national survey is required by law to report at intervals to Congress on the state of the nation's soils.

Soil monitoring should become one of the basic activities of soil survey organizations.[21] It is manifestly desirable for governments to know which land use practices maintain soil fertility and which degrade it. Projects to improve soil management could then be concentrated where they were needed. As in other fields, international organizations are dependent on good national data. Institutional strengthening will be required to do this: a recognition by governments of the need, provision of funding (with initial support from foreign aid), and retraining of staff. Recognition as the body responsible for monitoring soil change and advising on required changes in management practices would greatly raise the status of soil survey organizations. Unless countries obtain field measurements, attempts to monitor soil degradation cannot succeed. This is the most serious of all present shortfalls of soil information.

Climate, water, forests, and rangelands

Parallel advances have been made in methods for the survey of landforms, climate, water, forests, and pastures. Landforms are generally surveyed jointly with soils although, on occasion, geomorphological mapping has been used as the basis for land resource assessment. In the cases of water and forests, valid methods for survey and assessment are available, but are often not sufficiently applied.

Agroclimatology

Basic climatological data come from national meteorological services and, by comparison with other environmental factors, records are good. Climatic classifications have been available since the 1930s, the simple description classification of Köppen being the most useful for broad descriptive purposes.[22] In agro-climatology, the primary need is for assessment of water availability to crops. Methods for calculation of evaporation and plant transpiration, and their application to the soil moisture balance, were developed in the 1950s. Methods are known for calculating optimum planting dates, but in this respect the judgement of farmers is likely to be superior. The main application of the soil water balance is in assessment of irrigation requirements.[23]

A second demand for agro-climatological information came in assessment of the food-producing potential of developing countries. It became apparent that, when working at a global or continental scale, climate had a greater influence than soil on land productivity. After making allowance for temperature, the broad lines of potential crop yield are determined by the length of time during which there was water in the soil within reach of roots, called the growing period.[24] Lines of equal growing period were used to construct agro-ecological zones. With more than 270 days available, two or more crops can be taken; with less than 75 days, non-irrigated crop production is impossible; whilst between these limits, plant photosynthesis, and therefore potential crop yield, rises steadily with increase in growing period. The agro-ecological zones system of climatic information is now available as maps, databanks of many thousand stations, and computer programs for deriving growing periods. Much agricultural information has been built upon the agro-ecological zones scheme, which has become the standard basis for applications.[25]

Water resources

Most of the earth's water is the salt water of the oceans, and much of the fresh water is locked away in icecaps or deep in bedrock. For practical purposes, water resources consist of the 1% of the earth's water that is cycled as rainfall, soil moisture, evaporation, groundwater, and rivers, the hydrological cycle. There are two interlocking cycles, both starting with evaporation from the seas to the atmosphere. The first, shorter, cycle is from rainfall into the soil and then as evaporation and plant transpiration back to the atmosphere; this is sometimes called 'green' water. The second cycle, or 'blue' water, follows the longer path from rainfall through soil moisture, groundwater and rivers to the sea. By convention only 'blue' water, from groundwater and rivers, is considered as water resources, the direct use of rainfall in rainfed agriculture being treated as a climatic resource. About one fifth of the world's river water is considered inaccessible, meaning that it is found in places where it not needed, such as the Amazon, Zaire (Congo), and rivers of the Arctic.

The distinctive features of water resources are multiple use, the significance of quality as well as quantity, and sharing across national boundaries. The three major uses are domestic water supply, industry including power generation, and agriculture. It comes as a surprise to many people to learn that two thirds of the water withdrawn from rivers and groundwater is used for irrigated agriculture. If the 'green' water used in rainfed agriculture is added, the proportion is close to 90%.

This means that water resource planning cannot be confined to one sector, but must take account of competing demands: if supplies run short for one purpose, this will react upon the others. Water quality suffers through pollution, again with an interaction between sectors, agricultural wastes affecting quality for domestic use. A third feature is that the water of major rivers is frequently common to several countries.

Alternative supplies from desalination of seawater have some potential for domestic supplies where need is great. The cost, 10–20 times higher than supplies of fresh water, renders this source out of the question for the large volumes needed for irrigation.

There are no serious technical problems in the survey of water resources. From boreholes, changes in level of the groundwater table can be monitored, and quality assessed. Methods for gauging river flow were developed for the purpose of irrigation planning. However, in developing countries, nothing like enough data are available. 'There are considerable errors in the majority of assessments of water resources, at virtually every scale . . . The consequence is that statements of world water resources and water use are almost certainly seriously in error . . . Despite the forthcoming water crisis, only educated guesses can be made about the dimensions of the world's water resources.'[26]

Forest resource assessment

Very much the same situation applies to assessment of forest resources as to water: methods are available, but their application in developing countries is often inadequate. There are three stages to forest resource assessment: forest cover, inventory, and utilization. For survey purposes, forests are defined as land with a minimum 10% crown cover by trees and shrubs, provided it is not subject to agricultural practices (in Europe the limit is taken at 20%). Hence large areas are included which non-specialists would call savanna or thorn scrub. What in ordinary language is called forest is classed as closed forest, the remaining formations being open forest. The earlier international forest resource assessments were on the basis of reports from national forestry departments. For the 1980 and 1990 assessments, these were extensively supplemented by sample surveys from remote sensing.

Forest inventory is based on measurements of tree species, their size and form, for sample areas, linked with mapping of forest cover. Its original purpose was to assess the timber resources of a region, and a number of national forest inventories have been completed. The same method now provides a means to assess forest degradation, the depletion of resources by selective removal or overcutting.

Forest utilization can only be surveyed from field observation, using where possible records kept by forestry departments. The primary distinctions are between production forestry and protection forestry, and between natural forests and plantations. Mapping of forest utilization is less straightforward than might be supposed. A forestry department may intend that an area should be used for timber production when the trees are mature, but clearance for shifting or permanent cultivation renders this false. Multiple-use forest management has also blurred the distinction between production and protection objectives. In developing countries, economic and social functions such as collection of minor forest products and forest grazing are important but not easily quantified. There is a further survey problem in that designated, legal, forest reserves may no longer be under forest.

But the main problem is that the forestry departments of most developing countries do not have the staff and finance to conduct forest inventories. The status of information is best in the Asia–Pacific region, intermediate in Latin America, and poorest in Africa. The most recent data are, on average, 10 years old, a situation which makes action to improve conservation and management impossible. 'No [tropical] country has carried out a national forest inventory . . . to generate reliable estimates of total woody biomass volume and change. Only a few countries have reliable estimates of actual plantations, harvest and utilisation although such estimates are essential for national forestry planning and policy making.'[27]

Surveys of vegetation and rangelands

An early classic among natural resource surveys was a vegetation-soil map of Northern Rhodesia (Zambia). This was completed in the 1930s, when most of the country was covered with natural or semi-natural vegetation, variations in which provided a sensitive indication of soils and agricultural potential. There is a large body of information on vegetation mapping, more than one international classification system, and maps at continental scale have been produced.[28] In recent years, however, apart from specialized purposes of nature conservation, vegetation surveys have not been widely used in development, mainly because so little semi-natural vegetation remains. In the hands of a skilled ecologist, vegetation can be a sensitive indicator of agricultural potential, and both farmers and soil surveyors make use of indicator plants for shallow soils, impeded drainage, and other soil conditions.

Pasture resource surveys, to assess the grazing potential of open rangelands, have been produced for a number of countries, for example Zambia.[29] The extent of pastures and their plant composition is the main determinant of grazing capacity,[30] whilst the condition of plant communities is the primary indication of rangeland degradation or desertification. A satellite-based remote sensing technique, the normalized difference vegetation index, provides a measure of total green matter reflectance and is employed in an 'early-warning' monitoring system for the African sahel zone.

Surveys of the condition of grazing lands, showing how they are responding to grazing pressures, are of practical significance. It is not the short-term condition of the rangeland which matters – at the end of a long dry season this can appear catastrophic to the unpractised eye – but whether its capacity to recover during the rainy season, or in a later year of higher rainfall, has been impaired. However, systematic surveys of rangelands are not regularly carried out. One reason is that pasture management specialists do not appear to regard systematic surveys as a useful management tool. Another reason is the absence of appropriate national organizations with this responsibility. In semi-arid countries, rather than setting up a separate pasture resource survey unit, this could become a section of the soil survey organization, enlarged and renamed land resources survey. Regular monitoring, based on field transects linked with satellite imagery, could provide evidence of the extent and severity of desertification which is so lacking (chapter 7).

Remote sensing and computerized information systems

Four groups of technical aids have been applied to all types of land resource surveys: air photographs, satellite imagery, geographical information systems, and databases. Computerized techniques, including digital processing of imagery, have resulted in a data explosion in natural resource surveys. There are many excellent accounts of the nature and potential of these techniques, and their application to land resources.

Air photograph interpretation has been a mainstay of soil survey for more than 50 years, and for surveys at medium and large scales it remains far more valuable than satellite imagery. For getting to know a region, there is nothing comparable to photo-interpretation followed by field transects with the photographs in hand. Recent air photographs are better than maps in showing what is actually present on the ground, for example, whether a supposed forest reserve has been subject to clearance for agriculture. Farmers in developing countries readily recognize their surroundings on air photographs.

When satellite imagery first because available from 1972, it was sometimes assumed that because it was technically more complex it must be superior. Multispectral satellite images, including the familiar false-colour composites, are valuable for special purposes, notably mapping vegetation and land use. They allow rapid surveys at small scales; a land resource survey of the whole of Ethiopia was completed in five years using 70 images, which could never have been accomplished with air photographs.[31] The first reliable maps of land use in the tropics are now being obtained by this means. Particularly valuable is the repeat coverage available from satellite images, allowing monitoring of forest clearance, the condition of rangelands, and all forms of land cover, by farming, grasslands, forest, and settlements. Regular surveys of changes in land use would greatly aid planning both at national and district levels.

Geographic information systems (GIS) allow the overlay of a series of mapped sets of information, in the form of digitized maps, and the production of composite maps by calculation. Digitized soil maps, for example, can be combined with climatic information and crop requirements to produce crop suitability maps. Predictions of future changes according to specified assumptions can be made, and presented in the form of 'flying over the country' to show what it will look like in the future if nothing is done, or if policy measures are taken. Computerized databases are being employed to supplement information in map form, as in the *World soils and terrain digital database* (SOTER).[32] Among many aids to the assessment of land potential is a database of crop environmental requirements, ECOCROP.[33]

All of these computerized data sources are now becoming more widely accessible through the internet, on which all the major international agencies operate sites. There is little or no such information on the internet which is not also found

INFORMATION ON LAND RESOURCES

Extent and reliability of mapping and statistical data for less-developed countries

Climate	Good. World network of stations with long-term records. Converted to agro-ecological indices of temperature and moisture. Available as maps, tables, database.
Water	Adequate. Groundwater levels recorded, but many countries have few river gauging stations.
Soils	Uneven. World soil map at 1:5 million scale, and world soil reference database. Reconnaissance maps for most countries, but less than 25% coverage at medium scales. Systematic mapping has sometimes been discontinued.
Landforms	Not detailed. At world scale, data only as three slope classes; detail in course of compilation for soils and terrain database.
Forest	Moderate. Ten-year forest cover assessments provide benchmarks for often unreliable national data. Limited information on forest condition.
Rangeland	Poor. Systematic surveys of the extent and condition of rangeland are rarely undertaken.
Vegetation	Maps of natural/semi-natural vegetation available at continental scale; not generally used in planning.
Biota	Good. Data available from conservation organizations on plant and animal species, including endangered species, and on ecosystems, genetic resources, and protected areas.

elsewhere, but it is an aid to accessibility. In principle there could be applied interactive applications, such as requests from countries for information transmitted to a central agency, return of the data, and enquiries and responses on problems of analysis. A current need is for training of national staff in the use of such data sources.

The quantity of information obtainable from satellite imagery, geographic information systems, and databases is so great that practitioners of these techniques sometimes wonder why they do not solve all problems. A danger in their use is that it may lead to neglect of field observations, relegating them to a subordinate role referred to as 'ground truth'. In the early years, large numbers of studies were produced to demonstrate potential rather than to put the techniques to practical use, and there is still a tendency to use techniques for their own sake. On thing that remote sensing methods cannot do is to talk to farmers, and if a survey becomes too technique-orientated it will tend to neglect social aspects. Properly employed, however, as tools to be employed in the attainment of objectives, all of these methods can contribute greatly to mapping of resources and monitoring of change.

Land evaluation

Land evaluation originated from the justified criticism that soil surveys did not give the information needed for development planning. To a soil scientist, to say that 'the dominant soils on the north-eastern plain are ferric lixisols, often plinthic phase, with subordinate gleyic luvisols' conveys a great deal of information in compact form. It does not tell a project manager whether coffee will grow there, or what the yields are likely to be; and, if these are not known, the economic returns from a project cannot be predicted.

It was to provide information of this kind, directly relevant to development planning, that land evaluation was devised. The principles were set out in 1976 in *A framework for land evaluation*,[34] one of the most widely used FAO publications. Detailed guidelines were subsequently prepared on land evaluation methods for rainfed agriculture, forestry, irrigated agriculture, and extensive grazing.[35] The approach has been in regular use for 20 years at international, national, and district scales.

The essence of land evaluation is to match kinds of land use with types of land. It answers questions of two kinds:

1. We have an area of land; what is the best use to which it can be put?
2. We wish to expand a kind of land use; where are the best areas on which to do it?

Questions of the first kind arise when the planning objective is to improve the living standards of a region as a whole, by whatever kinds of development are found to be most appropriate. It can also be used at national-level planning, to identify priority areas for different kinds of development. Questions of the second kind apply where the objective is to find areas for development of specified kinds of land use, such as smallholder tea cultivation, softwood plantations, or fish farms. In these circumstances, the need is to find which areas of land will support such uses, and those where they would be likely to fail.

In view of these objectives, the focus of land evaluation is to identify the suitability of land for specified kinds of land use, each one separately. This is an advance on an older system, called land capability classification, in which there was an in-built order of priorities for use, agriculture taking precedence over forestry, and both over conservation. In land evaluation (distinguished where necessary as land suitability evaluation), nature conservation could be one of the defined kinds of land use, in which case rocky cliffs which provide breeding sites for rare birds could be the most highly valued land. In surveys at a reconnaissance scale, the kinds of land use may be very broadly described, often being simply the growth of a single crop or tree species. At larger scales, the land use is specified in more detail. For example, land suited to maize cultivation by smallholders using animal-drawn machinery, with low-level inputs, would not be coincident with land suited to large-scale mechanized cultivation of maize with high inputs.

The key requirement for land evaluation is to identify and describe the types of land use to be compared.[36] These will begin with the existing uses, the farming systems, extensive grazing, or whatever it may be. Potential uses, those which are plausible objectives of development, are then added. These may be complete changes in major kinds of use, such as forest clearance for agriculture, or introduction of new crops. On the other hand, they can also refer to the same products but with improved management. If the present land use is maize production by traditional methods, with no external inputs, then maize production with specified levels of fertilizer, and perhaps soil conservation measures, would be a second land use type taken as the subject for evaluation. This is the key to the approach of land evaluation: give equal attention to formulating the land use as you give to describing the land.

In idealized form, the evaluation proceeds as in Figure 3, although in practice there is much interaction between the stages. The early procedure, 'diagnosis of land use problems', did not form part of the original method, but was added to reconcile the sequence with the approach of diagnosis and design (p. 75), strengthening the diagnostic element of land evaluation.[37] By supplying information on the existing problems, it helps in the formulation of improved forms of land management.

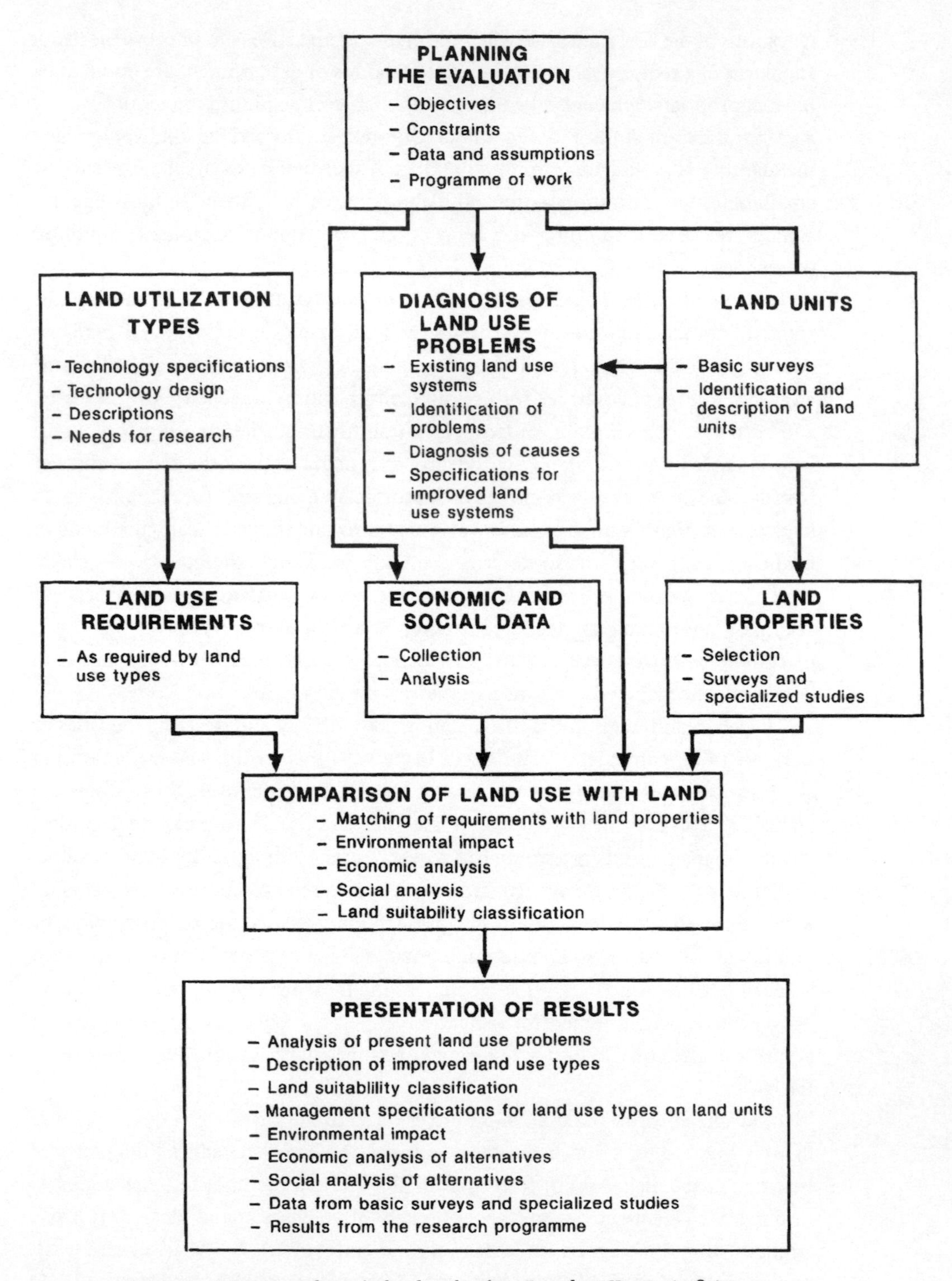

FIGURE 3 Procedures in land evaluation. Based on Young (1985).

Having identified and described the land use types, the next step is to ask what features of the environment are necessary or desirable in order to practise them; these features are called the requirements of the land use. They might include an adequate growing period for the crop concerned, gentle slopes and absence of rocky ground if cultivation is to be by mechanized means, or the presence of mature forest with good timber species if evaluation is for production forestry.

If surveys of land resources have not already been completed, they form an early stage in land evaluation. Advance consideration of probable kinds of land use, however, allows selection of which types of natural resource information are needed and in how much detail. With this information available, the requirements of the land use are then compared with the actual characteristics of the land, not by a static comparison but by a process of successive adjustment known as matching. If, for example, the land use as first defined did not involve use of insecticides, but there is found to be an endemic pest in the area, then the land use type is redefined to include appropriate control measures.

Having matched land use to land, criteria are set up for evaluation. For each kind of land use separately, criteria are established to assess each of the mapped units of land as suitable or not suitable to each defined type of land use. There is usually a further grading into highly, moderately, or marginally suitable. Evaluation includes from the start an assessment of whether there are dangers of land degradation, and thus if the land use is sustainable. The output at this stage, known as physical land suitability evaluation, consists of:

- definitions of land use types;
- land suitability maps, showing for each land use type areas on which it is suitable;
- management measures that will be needed to adapt each type of use to different areas of land, for example conservation practices, cropping sequences, inputs;
- an estimate of the benefits from each use, as production, conservation, or other services.

Although the examples above have been taken from agriculture and forestry, evaluation can be applied to all types of land use, for example land suitability for nature conservation or waste disposal.

In practice, a high proportion of land evaluation studies have ended with evaluation in physical terms. There is a further stage of economic suitability evaluation. This is not a full economic analysis of a project, but a basic input–output analysis at farm-gate level (or forest plantation, etc.). A set of background economic assumptions is first made, including costs of inputs, prices of outputs, and, where necessary, an economic valuation of off-site impacts and intangible benefits. Taking the above estimates of inputs and labour needed, and outputs predicted, standard economic analysis of costs and returns is made. What is distinctive to land evaluation is that this will differ not only between types of use but between the different classes of land on which these are practised. Sloping land, for example, may have been rated suitable for cereal production by defining the land use type to include terracing, but the cost of the latter may prove to be unacceptable. This procedure corresponds with a situation often found in practice, where the objective is to expand the area under a particular use; thus if the objective is sugar cane production, there may be low-lying land which is too wet but which could be made suitable by drainage works, the cost of which will lower the profitability. For forestry plantations, sites where tree growth will be slower, giving a longer rotation between harvests, give lower economic returns.

A new class boundary can be added, separating areas which are suitable, or economically viable, from land which is 'currently not suitable', meaning that on this type of land the use would be possible from a physical point of view – the crops would grow – but is not economically viable at current costs and prices. These areas might become suitable in the future if costs fell or prices rose. There is a computer program, the Automated Land Evaluation System (ALES), which not only produces economic evaluation but shows the physical causes, or land limitations, which have led to differences in profitability between areas of land.[38]

More recently, a system has been proposed specifically for evaluating sustainable land management.[39] In its draft form, this is unduly complex and would be difficult to implement. It is of value for drawing attention to indicators, threshold values, and off-site effects. However, sustainability has been incorporated into land evaluation procedures from their inception; in the original statement in *A framework for land evaluation*, Principle 5 is, 'Suitability refers to use on a sustained basis.'[40]

A large number of physical land evaluation studies have been carried out, at national, district, and local scales; the method has also been applied at international level in assessing population and food requirements (chapter 13). A number of countries possess, or are in course of developing, national evaluation systems, to provide a basis for converting survey results into information on land potential. At national level, land evaluations provide a basis for assessing development priorities, acting as a channel of communication from resource scientists to development agencies. A wide range of evaluations have been completed as a basis of national planning in Kenya, covering crops, soil conservation requirements, livestock, and fuelwood productivity.[41]

At the district or project level, there is a problem over what can be done to implement the results. An evaluation may lead to the recommendation that a particular area is suited to, say, fruit trees, but not suitable for annual crops; but no one is able to compel farmers to grow the former or forbid the latter. Such results can be used, however, in more flexible ways, for example as a guide to locating fruit-marketing facilities. Bangladesh is a densely settled country, and possesses an unusually comprehensive land evaluation, built upon soil survey; when improved crop varieties were introduced, this was employed as a guide to distribution sites for differently adapted varieties, tolerant of salinity, flooding, etc.

Economic land evaluations have been carried out as academic studies, but much less frequently as an aid to practical development planning. The reasons are partly institutional, many evaluations being carried out in soil surveys and other institutions which did not employ economists. A common practice is to present the results of physical evaluation, saying that these can now be taken up by economists, but this does not happen in practice. For a thorough-going economic land evaluation it is necessary to secure from the start the full commitment of economists in a team to its objectives and methods. One means would be to recruit economists into soil survey organizations. Land evaluation provides a means to avoid the 'standard package' approach to development planning, by demonstrating how local variations in environmental conditions influence land use inputs, outputs, and hence profitability.

Making use of survey results

Natural resource surveys and their interpretation, as land evaluations, are carried out for a purpose: to help people to make the best possible use of their land. They help to avoid costly planning mistakes, and to ensure that scarce funds are spent efficiently. Three questions may be asked:

Is there an adequate range of methods, scientifically sound and relevant to development needs?

Have surveys been carried out in sufficient number and at appropriate scales?

Have the results been properly applied?

With one exception, there is no doubt that the answer to the first question is yes. Valid methods exist for surveys of climate and water resources, soils and landscapes, forests and rangelands, and for interpreting the results of these surveys in terms relevant to land resource management and development. The exception is that surveys have taken too static a view of resources, whereas, in the light of land degradation, there is now a need to find ways of measuring change. Methods for soil monitoring, measuring changes in soil properties in response to management, need to be developed further.

It is possible to view the second question, whether enough surveys have been completed, in two ways. On the one hand, knowledge of land resources has advanced enormously over the last 50 years. For most countries, there is now some systematic knowledge of the distribution and potential of the soil, water, and forests resources; some have systematic coverage at reconnaissance scales, in others, knowledge is based on collation of local surveys made for specific purposes. Most land development projects begin with reports on land resources, in some cases running to several volumes. An outstanding example is the collection of Land Resource Studies produced by the UK Natural Resources Institute, not a uniform set but one which includes forest inventories, pasture resource surveys, land evaluations, and studies which integrate economics with resource surveys, all produced in response to development needs.[42]

But, conversely, there remain large deficiencies in the adequacy of information. Many countries still lack a basic national reconnaissance survey on a land systems or similar basis. This is primarily due to the inadequate funding of national resource survey organizations, dependent in turn on inadequate recognition of the functions they serve. Much progress has been made in the education of scientists from developing countries, but they are not given adequate working budgets. Unfortunately, it has become out of fashion to conduct resource surveys except where there is an immediate development purpose, and even then to proceed if possible on the basis of existing information. This attitude leads to development investment being founded on land resource information which may be unreliable, out of date and at an inappropriate scale.

The general answer to the third question, whether the results of surveys have been put to proper use, must be no.[43] Surveys are put to use in the initial stages of planning, such as in deciding the limits of a development project, and the broad layout of land settlement schemes; many irrigation schemes are planned by comparing water resource potential with land suited to irrigation. But the use made of natural resource information is often confined to the early stages: 'The soil surveyors tell us where the hills and swamps are', as a project manager once expressed it. At the later stages of planning and during implementation, information from land resource surveys does not usually play a large role. The post-investment activity of monitoring and evaluation is conducted mainly in economic terms, effects on the land being observed only indirectly through crop yields.

Three reasons can be given for this situation:

failure to collect the information most needed;

a failure of communication by natural scientists, in not translating the implications of resource aspects into terms relevant to development and land management;

insufficient links between resource surveys and the approach of participation by people.

The last reason, neglect of the human dimensions of land development, is discussed in chapter 5. The first originates in the terms of reference of surveys. These should include a clear statement of purpose: what are the objectives, and what information is needed to meet these. A starting-point is to ask critical questions: what are the land resources of the area, how are they currently being used, is this use sustainable and, if not, how can it be modified?

The failure of communication is fundamental. The natural scientists have not succeeded in conveying to planners and decision-makers the full implications of the information they have acquired. They should participate throughout the process of development, including in the important stage of modifying plans in the light of ongoing experience. The forward translation of knowledge about land resources involves far more than simply converting results into economic terms, although this is an important element. The combination of efficient production with resource conservation, or sustainable land use, is so central to land development that it should permeate the whole fabric of decision-making.[44]

There are educational and institutional causes for this lack of communication. On the one hand, natural scientists have not been taught how to work out the implications of their findings for planning and management; along with their specialist training, they should receive education in the wider context of planning and development. Conversely, economists, planners, and decision-makers are not sufficiently aware of the usefulness and relevance of natural resource information. 'A range of initiatives is needed to develop land literacy and awareness of natural resources, the consequences of their mismanagement, and the value of natural resources information . . . amongst policy-makers and task managers in both governments and aid agencies.'[45] The situation is much improved where the decision-making team includes resource scientists. Institutionally, communication would be improved if resource survey organizations included agronomists and economists on their staff, charged with linking information to land management and development.

4 Competition for land

In addition to its main function for agricultural and forestry production, land is needed for many other purposes, including regulation of the atmospheric and water cycles, mineral supply, nature conservation, settlement, and waste disposal. There is often conflict between competing uses. Some major issues in policy and planning centre upon land use: the supply of arable land, loss of forest cover, and land needed to meet the urban expansion which is a marked feature of the developing world. Land use, the functions which it performs, is distinct from land cover, by plants or built structures. There is a need for common classifications of land use and cover, as a reference base for conversion between national systems. Means are available to survey land use, but in many less developed countries scarcely any reliable surveys have been carried out. As a result, data on land use are among the least reliable of all kinds of international statistics. Until information on land use is improved, development planning will remain without an adequate basis on many major issues. International conventions, such as the World Soil Charter, are of value mainly as a guide to policies at national level. All countries need a national land use policy, under the guidance of a committee at ministerial level, to co-ordinate and reconcile the many sectoral interests.

Land use is a basic element in human activity. Most things that we want to do – grow food, manufacture goods, travel, live in houses, take recreation, preserve nature – requires land on which to do it, often land of a certain kind or in a particular place. For the most part, the same piece of land cannot be used for more than one purpose at the same time, a wheat field as a forest, for example, or either of these as a motorway, added to which, land is almost universally in short supply. As a result, competition between different kinds of activity is the theme which runs through any consideration of land use.

The use of land, and government policy and action in controlling or directing this use, is fundamental to planning at national, regional, and local levels. Any form of development planning is going to require land. In the 1950s and 1960s, there was sometimes land which was (or was thought to be) 'spare', in formal terms classed as not used. This is now rarely the case. Even where land is neither settled nor under any kind of productive use, it is serving essential functions such as water supply or nature conservation. Development therefore calls for land conversion, or change in use; a gain to one form of activity is a loss to another, in a zero sum game.

Three elements of land use are of key importance at all scales, whether for international policy, national planning, or for purposes of a development project:

Arable land How much land is cultivated? Most basic foodstuffs are produced from arable land. We need to know the area presently under cultivation to compare it with the potentially cultivable land, in order to assess what potential remains for expansion.

Forest cover The area under forest is of much international concern for environmental reasons. At country level, a continued decline in forest area will inevitably lead to shortage of wood products and non-sustainable overcutting.

Urban land The trend towards urbanization in the developing world is strong and will continue. Much of the expansion of the built-up area unfortunately takes place on prime-quality agricultural land. A balance must be struck between conflicting interests.

Besides its major role in planning, land use is relevant in two other respects. First, the state of world land use, particularly but not only forest cover, is a matter of concern for environmental issues: regulation of atmospheric carbon, and preservation of nature and genetic resources. Secondly, in assessing potential for food production, estimates of the land requirements for other kinds of use are needed.

The state of knowledge about land use, and the degree of government control over it, differs greatly between developed and less-developed countries. In the developed world there is generally good statistical information on the use of land. There is intense competition between uses, and planning controls over changes are, by and large, enforced. This applies not only to the densely settled regions, such as Europe and eastern United States, but also to sparsely populated areas such as much of Canada and Australia, on which proposals to change the type of land use must meet legislative controls and requirements. Even completely empty regions may be legally designated wilderness sites.

In the less-developed world, by contrast, information about land use is often extremely poor. Empty land, not under any kind of use, existed formerly but is rapidly being reduced, and now mostly confined to arid regions and steeplands. Competition between uses is growing. Legal constraints, such as forest reserves and urban planning restrictions, exist but are poorly enforced. One of the least enforced of land use controls is found in countries which legally forbid cultivation of land of more that a certain steepness.

Functions of land

In considering land use we think first of agriculture and forestry, together with urban settlement. But there is a wider range of functions which land offers to human society. Putting the production first, these functions are as follows:[1]

Production based on plant growth: production of food, animal fodder, fibres, timber, and fuelwood, by means of agriculture, forestry, and freshwater fisheries.

Regulation of the atmospheric and hydrological cycles.

Conservation of biodiversity and habitats: ecosystems, plant and animal species, genetic resources.

Storage and ongoing supply of non-renewable resources: fuels, minerals, and non-biotic raw materials.

Functions related to human settlement: housing, industry, transport, recreation.

Waste disposal: receiving, filtering, and transforming the waste products of settlement.

The heritage function: preserving natural sites of interest and beauty, and evidence of cultural history.

Some of these functions may appear to be mutually exclusive, but even in apparently single-purpose uses there is a degree of multiple functions: cereal crop production contributes to atmospheric and hydrological regulation, urban areas include appreciable areas of trees, grass, and crop or animal production. In some cases there are more substantial overlaps. It is now recognized that agricultural land has an important role in wildlife conservation. The heritage of ancient and medieval Rome coexists with the functions of the modern city. The former concept of forests being managed exclusively for wood production, with 'keep out' regulations to outsiders, has largely been replaced by systems of multiple-use forest management.

There is another 'use' of land, if it can be so called: its value for investment. In countries with unstable financial institutions and an ever-present danger of inflation, land gets bought as a store of wealth and speculative gain. It may simply be held unused, awaiting the arrival of the housing estate or office block, or let out to tenant farmers, one of whose duties is to clear the forest and so increase the value. On balance, one must hope that, after a period of being held idle, the value of such land to society is increased.

Information on land use and cover

Classification: towards a common reference base

To an extent which, viewed in retrospect, is remarkable, methods for the collection and analysis of land use data have lagged behind those for natural resource survey.

Any planning or development activity which tried to compare the potential of land with the use to which it was put has had to operate from an unbalanced basis: on the one hand, there were surveys of soils and other natural resources which were conducted on an internationally agreed basis, but, on the other, there were no standardized systems for recording land use. A prerequisite for mapping and statistics is an agreed classification system, but, in the early 1990s, the situation with respect to land use classification was comparable with that for soils in about 1950: a large number of systems devised for national use, with no guidelines for comparison.

This situation has arisen at a time when requirements for comparable and reliable land use data are increasing. At international level, information on agricultural land use is co-ordinated through the World Agricultural Census, not in fact a single operation but an attempt to provide standardized guidelines for national censuses. Forest cover is surveyed at 10–year intervals at the beginning of each decade (1990, 2000, etc.) through the Forest Resources Assessments.[2] The UN Earthwatch programme, largely operated through UNEP, is concerned with the effects of world land use changes on climate and atmospheric pollution, and the scientific community is conducting a research programme into the consequences of land use changes through the International Geosphere-Biosphere Programme (IGBP). At national level, many countries are now seeking to monitor land use change as a basis for policy guidelines and action.

The ideal would be a standardized classification system agreed world-wide, with multiple levels adaptable to surveys at different scales. It is most unlikely that this can be achieved. Instead, efforts are being made (co-ordinated by FAO and UNEP) in three directions:[3]

- a common classification system to serve as a basis of reference (comparable with the World Reference Base for Soil Resources);
- 'translation systems', to permit conversion of data between one national system and another;
- the description of land use by a set of standardized multiple attributes (comparable with soil properties), so that anyone who wants can place a site observation within their own, purpose-built, classification.

There is a fundamental distinction between land use and land cover. Land use refers to the human activities which are directly related to land, making use of its resources and having an impact upon it. Crop production, a pulpwood plantation, and residential settlement are forms of land use. Land cover refers to the vegetation (natural or planted) or human constructions (buildings, etc.) that cover the earth's surface, for example, forest, grassland, agricultural crops and houses. Land cover is a simple observational fact of what is there. Its causes are, first, by the natural vegetation, and secondly, by the human actions that have been applied to the area, such as vegetation clearance, ploughing, planting, or building.

Confusion has arisen because, in certain cases, use and cover are closely equivalent; the land use 'production from perennial crops' corresponds largely with a land cover of perennial crops (but this use includes areas in rotation which have a different cover); industrial production usually means a cover of industrial-type buildings (but production can also be from cottage industry). In other cases, there is no such correspondence. A land cover of grassland can be under grazing or recreational use, whilst a cover of forest may be for wood production, catchment protection, nature conservation, recreation, or a combination of these. In principle, land use and land cover are quite distinct, and should be kept separate in classification, mapping, and for purposes of analysis. In practice, cover is widely employed to diagnose use, whether from ground observation or aerial survey. One can often guess the uses of areas of grassland and built-up sites, but with forest this is not easy. For the tropics, the 1990 Forest Resources Assessment was concerned only with forest cover. For Europe, together with the USA, an attempt was made to identify land use as well, by asking forestry departments to grade forests according to the relative importance of seven functions: wood production, protection, water, grazing, hunting, nature conservation, and recreation. Wood production proved to be the most important, followed in the USA by protection, but in Europe by hunting (for recreation, not survival).[4]

One of the best of the earlier classification systems, devised for the USA, explicitly mixed use and cover, for the practical reason that it was intended for application to data obtained from remote sensing.[5] Statistical data obtained from ground censuses, such as questionnaire returns by farmers, refer mostly to use. There is now a strong pressure towards making maximum use of cover, because of the directness with which it can be obtained from satellite imagery; for example, the European Community survey, CORINE, is based upon land cover. A need still remains for data on land use, however, for all purposes related to comparison of benefits and competition between uses.

There is a further distinction between functional and biophysical land use. Functional land use is the purpose for which the land is used, or the benefits obtained from it; these benefits may be products (e.g. crops, wood) or services (e.g. conservation, recreation). Biophysical land use refers to the sequence of operations carried out on the land in order to obtain benefits, operations such as vegetation clearance, ploughing, grazing, application of fertilizers, or erecting buildings. Functional land use is the focus of interest when discussing the outputs obtained, whether, for example, a national park or forest reserve should be turned over to agricultural production. Biophysical land use is particularly relevant in analysis of the impact of use upon land, for example, whether land degradation is occurring or

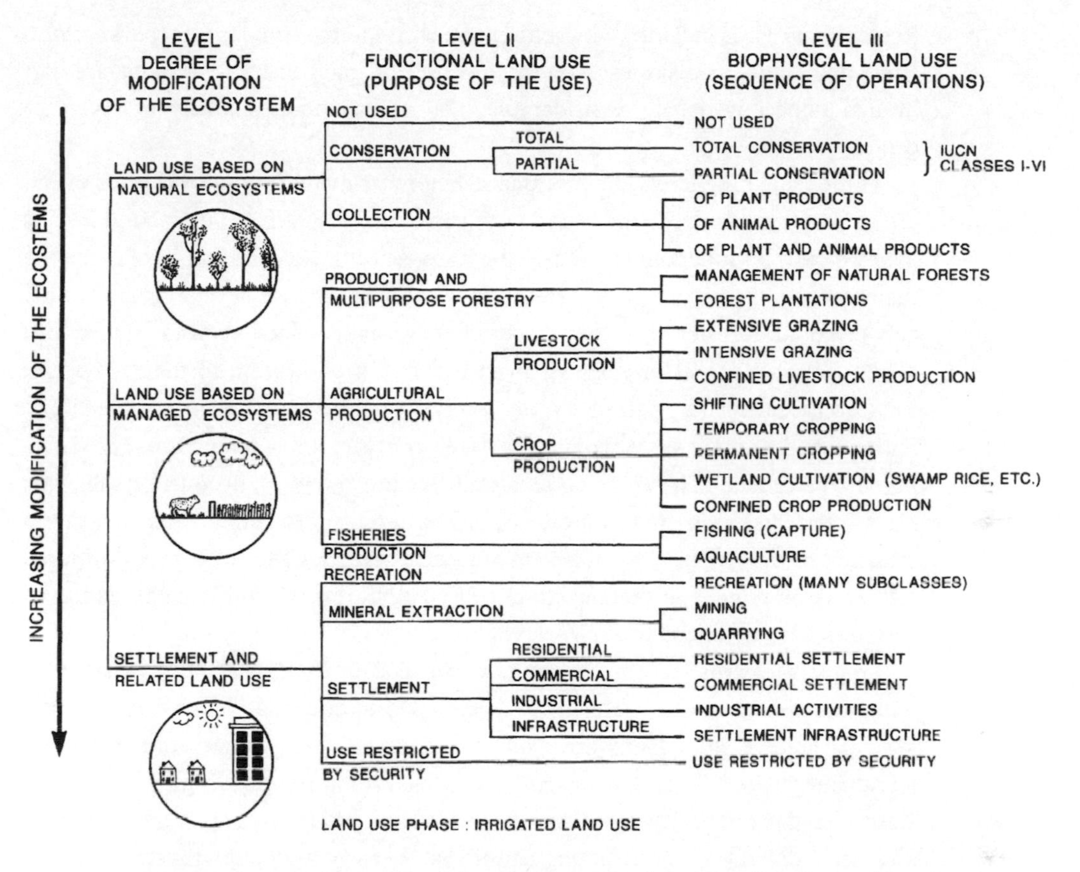

FIGURE 4 An international classification of land use. Based on Young (1994c).

likely to occur. Both aspects need to be included in a classification intended for a wide range of purposes.

Figure 4 gives an international classification of land use. It commences with 'not used', a class that is present at the limiting case in principle although rarely, nowadays, in practice. There is an overall gradation, from top to bottom of the table, based on the degree of modification of the natural ecosystem. The three groups at the highest level are distinguished by this feature:

natural ecosystems, rarely untouched but not planted;

managed ecosystems, based on plant production but with the plants either replaced, as in agriculture and plantation forestry, or recurrently altered, as in grazing and production from natural forests;

settlement, in which buildings replace a plant cover over substantial parts of the land.

Recreational use is included with settlement, even though some forms (parks, camp sites, golf course, ski slopes, etc.) have a cover of grass or trees; reasons are that human impact is usually considerable, and such land is unlikely to revert to productive uses.

At the remaining levels, a conscious attempt was made to employ classes which are generally recognized, and found widely in existing systems. The second level is based mainly on functional land use, the purpose of the use. At this level come the major groupings employed at international and other high-level classifications: conservation, forestry, livestock production, crop production, and settlement. Where possible the listings are again in order of modification of the ecosystem: total conservation (e.g. strict nature reserves) before partial conservation (e.g. national parks), management of natural forestry before forest plantations, extensive grazing before intensive (grazing plus stall-feeding) livestock production. The last class listed, 'use restricted by security', is a euphemism for military uses, weapon sites and the like; these are unlikely to appear in statistics and were once non-land, left blank on maps and literally cut out of air photographs, but the spy nature of satellites has put an end to such secrecy.

Progress towards securing agreement on a classification is bound to be slow. As with the case of soils, countries with effective national land use classification systems neither should nor are likely to abandon them. However, the present *laissez-faire* position is far from satisfactory; it is one cause of the highly inadequate state of land use data. The two elements needed are, first, a common reference base, and, secondly, rules for translating national and local classification systems into it.

Survey of land use and cover

Information on land use and cover can be obtained by four methods:

Census data Data are based on questionnaires filled in by farmers, other land users, or enumerators. Coverage may be total, or by means of an area sampling frame. This permits the direct survey of land use, both functional and biophysical. It supplies statistical data but not detailed maps. Examples are national agricultural censuses.

Remote sensing Data are obtained from satellite or aerial imagery, with control by ground observation of sample areas. This is primarily a source of information on land cover, with land use being inferred. It supplies maps but not, directly, statistical data.

Ground observation Data are directly observed on the ground, field by field. Observation can be both of land use and cover.

Administrative data This method involves the processing of information on areas of land assigned to specific purposes, such as legally designated conservation areas, forest reserves, or areas licensed for the growing of specific crops. A problem is that the designated function may not correspond with the actual use, for example through illegal agricultural incursion into designated forest areas.

Census data in developed countries are often based on annual returns filled in by farmers. These are generally assumed to be fairly reliable, although it is recognized that farmers regard their completion as something of a chore, and may falsify information to obtain subsidies. In less-developed countries this work is carried out by government enumerators who, it is to be hoped, visit the area, sit down with local elders under a shady tree, and decide what figures will, on the one hand, be regarded as plausible, and, on the other, be in the interests of the village.

Remote sensing initially produces data on land cover in the form of maps; this can be converted by overlay with the boundaries of administrative units and area measurement. Fortunately, three of the elements of greatest interest, forest cover, arable land, and built-up areas, can be directly identified. Rice fields, olive groves and some other individual crops can be fairly reliably identified. Identification can either be by spectral signatures, training a computer to recognize patterns of colour and density, or 'eyeballing', doing the same thing yourself; which of these is preferable is a point of dispute among specialists. The availability of images at a wide range of scales is a further advantage. It is essential that image interpretation should be systematically and regularly checked against field observation of sample sites.

Remote sensing can be a powerful method for monitoring changes in land cover and use. An example is a land cover atlas of Cambodia which compared Landsat images from two periods, 1985–7 and 1992–3.[6] Changes were obtained by overlaying cover at the two periods, using a geographical information system, in this case showing primarily a decrease in forest and an increase in rice cultivation. What is so informative is that land use changes were obtained first as statistical data by provinces, and, secondly, in map form showing the actual sites of conversion. This is the primary data source used by CORINE, the land cover survey system of the European Community.[7]

It has occurred to many research organizations that monitoring of land cover change is possible for sub-continental regions or, in principle, for the whole world, and a number of exercises are in progress to do this, for example in eastern Africa and the Amazon Basin. The effort involved is immense and the cost considerable, bearing in mind that regular and repeated ground checks, visiting sites in the field, are essential. Remote sensing has been the primary source for data on deforestation. It is to be hoped that a comparable effort will be put into monitoring changes in urban and agricultural land.

Ground observation, actually going to the site and mapping the land use field by field or other mapped boundary, is incomparably the most reliable method of mapping land use. It was the method used in the first land utilization surveys of the Britain, conducted in the mid-1930s on a minuscule budget using voluntary staff. This has since been repeated twice, once with complete mapping coverage again and currently on a kilometre-grid sampling basis. A remarkable exercise is carried out in France. Many years ago a statistically controlled network of points, one-metre squares, was established by a combination of map grid and sample air photographs. Each year, local officials are required to visit each point and record the use. Of course, in a large proportion of cases this has never varied, but on a national basis it records changes such as growth of towns and clearance of olive groves. There is less scope for ground observation in developing countries, because the map base may be poor, there are fewer reference lines such as fixed field boundaries, the areas to be covered are large, and the patterns of use may be highly complex with uses covering less than one hectare. Field transects provide a possible means. As geographers have long maintained, there is nothing to match field observation as a way of finding out what is actually going on.

In practice, census-type statistical methods and remote sensing are the two principal sources of data, each with different strengths and weaknesses. Data of these two kinds can be made compatible where:

- both types of survey collect data on the basis of common classification systems;
- census data are collected for geo-referenced areas (i.e. farm returns state within which administrative units – or better, grid squares – they lie);
- in remote sensing surveys, the boundaries of administrative units are transferred to maps, and quantitative data abstracted for these.

Integration of this kind has been achieved in Canada, through the use of geographical information systems. The practical problems of achieving it in developing countries are considerable.

A principle in land use survey is that reliability is best achieved by combining data from two or more sources. Remote sensing has great advantages of speed and cost, and, as would be expected, remote sensing scientists themselves strongly promote its use. Their arguments are largely valid, provided that the need for substantial and repeated comparison with sample field survey is actually implemented and not merely given lip service.

Survey of land use and cover, repeated at intervals to monitor change, should become a fundamental activity of national resource survey and land use planning organizations. The effort that was once applied to geological and soil surveys ought to be directed towards this focal piece of information, how the land of a country is being used. Of all the deficiencies in knowledge in the area of land resources this, as will be seen, is one of the greatest.

Data on land use

Information on land use in developing countries is probably less adequate, and certainly less reliable, than for any other kind of statistics on the environment. This is the more regrettable in that such information is widely used in analysis, for policy, research, and planning. The inconsistencies of data, revealing how little is known of the true position, are so great that at times it is hard to resist being satirical.[8] The examples given refer mainly to data for major land use types at national level, but there is no reason to suppose that it is any better at other scales.

Each year, Table 1 in the FAO *Production yearbook* gives the latest country statistics for what is called in the table itself 'Land use', but in recent years has been referred to in the notes as 'Land use (land cover)'. These are the data largely used as the basis for the table on land area and use in the biennial *World resources*, an influential and authoritative source, being a joint publication of the World Resources Institute, UNEP, UNDP, and the World Bank. These two sources, the one direct, the other derived, are in turn very widely quoted in other tables of statistical data. It terms of order of listing, land use is generally treated as the second most fundamental set of statistics, following population. These values are then very widely reproduced in other data sets, and used as a basis for discussions of the most fundamental kind, including of world food supplies and deforestation. It is safe to say that these national land use figures are used very much more often than, say, statistics on world soil types. They are all obtained from this one basic source, FAO's 'attempt to bring together all available data on land use (land cover) . . . throughout the world'.[9]

FAO are at pains to point out the difficulties under which they operate. With some exceptions (notably the forest resources assessment) they do not themselves carry out data collection, but are dependent on what governments supply; even where they believe such data to be wrong, it is difficult for them not to print it owing to their official status as the servant of member governments. 'It should be borne in mind that definitions used by reporting countries vary considerably and items classified under the same category often relate to greatly differing kinds of land' states FAO, whilst *World resources* comments that, 'FAO often adjusts the definitions of land use categories and sometimes substantially revises earlier data . . . [consequently] apparent trends should be interpreted with caution.' They should, indeed.

Table 1. *Land use statistics: a comparison between 1992 and 1993 data*

	Area, thousand hectares			
	Cropland		Forest	
	1992	1993	1992	1993
Kenya	2450	4520	2300	16800
Madagascar	3105	3105	15450	23200
Somalia	1038	1020	9040	23200
Togo	669	2430	1449	900
Zimbabwe	2846	2876	19000	8800
Brazil	59000	48955	488000	488000
Haiti	910	910	35	140
Jamaica	219	219	184	185
Peru	3730	3430	68000	84800
Lebanon	306	306	80	80
Laos	805	805	12500	12500

Source: FAO Production Yearbooks.

The classes given in the table are:

Arable land Land under annual crops; includes rotational grassland, but (nowadays) excludes fallow following shifting cultivation.

Land under permanent crops Includes crops such as coffee, rubber, fruit trees, and vines, but excludes trees grown for timber.

Arable and permanent crops The sum of the two above classes.

Permanent meadows and pastures Includes both improved pastures and semi-natural pastures under extensive grazing.

Forests and woodland For the most part follows the foresters' definition, land with more than 10% cover of trees and shrubs, though some countries use 20%.

Other land All land not listed above. This is obtained purely by subtracting the sum of the above classes from the total land area. It thus includes both towns and barren wastelands.

In the following comments, 'arable and permanent crops' is abbreviated to cropland, and 'forest and woodland' to forest.

Let us consider data for cropland and forest for two successive years, 1992 and 1993, obtained directly from the primary source (Table 1). The sample of countries shown is selective, although it by no means exhausts cases of sudden change. In Kenya, cropland almost doubles between these years, whilst the forest area increases by 630%. From a forestry point of view, 'the good news' is that the forest area is larger for Madagascar, Somalia, Haiti and Peru; 'the bad news' is that it has fallen in Togo and dropped to less than half in Zimbabwe. Cropland has fallen in Brazil, but risen by a spectacular 260% in Togo. In Jamaica, by contrast, figures for the two years listed are entirely consistent; the 1991 value, however, had been 24% higher. In Laos and Lebanon, a no-change situation is reported by FAO. *World resources*, however, reports the average forest cover for 1991–3, in thousand hectares, as 703 for Laos and 2819 for Lebanon, respectively 18 times lower and 35 times higher.

Does this mean that drastic changes have taken place in these countries within a year? It is true that some of the countries listed have experienced internal political strife, but that is not the reason for the large and abrupt alterations. They are brought about by revised estimates, which by coincidence were applied to both types of data in the same year. The study, *World agriculture: towards 2010* needed estimates of present cropland as a basis for assessing the potential for its future expansion. An inconsistency was detected, by summing the reported harvested areas of individual crops and comparing these with the reported totals for cropland (they are well aware of the practice of multiple cropping and no doubt made allowance for it). Present arable land for the base period 1988/90 is therefore reported as the original data and 'adjusted', for example:

	Cropland, 1988/90, million hectares	
	Original	Adjusted
Sub-Saharan Africa	140	212
Latin America and Caribbean	150	189

For 42 of the 91 countries in the study, the adjustments exceeded 20%. Truly, 'existing historical data . . . do not constitute a sufficiently reliable basis for analysing the historical evolution of this variable.'[10] Results from the 1990 Forest Resources Assessment were finalized by 1992–3, and are the reason for the 'adjustments' to forest area.

These data are available on computer diskette, showing land use for every year from 1961 to the present.[11] Can we therefore expect abrupt discontinuities in the statistical series? Inspection shows that this is not the case; most values either remain totally constant year after year, or change gradually. Furthermore, the *Production yearbook* gives data not only for the current year but 5 and 10 years previously. Taking some of the countries in Table 1, we can compare estimates of the same type of land use for the same year, 1988, the one taken from the yearbook when they first became available, that for 1989, the other being the historical data given in the yearbook for 1994. Some selected comparisons are:

	Land use, 1988, thousand hectares	
	As given in FAO Production Yearbook for	
	1989	1994
Kenya, cropland	2 425	4 495
Madagascar, forest	14 580	23 200
Somalia, forest	8 750	15 000
Togo, cropland	1 439	2 360
Jamaica, cropland	269	220
Brazil, forest	78 550	55 000
Brazil, permanent crops	12 050	8 000

In the interests of statistical continuity therefore, or, as some would say, in the light of improved information, the historical data are rewritten, truly a George Orwellian '1984' situation. A statistic drawn from this source must therefore be quoted in the form, 'The cropland area for 1980, *as estimated in* 1995 . . .'

The position is quite different with respect to pasture. Consider, for example, the pastures of Mexico. In 1949, these were given as an honest approximation of 100 Mha, continuing until in 1954 a figure of 67 296 000 ha was obtained from the 1950 census of agriculture. The latter area remained unchanged every year until 1961, when it inexplicably rose to 75 156 000 ha *based on the same* 1950 *source*. In 1978 the estimate became 74 499 000 ha; this has proved to be a landmark figure, for it has been quoted every year since (except 1985, when someone had the statistical decency to round it up to 74 500 000, reverting to pseudo-exactitude the following year). Furthermore, the 74 499 000 hectare figure is now quoted as the estimate for every single year back to 1961, showing no change in area for over 30 years. These pastures are by now widely and severely degraded, but there is clearly not the slightest reliable knowledge of area on which to base any kind of analysis of livestock densities and fodder production.

This is not an isolated example. Data for percentage change over the past 10 or more years show that about a quarter of developing countries simply never change the figures reported. National organizations are not interested in the areas under grazing, and one finds also that international grassland management experts have a low regard for the value of pasture surveys.

One can hardly refer to this situation as one of data inaccuracies; it is more nearly a case of absence of knowledge, in an area where good data are essential to forward planning. Uncertainties of the order of 50–100% plus or minus would never be tolerated in economic or social data. Here it is easier to cite examples at international level, one of which, the FAO study of world agriculture to the year 2010, has already been noted. Another example can be taken from research. A population specialist came across the computerized version of FAO land use statistics and must have been excited by the prospect of finding correlations between population changes and land use. Using 180 countries, he tried plotting annual percentage change in population against annual change in arable land. The points on the graph are scattered almost completely at random. The same was found when change in arable land was plotted against either population density or gross national product, and when change in forests was compared with population increase. He concludes that, 'We can find scant first-order correlation between demographic variables and FAO's land-use categories . . . Admittedly, this . . . does not prove that there is independence between population and land-use change. *Flaws in the FAO data might explain some of the results*'.[12]

Data founded on such uncertainty are taken as the basis for analysis at the highest levels. For example, the technical background documents of the 1996 World Food Summit quotes two figures for the arable area of developing countries, 669 and 757 Mha, before and after 'adjustment to correct inconsistencies of data'. It gives a graph of 1970–1990 changes in percentage of land under permanent pasture by agro-ecological zone which, not surprisingly, shows very little change in any zone.[13]

INFORMATION ON LAND USE AND COVER

Extent and reliability of mapping and statistical data for less developed countries

Land use Poor. The widely used national data are highly generalized and of doubtful accuracy. Lack of standardized definitions and methods of mapping. Data on irrigation relatively good, on urban land inadequate.

Land cover Fragmentary but improving. Remote sensing provides methods for survey and monitoring at all scales. Examples show that these methods are practicable, and more surveys are being undertaken.

Questionable reliability for information on three aspects of major importance: arable land, forest cover, and settlements.

These data problems are serious as a basis for analysis at international level. However, the most important need, as has been emphasized repeatedly in other contexts, is for policy and planning at national level: in plain language, how will a country meet its future needs for cropland, grazing land, and productive forest, whilst protecting fragile environments and conserving biodiversity. National resource survey organizations should be systematically monitoring land use, analyzing the results, and reporting the present state and future consequences to policy makes at the highest level. Developed countries recognized the need for land use data from about the 1940s or 1950s. For most developing countries at present, policy and planning rest on a situation of severely deficient knowledge. As expressed in the title of a World Resources Institute review of data, we are walking '*Eyeless in Gaia*'.[14] (See note on land use statistics, p. 301.)

Urbanization and urban land use

In about the year 2000, a watershed in the evolution of human society will be reached. Throughout all human history up to then, more than half the population has been rural, that is, has lived mainly in villages and small towns, a few in isolated farms. From 2000 onwards into the foreseeable future, more than half of us will be urban, will live in cities and large towns. The UN data will show this happening around 2005, but this is an accident of the definition of what constitutes 'urban', which varies between one country and another. Future generations will link this milestone in history with the new millennium.[15]

One third of the population lived in cities as recently as 1975, and it is fairly likely that two thirds will do so by 2025. Most of the Western world is already about 80% urban, 75% for Europe, slightly higher for Canada and the United States, and over 85% for Australia and Argentina. These proportions are changing only slowly. It is in the developing world that the transformation is occurring. Africa and Asia have both moved from 25% urban in 1975 to 35% in 1995, and there is every likelihood that a growth rate of nearly 1% per year will take the urban population past 50% within a generation. Most countries of Central and South America have already passed that crossover point. Much of this urbanization takes place into cities of 1 million or more population, and more than half the 18 agglomerations of over 10 million are in the developing world, including Mexico City, Bombay, Shanghai, Beijing, Calcutta, Seoul, Jakarta, Lagos, Delhi, and Karachi.

Third-world urbanization creates formidable environmental problems of water supply, health, pollution, waste disposal, travel, congestion, and crime, which lie outside the scope of the present discussion.[16] To the eyes of Westerners, life in shanty towns appears unattractive and frightening, and if they had to be poor they would rather live in villages. That this is not the view of developing country citizens, taking economic realities into account, is evident from the way they vote with their feet. There is some evidence, however, of an element of return migration to rural homelands.[17] As there is an in-built urban bias, the problems of urban areas take up well over half the budgets of most developing countries, and a substantial proportion of foreign aid.

In the context of national land use, the focus of interest is on the land taken up by urban settlements. In this respect, the paucity of data is quite astonishing. One might have expected 'Settlements' to be a class in the long-term statistical series of national land use, but instead these are combined with mountains and deserts as the infamous 'Other land'. FAO staff give absence of data as the reason. Nor can tables of urban land areas be found amid otherwise comprehensive summaries by international organizations, or academic studies.[18] This is the more surprising since settlement, that is, areas covered by built structures, shows up so clearly on false-colour satellite imagery.

Some 40 years ago, Western countries became scared that uncontrolled urban expansion would take up 'good agricultural land', not only as housing and industry, but as transport structures, especially wide modern roads. Studies were made, with results that were broadly reassuring. Given the slow pace of urbanization, in places static or negative, and the setting up of planning controls over building, it is generally considered that this threat is being held in check. In the UK, for example, buildings cover 10% of the surface and communications a further 3%. The proportion of the best grades of agricultural land is much higher since, by their nature, towns and roads occupy such land disproportionately. London's Heathrow Airport occupies superb soils developed from Thames river terraces, formerly under market gardening. Similarly in the US, urban expansion is strongly biased towards prime agricultural land.[19] So strong are the economic and social pressures that it is only with difficulty that planning controls can limit such encroachment.

Current estimates of built-up land in developing countries have been made indirectly, by seeking a regionally typical value for land per person and multiplying by the population. There is a distinction of environmental significance between built-up areas as a whole, which include parks, gardens, sports fields, etc., and that actually covered by bricks and concrete. An estimate using both Western and third-world cities gave 250 m^2 per capita, or 40 people per hectare, for built-up areas (of which 25 m^2 per capita is covered by structures).[20] Multiplied by six billion people this is 150 Mha, just over 1% of the world's land area – but equivalent to 10% of its present cropland, on which most urban areas are built.

The study of *World agriculture: towards 2010*[21] needed a value for urban land in developing countries in order to know how much of the potential cultivable land this would take up. Their analysis begins with a statement which sums up the state of incertitude, 'As far as some speculative estimates could be made, perhaps some . . .'! They estimate that 94 Mha of land are presently occupied by settlements in developing countries (excluding China), or 30 people per hectare. Future increase is assumed to be at a higher density, 50 people per hectare. Of the present area, 50 Mha is probably included in the 'land balance', the area they believe to be cultivable but not yet cultivated, and this would occupy only 4% of such land by 2010.

One does not need to think of many cities of the developing world, Delhi and Bangkok for example, to realize that these, too, are largely built on flood-plain, river terrace, or other gently sloping land which would otherwise have a high crop-production potential. It is true that some suburbs, both shanty towns and high-class residential areas, may clamber their way up surrounding hillsides (e.g. Mexico City), but this is the exception. Some occupy the best agricultural land in the country; Lilongwe, the new capital city of Malawi, is built on the best soil in the country.[22] There are good reasons, historical and contemporary, why they should do so, such as sites on river crossings or estuaries in the most developed part of a country. All the urban development in Jamaica is taking place on the narrow coastal fringes which form the only non-steepland areas of the country.

Two factors will lead to further urban encroachment. The first is the difficulty of enforcing effective planning controls amid the corruption which affects many city and local authority governments. The second is the very real need for better housing, to replace the high-density slums in which new settlement is first concentrated, as well as for industrial development and better transport facilities. The price of land developed for urban purposes is so much higher than its sale price for agriculture (although this is not thıe economic value, see chapter 10) that there is bound to be a continuing loss of good agricultural land to settlement.

Urban interests can be compatible, but are often conflicting, with those of rural areas.[23] Urban populations want cheap food, farmers a high price for their crops. Urban areas need water, leading to flooding of valley land by reservoirs. The urban demand for fuelwood can easily lead, through market forces, to overcutting of woodlands. Cities and industrial activities produce wastes, potentially polluting the atmosphere and rivers, and taking land for solid waste disposal. These are in addition to straight competition for flat, fertile, and therefore easily built-upon land.

The transformation of the earth's surface

In writing of the magnitude of change in global land use we meet the dilemma that will be encountered more than once in this book: on the one hand, to maintain that the necessary data are highly incomplete and inaccurate, and on the other, to assert that considerable changes have occurred. One can only justify this two-faced attitude by saying that first, there is some plausibility in the broad magnitude of the statistics, and secondly, the trends described are supported by what you see as you travel around. In the case of land use there is a graphic additional source of evidence, in that on looking at a satellite photograph of areas other than mountain and desert, much of the pattern that can be seen is manifestly the result of human activity.[24]

Table 2. *World land use change, 1750–1980*

		Million hectares				
		Area 1700	Change 1700–1950	Area 1950	Change 1950–1980	Area 1980
Croplands	Developing	157	+411	568	+302	870
	Developed	108	+494	602	+29	631
	World	265	+905	1 170	+331	1 501
Forest	Developing	3 564	−523	3 041	−329	2 712
	Developed	2 651	−303	2 348	−7	2 341
	World	6 215	−826	5 389	−336	5 053

Source: Turner et al. (1991).

An attempt has been made to obtain a broader historical view, by estimating world land use changes at intervals of half a century, using historical data for the earlier periods. The title of a massive symposium on the topic, *The earth as transformed by human action*,[25] rightly expresses the magnitude of change. Table 2 shows the changes in croplands and forest for three periods, which may be called early historical (to 1700), late historical (1700–1950), and recent (1950–1980).[26] Relative to a world land area of 13 000 Mha, about half of which is not polar or desert, the land transformations in the early historical period, the whole of human history to 1700, were quite small (assuming forest loss as approximately equal to cropland gain). Changes were much greater in the ensuing 250 years of the late historical period, converting the use of something like a further 1000 Mha; this transformation took place equally in what are now the developed and less developed worlds, although, of course, at that time the 'frontier of settlement' areas such as in North America were the developing world in a different sense. In the modern period, land use change in the developed world has virtually stood still, and, viewed on this macro-scale, is likely to continue to do so. But the less developed countries have experienced almost as much land transformation in these 30 years as in the previous 250, and currently there are no obvious forces likely to arrest continuation of these trends.

To the present, therefore, human impact has converted 1500 Mha to cropland, over 10% of the earth's land surface and 20% of the non-desert, non-polar regions; and has cleared at least 1000 Mha of forest, over 25% of the likely original cover. In addition, some 250 Mha have been transformed by irrigation, and perhaps 150 Mha built upon. All such transformations affect the global atmosphere, and local and regional hydrologic cycles, and substantially alter the properties of soils. Whatever the arguments about the nature and magnitude of these environmental effects, there is no denying that the human species has massively transformed the nature of the earth's surface.

Issues in land use

Competition between alternative uses, to be resolved by some combination of economic forces and government control, underlies all the major issues in land resources and their management. This is true at all scales, global, national, and local. Examples of land use aspects of these issues, discussed in later chapters, are:

Out of the limited area of land with a potential for food production, how much must or will be taken up by alternative uses: settlement, forestry, and conservation?

To what extent can livestock production, and the livelihood of peoples dependent upon it, be protected from loss of the best grazing lands to cultivation?

Can the rate of forest clearance, with its implication for the global atmosphere, catchment protection, and supply of wood products, be checked?

Can ways be found to direct the, very necessary, expansion of settlement and the urban infrastructure along paths which will lead to a better acceptable quality of urban life?

Can the most productive agricultural land be protected from permanent loss to alternative uses, especially urban encroachment?

How much land can be left for nature?

Given the importance of such issues to welfare and development, the failure of governments in developing countries to collect land use data, as the starting-point for policy and planning, can only be described as negligent.

Land use issues are by their nature intersectoral. In any one instance there is usually a sector – agriculture, forestry, conservation, etc. – whose interests are dominant, but which must be matched against those of competing demands for the land which that sector formerly controlled. In regional development projects (not the easiest form of development planning), all of these interests must be taken into account. In addition, production and resource conservation must be considered, the needs of local communities set against national welfare, and urban needs matched with rural. Balancing all of these interests is the problem faced by land use planning.

Soils and land use policy

International conventions

Conventions, charters, and policy statements are intended to secure long-term commitment of the parties which sign them to maintaining a set of policies. At international level, they provide guidelines to national governments as to what is expected of them. They can also be used in a mildly coercive way, through the granting or withholding of aid funds to countries which have or have not signed formal agreement. At national level, government departments can seek budget allocations by showing that their activities are in accord with national policy. Of course, policy statements in themselves are no more than forms of words, which can be ignored at will. It is easy to give lip-service to statements in the form of, 'will promote, encourage, conserve, etc.' without taking action to do so, but even these serve as a guide and reminder. At national level, policy statements can be given more bite by references to specific issues.

The *World soil charter* was adopted by FAO in 1981.[27] It is commendably compact at 5 pages, short enough to be read by high-level officials. 'The Charter establishes a set of principles for the optimum use of the world's land resources, for the improvement of their productivity, and for their conservation for future generations' says the Foreword, anticipating the concept of sustainability. There are 13 principles, followed by guidelines for action by national governments and international organizations. Many of the principles are inevitably couched in terms which sound platitudinous, as in, 'Recognizing the paramount importance of land resources for the survival and welfare of people . . . it is imperative to give high priority to promoting optimum land use . . . and conserving soil resources.' But, even at this highly generalized level, there are statements which could help to promote action: needs for proper incentives to farmers, a sound technical, institutional and legal framework, and educational programmes. Under the guidelines, governments are asked, for example, to develop an institutional framework for monitoring soil conservation, and conduct research programmes; such statements are used by under-staffed survey and research departments as an argument to obtain aid to strengthen their capabilities.

There are many similar international agreements. The UN *convention to combat desertification*[28] contains, as would be expected, a large number of statements of similar effect to the soils charter. It is a longer document of 40 'Articles' (including the admirably compact, 'Article 37: Reservations. No reservations may be made to this Convention.'). By ratifying this agreement, countries commit themselves to checking land degradation in dry zones, and, in return, are likely to gain prior access to aid support for this purpose. There are a number of similar conventions in the area of nature conservation. The *Tropical forestry action plan*[29] begins with policy statements, to be translated into national plans, and there is a 'Statement of forest principles' in *Agenda 21*,[30] the 1992 Rio declaration on environment and development. More generally, *Agenda 21*, by now widely ratified by governments, reinforces many sectoral conventions.

Policies at the national level

Action Number 1 in the *World soil charter* calls on governments to 'Develop a policy for wise land use'. Because demands for land are so competitive, there is a need for government to exercise a degree of control, to act on behalf of society to moderate an otherwise *laissez-faire* situation. This can be done on a prescriptive basis through legislation, requiring or forbidding activities and management measures in certain areas; and, on an indicative basis, by providing information, technical advice, and encouragement. As applied to public lands – national parks, forest reserves, etc. – government has a more or less free hand in management. For private or communally owned lands, zonation of permitted or forbidden uses is a common method, although subject to difficulties of enforcement. Major land use decisions, for example to create or excise a national park, are taken at the highest level. They should be based, first, on a sound basis of information, and, secondly, on a set of consistent policies.

What could go into a statement of national land use policy? It would commence with some objectives and statements of principle, for example:

One or more objectives, such as 'a healthy and productive life for the people, based upon sustainable use of natural resources and harmony with nature'.

A specific statement about conservation of the land resource heritage for future generations.

An intention to increase production, based on land suitability for different kinds of use and the needs of the country and its people.

A policy to give fair opportunities to all sections of the community, recognizing the particular value of land resources to the poor, and (specific) minority ethnic groups.

The government, on behalf of society, will exercise a degree of control over types of land use, through zoning, etc., whilst seeking not to hinder economic incentives to better use. Such allocation will be based on the qualities of the land and the interests of the people.

Set out in this generalized way this sounds platitudinous, but such statements acquire solidity and force when applied to specific national issues. Under UNEP and FAO guidance, a number of countries have formulated what are called national soils policies, although these necessarily extend into wider aspects of resources and land

A NATIONAL SOILS POLICY FOR JAMAICA

Jamaica is an island nation divided into two zones, with a marked contrast in farming systems: a central core of hill lands dominated by small farms of a mixed subsistence-commercial nature, and a fringe of coastal plains with large commercial farms. The original forest cover has been extensively cleared. For many years, soil erosion has been a problem on the hill lands; past attempts at soil conservation based on physical conservation works were not successful. Other problems are rehabilitation of land after opencast mining for bauxite, including a disposal of waste ('red mud'); urban encroachment onto the limited areas of good soils on the plains; and control of land required for the tourist industry.

The national policy was founded on a division of the hill area into two zones, the very steeply sloping upper watersheds, and the remaining areas termed the hill zone. Elements of policy were:

Steeply sloping upper watershed zone The policy is to preserve this land for water supply purposes, conservation, and tourism. Soil conservation will be achieved through maintenance and improvement of the natural forest cover. Existing agriculture will be restricted as far as possible to valley floors, and further agricultural settlement discouraged.

Hill zone Here, the policy is to make productive use of the natural resources supplied by a high rainfall and often fertile soils, to benefit population of the zone and the national economy. A major effort will be made to develop soil conservation methods which are economically and socially acceptable, primarily through maintenance of a soil cover.

The coastal plains and interior basins The main objective of the policy is to preserve the fertile soils of this zone for agricultural use. This will require controls of expansion of urban land and tourist facilities, primarily through a system of land zoning.

The policy also covers rationalization and strengthening of institutions, and reform of environmental legislation.

use. These policy statements include technical, socioeconomic, institutional, and legal elements.

The national soils policy for Jamaica may be taken as an example.[31] Being a relatively small country, the land use issues are relatively clear cut. The policy objectives for the three major land resource zones were founded on a consensus of opinion (see Box above). The interest and backing of ministers and senior civil servants was critical to the formulation of the policy.

Two common situations were found in Jamaica at the time of preparation of the policy: a multiplicity of institutions with interests in land use, and legislation which was not effectively enforced. Responsibility for soil conservation was divided between agricultural and forestry departments, on their respective lands, together with an environmental conservation authority. Statutes in force affecting land use included a forest act, watersheds protection act, and town and country planning act, but these were difficult to enforce; for forest protection, for example, the number of forest rangers was inadequate, and even the legally permitted maximum level of fines for contravention had been rendered meaningless by inflation. Among the actions required to implement the policy were institutional strengthening and rationalization of responsibilities, and revision of environmental legislation.

A similar although more complex policy was drafted for Indonesia.[32] This called for the establishment of an integrated land use plan, based on the soil potential of the extensive and varied lands of the country and the anticipated future needs. Successful implementation would require an authoritative national land use planning body.

Somewhat on the same lines are national environmental action plans (NEAPS), set up at the instigation of the World Bank. More than 60 of these have been completed, mainly by developing countries, but also by a number of developed countries in receipt of aid, for example Cyprus and Poland. National environmental action plans cover pollution and waste aspects as well as resource conservation. Experience of success or otherwise in leading to action has been variable. Early findings are that good analytical work is essential to determine priorities, public involvement is crucial, and 'A plan is more likely to be effective when its preparation involves those responsible for economic, as well as environmental, decision-making.'[33] These plans cover all three aspects of the environment: resources, pollution, and conservation. The first step after preparing the plans is to strengthen the capabilities of the relevant national institutions to monitor environmental change and to implement the plan, and a number of World Bank projects to do this are in operation. After that, one hopes that action will be taken.

The primary need for national land use plans is to co-ordinate the many, and partly conflicting, sectoral interests. Long-standing (and generally good-humoured), differences are found between agricultural and forestry departments, the latter resenting the fact that they are relegated to the poorest land, because conventional cost–benefit analysis attributes a low value to forest products. Urban planning authorities are far removed from rural in governmental structures. There are often newly founded departments with responsibility for the environment, generally lacking in budgets and political weight as compared with the production sectors. The institute with nominal responsibility for land use is generally restricted to conducting surveys and evaluation studies.

A national land use policy, possibly including some form of master plan, is essential to the achievement of rational and consistent development. The technical land use authority plays a vital part in collecting information and analysing consequences of different courses of action, but such bodies inevitably lack political weight. There needs to be a high-level land use co-ordinating committee, at ministerial level, with advisory powers to government, if a rational and consistent policy is to be achieved.

5 Working with farmers

Only recently has a fundamental fact been realized in development planning: that whatever may be achieved by development projects, ultimately it is people who manage land resources: farmers, other land users, and local communities. Farm systems research was only a stage along the route to this realization. A true participatory approach is based on the sharing of knowledge and views between advisers and farmers, linking scientific advances with the store of local knowledge. Participatory projects by non-governmental organizations have achieved striking local successes, particularly in the area of soil conservation. But this approach can never be scaled up to reach the rural community at large until it is implemented by countries themselves, either through national advisory services or, if these fail, through self-help by farming communities.

If there is a single theme which runs through the success, or otherwise, of projects in natural resource management, it would be 'people matter'. This goes without saying in developed countries, where farmers and residents' associations are organized into strong lobbies, and will respond to any proposals for change. Continuing co-operation with local land users, from the first conception of a project to its implementation, is as important or more so in the typical situation of developing countries, in which land users are more numerous but less powerfully organized.

There is a set of approaches which have in common the interaction between farmers and developers – more precisely, between the local land users on the one hand, and the scientists, planners, developers, or other non-local people on the other. The developers meet the farmers, ascertain their problems, and ask them for their suggestions for improvement. Before any changes are made, they are explained to the farmers, and not implemented without their agreement. This does not merely mean securing agreement to the introduction of developments, but that the farmers are so convinced that these will benefit them that they will put in the work – especially, the maintenance – needed for success.

This may seem so evidently desirable as not to require justification. This was not always the case. During the era of top–down planning in the 1950s and 1960s, new technologies were frequently introduced without substantial consultation with those who would be affected, nor with their agreement. Not surprisingly, this was a major cause of project failures. It would have been considered revolutionary during the early part of that period, and later innovative, to have a sociologist in a planning team.[1]

The participatory approach, or working with farmers, is now recognized as an essential element in development. Rural sociologists, farm system specialists, or the like, are routinely employed in development teams. Research stations accept the need to maintain ongoing contacts with land users. A number of converging strands led to this change of attitude:

Failures of development projects through non-participation of land users. The necessary work (e.g. planting of trees, maintenance of irrigation channels) simply was not done.

Specifically, failures of soil conservation schemes, through failure to maintain capital works.

A prior need for land tenure reform, if land users were to have a stake in improvements. This was the focus of the World Conference on Agrarian Reform and Rural Development (WCARRD) in 1980, and subsequent national programmes.[2]

Problems with the adoption of supposedly 'improved' technologies; for example, some early high-yielding crop varieties were not disease-resistant.

A move, particularly in Asia, towards decentralization of planning activities, as multi-level planning.[3]

Recognition of the value of local knowledge of the environment and land resource management.

Approaches and methods

A confusingly large number of methods have been developed under titles such as 'the grass roots approach', 'working with farmers', 'farmers first', and 'participatory rural appraisal'. One compilation lists 29 approaches, all known by acronyms, 10 including the word 'rapid'![4] Some are different in objectives, others only in name. These methods are overlapping, but will be summarized in three groups: farm systems research, participatory rural appraisal (including diagnosis and design), and making use of local knowledge.

Farm systems research

The first attempt to reach 'grass roots' level was farm systems research.[5] It became an established part of the international agricultural research centres. These centres are focused on single elements in agriculture, such as plant breeding, pest control. It was realized that these elements interacted on farms, and that a change in one might have knock-on effects upon another. One therefore needs to understand the functioning of the farm as a whole.

Initially, the need was to find out basic facts. There are three elements to a farm system: the family or household, the farm itself, and off-farm activities of the household.[6] Data such as area, people supported, crops grown and livestock, inputs and labour, costs and revenue, were collected by sample questionnaires, with the results tabulated. Some of these could be instructive to researchers and developers; often, for example, there are two distinct farming systems in the same region, with and without livestock, with different needs and resources. However, data collection was apt to become an end in itself, and the value of the approach was disputed. One might also note that farm systems studies are conducted largely by social scientists; it is rare to find data, by farm, on such matters as perceived soil deficiencies, water availability, pest attacks, tree species planted, or other aspects of physical resources.

Farm system studies were based on the development advisers finding out what was happening on farms. This is not interactive, nor truly participatory. These studies were of value educationally – for the self-education of the advisers – and as a necessary background for planning.[7] Specialists need to appreciate the constraints under which farmers work. If a family of seven is supporting itself on one hectare, with no local availability of fertilizers, it makes a difference to development potential. New insights were gained, no doubt, as a result of farm systems work, but the data compiled was unselective. What was needed was a better definition of objectives, which would give reasons for selection of relevant information, together with a defined basis for subsequent action.

Participatory rural appraisal

A clearer focus to working with farmers is provided by participatory rural appraisal (PRA).[8] There can be two distinct objectives:

1. To find out the problems of farmers in order to design a programme of research to help solve these;
2. To find out the problems of farmers in order to help them with solutions, based on existing knowledge.

In principle these overlap, since the investigator does not know the extent to which existing knowledge is sufficient. In practice, however, the first type of study is conducted at the instigation of research stations, the second in the design of a development project.

There are many versions of participatory appraisal. All have the same philosophy, that of a mutual exchange of ideas between the person or team providing assistance and the farmer or other land user, but each has different procedures. These will be illustrated by the method of diagnosis and design (D & D), meaning diagnosis of the problems of the land use system, and design of ways to ameliorate these problems. The analogy is with a doctor, who must diagnose a disease before treating it. Diagnosis and design was originally devised for the first purpose above, to design a research programme, and specifically for research in agroforestry. However, it is equally or more relevant, to the second purpose, that of providing direct help to farmers; and it is applicable to land management changes of any kind – plant nutrition, irrigation, pest management, livestock management, as well as agroforestry. To avoid repeated qualification, 'farmer' means farmer or other land user, and 'adviser' refers to the development team or individual carrying out the study.

For this general case, the steps are outlined in the Box below.[9] Steps 1–3 are preliminaries, not part of the participatory process as such. From Step 4 onwards, all work is carried out with the participation of the farmers. Diagnostic survey involves identification of the problems of each farming system. These include problems of the farmer, such as shortages of food, water, fodder, or fuelwood, a

STEPS IN DIAGNOSIS AND DESIGN

REGIONAL SURVEY AND SITE SELECTION

1 **Planning the study** Objectives, area covered, identification of collaborating institutions, programme of work.

2 **Environmental description** Identification of natural land units, e.g land systems.

3 **Identification and selection of land use systems** Identification of farming systems or other land use systems, selection of priority systems, design of a stratified sampling programme.

DIAGNOSIS OF PROBLEMS

4 **Diagnostic survey** Land use problems and priorities.

5 **Diagnostic analysis** Causes of the problems.

DESIGN OF ALTERNATIVES

6 **Identification and selection of improvements** Identification of improvements in infrastructure or land management, selection of priorities.

7 **Design of improvements** Technical specifications and design of the selected improvements.

EVALUATION AND PLANNING

8 **Evaluation** Evaluation of the proposed improvements in terms of economic viability, environmental impact, and acceptability.

9 **Planning** Formulation of a plan to put the improvements into effect.

IMPLEMENTATION

10 **Implementation** Putting the plan into practice.

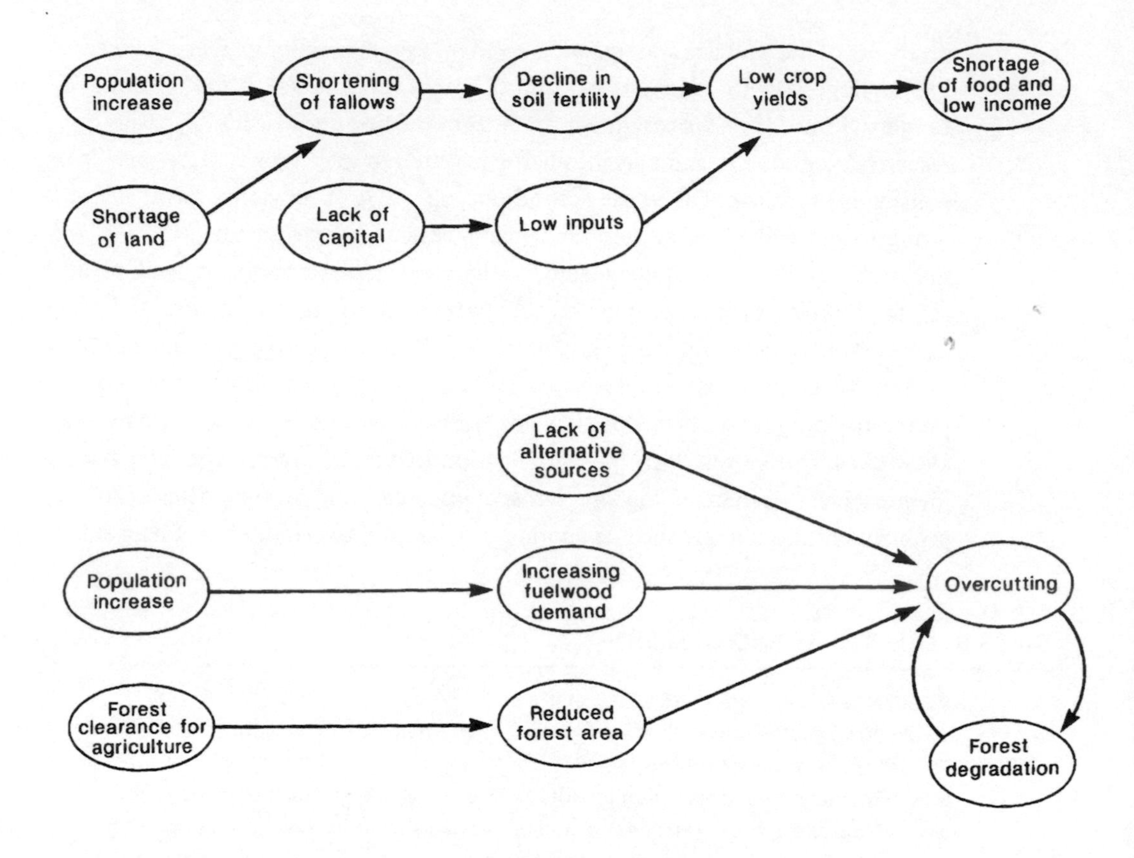

FIGURE 5 Chains of cause and effect identified in diagnosis and design.

low cash income, or distant and unreliable markets; and problems of the land, like erosion, weeds, pests, or poor condition of pastures. In practice, a low cash income is invariably identified as a priority. These are primarily the problems as perceived by the farmer. Diagnostic analysis means finding the causes, or chains of cause and effect, which lead to these problems. Some of these chains link problems of the farmer with those of the land, as in the examples illustrated in Figure 5. Some can be tackled through changes to land management; others are beyond the control of the farmer, as when markets are distant because there is no road. In seeking causes of problems, the local perceptions of the farmer may differ from the wider view of the adviser.

The selection of proposed improvements is carried out by mutual discussion. The farmer will probably have a clear idea of what is needed, the adviser of what is practicable. The detailed design of what is to be done may be initiated by the adviser, moderated by the farmer. This is the point at which a research offshoot may be needed, such as identification of nutrient deficiencies, or finding the best tree species to plant. Both farmer and adviser will carry out an evaluation, the farmer probably on a subjective estimate, the adviser in a more formal way. Following the planning step there may be a gap in time, to secure funding. Implementation may require action both at project level, for improvements to infrastructure, and at farm level, as regards land management.

The diagnosis and design procedure can be integrated with land evaluation.[10] An essential feature is that it provides a focus to the inquiries, and hence a reason why specific items of information are or are not needed. The farmers report a shortage of fodder, for example; so there is a need to find out the livestock types and numbers, and present fodder sources, in order to assess how these can be augmented. As originally described, diagnosis and design was one of the 'rapid' survey techniques, intended to be carried out by a team of external advisers in the space of a few weeks. Whilst practical circumstances may sometimes require this, it is far from ideal. The best time span is for the adviser to be permanently resident in the area, to continue interactions with farmers as the changes go ahead, and to live with the results; and this can only be fully achieved if the adviser is not a temporary visitor to the area, whether expatriate or national, but a permanent member of the local extension staff.

Considerable effort has been necessary to replace former attitudes, in which 'experts' told supposedly ignorant farmers how to improve their 'traditional ways', by a way of thinking based on real participation. Even the words 'adviser' and 'improvement' are regarded by purists as pejorative (who is advising whom, and from whose perspective is it an improvement?), although this criticism applies only to the wrong sort of advice. Some precepts of the participatory approach are:[11]

Rural communities have considerable local knowledge; they make rational decisions within the framework of external constraints in which they operate (agricultural services, inputs, marketing, prices, etc.).

Outsiders can learn from local people, as well as vice versa; farm families should be allowed to judge what is best in their situation.

To begin with, technical advisers have low credibility with local people; there needs to be self-critical awareness.

Data should not be collected for its own sake, but only that which is needed to identify, and help solve, the problems; the advisers are there to help with the business of farming, not to prepare a fine-looking project report.

As far as possible, local communities themselves should initiate, design, and implement changes, so that they feel responsible for the outcome.

Some of this may seem idealistic, and provoke the reaction, that if the farmers know best, what is the use of advisers? However, the role of farmers' local knowledge

FIGURE 6 Land resource scientists should also talk with farmers.

receives emphasis partly to counteract the older way of proceeding. Of course, the advisory team possesses technical knowledge which is potentially of use to the farmer, as well as better opportunities to influence the policy framework. What is central to the approach is that local communities and farmers should have the last word when it comes to implementation, since the success of any changes in land management depends critically on their active participation.

Making use of local knowledge

In a World Bank agricultural development project for the Gusau area of Sokoto State, northern Nigeria, in the late 1970s, an extension handbook was prepared. It largely covered single cropping. One point of emphasis was that, in view of the short rainy season, cotton should be the first crop planted since it had a long growing season. Advice give in the handbook was not much followed, so a study was made of local farming practices. This revealed a sophisticated system directed at minimizing food risk in a region where the early rains might be followed by a dry spell. At the first rains, drought-resistant food crops were planted individually by hoe at a wide spacing. If the rains then failed, this could be repeated without much loss. If the rains continued, then other food crops could be planted in the spaces, giving a multiple cropping system. Only after the food crops were established was cotton planted.[12]

In a long-established farming community, there is always a store of local knowledge, passed on through families and shared by communities. Risk-avoidance strategies, such as that outlined above, are common. There are frequently local names for soil types, coupled with a knowledge of what each should and should not be used for. Arising from times when fallowing was the principal means of maintaining soil fertility, there are recognized plant indicators of soil condition. There is a sophisticated knowledge of how to survive in a bad year, particularly among pastoral peoples. Long-established indigenous communities may view land not as a commodity, to be traded and exploited, but as a part of their being, often sacred. There are problems in maintaining traditional ways, brought about by population increase and outside pressures, but people want to create and control their own destinies.[13]

Local knowledge is by no means confined to traditional ways of farming. Views about whether and where to use newly introduced crop varieties can be quickly passed around the community, by what has been called the informal extension service. Farmers' associations can play an important part in sharing and updating local knowledge, and in integrating the traditional with the new.

There are opportunities for systematizing some of this local knowledge, and applying it on a project scale. Early examples were the compilations made by government ecologists along the lines of, 'Useful plants of . . .', which have proved valuable in the recent development of agroforestry. Soil surveyors do well to enquire about locally recognized indicator plants, and some national lists of these have been compiled. Indigenous methods of soil and water conservation have been collected, with a potential for transfers between regions.[14] Of course, some traditional knowledge, effective in its time, may be no longer applicable under modern land pressures and farming conditions.

Women's role

Many tropical farmers are women, sometimes as sole owners but more often where the men in the household have taken up non-farm work, locally or by migration. There are often gender divisions of labour, for example the women cultivating food crops and the men cash crops. There may be special roles for women's communal working parties. In many societies, women do much of the hard work of collecting water, fuelwood, and fodder, often walking long distances carrying heavy loads. It has been claimed that women are keepers of local environmental knowledge, and that they suffer disproportionately from land degradation.[15]

Given the distinct social position and role of women in developing countries, which can change only slowly, some degree of specific attention to their part in land management needs to be given. An elementary need is to include women among local extension workers, formerly a rare occurrence, now increasingly to be found. Within advisory teams, women advisers may be able to mobilize women's communal working groups in a way that men could not. In the diagnosis stage of participatory appraisal, women can be separately asked to identify problems, and will have different orders of their priority.

The taking of measures of this kind should not need pressures from the feminist movement, although these played a part in the recent recognition of gender roles in land use. Once these roles are recognized, it is no more than common sense to give them attention. A wider question is the generally lower educational levels of women, which is bound to affect the ease with which changes in land management can be introduced. In this respect, there is a link with wider questions of population and land, discussed in later chapters.

Potential and problems of the participatory approach

Some obvious truths underlie the participatory approach. It is people who manage natural resources. Developers may bring machinery and construct capital works, but farmers have to maintain them. Farmers know their problems, as should local extension workers, if there is a reasonable level of competence in the advisory services. It has been commented that if you need a questionnaire to find out the problems of farmers, then you are probably not the right person to be doing the work![16] But, just as obviously, farmers do not know all the answers; if they did, we should not need research. Advisory and development teams can influence policy and improve infrastructure; and if soil scientists and foresters do not have some useful and relevant technical knowledge to offer to farmers, then they have not been properly educated.

Even though self-help by farmers is the primary objective, diagnostic surveys should not be started without an operating budget to implement the changes that are proposed. To do so loses credibility. Over and above the staffing costs of the diagnostic survey, funds must be kept available for construction of agricultural service centres, for drilling tubewells, setting up forest nurseries, or whatever are deemed to be the development priorities. These lend support to on-farm management improvements such as composting.

Participatory studies have been conducted on occasion as part of UN or World Bank projects,[17] but they are the special province of work by non-governmental organizations (NGOs). Dedicated, usually young, expatriates, working at the scale of a small group of villages, have achieved some notable successes, particularly in the area of soil conservation where it has given rise to a wider concept known as land husbandry (p. 180). An early step is to take on and train local staff, who can become equally committed. In a degraded watershed in Tamil Nad, India, a 'cluster council' of 44 villages was established, composed of one person from each village employed by the project; with local agreement, measures were taken to revegetate the soil by control of grazing. The soil was upgraded and the land became visibly greener than adjacent areas. In Honduras and Guatemala, participatory soil conservation projects that had ended in 1991 were revised three years later; in the interval, adoption of conservation had increased and crop yields had been further raised. In a remote region of Haiti, one US volunteer with one Haitian successfully checked erosion by introducing contour hedgerows of *Gliricidia sepium*, which provided the stakes needed to grow yams. These examples could be multiplied many times, and are achieved at cost levels of 1–5% of the budget of capital-intensive development projects.[18]

Successes by non-governmental organizations raise questions. How can the transition be made between successful pilot projects and assistance reaching the rural community as a whole? How can one plan for withdrawal of external assistance, with continuation of the good work under local initiative afterwards? How can governments be made interested in improvements on a small and local scale, when the 'big money' comes from projects involving capital construction? What can projects at the local level do if the macro-economic environment is unfavourable, or the stability and effectiveness of government are insufficient to provide support? Above all, how can schemes that are successful for a small group of villages be scaled up to district level?

The scaling up problem is the greatest. It is instructive to fly over the Indo-Gangetic plains. Because of alignment along canal systems, the pattern of fields looks somewhat like a sheet of graph paper, stretching out to the horizon. Here and there are darker squares, in each of which a thousand souls are spending most of their lives. In a very small number of these villages, lying near roads or cities, schemes are in progress which will improve production and conservation, at least in the short term. Is it possible to consider the same being repeated in all of them?

A situation could be imagined in which funds of $100 million, which a development agency initially intended to spend on a dam, road or other construction project, were instead diverted into a farmers-first project covering about 1000 villages; and it would be instructive to compare the respective cost–benefit analyses. But neither is this plausible nor is it the best way of proceeding.

At present, the small-scale projects run by non-governmental organizations are doing useful work, at low cost, and gaining experience; but it will not enable the 'grass roots' approach to reach a higher proportion of the population. For this to happen, participatory adviser–farmer projects need to be conducted by officials of the country itself. In view of the poor status and funding of many national advisory services, this will not be easy to achieve, but it is where the future of the approach lies.[19] Many difficulties have been encountered in top–down advisory services, such as on the 'training and visit' system.[20] Local extension staff, speaking the language and coming from the same area, should be better fitted than outsiders to conduct participatory extension advice, adapted to local conditions, once the appropriate type of training is given.

Some experienced outside advisers are now running participatory training courses, very different in nature from conventional courses since they are themselves based on sharing of experience between 'instructors' and participants. Unless and until governments are prepared to give adequate support to their national advisory services, the participatory effect cannot multiply in this way.

There is a further stage, that of farmer-to-farmer extension. This self-help approach is growing in a number of countries, by informal contacts and through farmers' associations or similar institutions. It overcomes the inadequacy of government services through bypassing them. This is the ultimate and the only fully effective stage of a community-based, people-centred, approach to the management of natural resources.

In the final analysis it is people who manage land resources – farmers and local communities. If they do not do so sustainably, then no one else will. Aid projects based on the participatory approach, the particular although not exclusive province of non-governmental organizations, have achieved some striking successes in mobilizing local action. The first major problem that they face is that of scaling up. Where these are implemented by external organizations, they can never hope to reach the mass of the population. This can only come about if the same type of self-help activities are carried out either through the national advisory services or, if these cannot be mobilized, ultimately through mutual aid between farmers themselves. The second problem is that local communities must act within the wider context. They can do little to remedy deficiencies in government services or economic policy, and they face the inexorable pressures brought about by population increase and land shortage.

6 Land use planning

Resource surveys and land evaluation studies are only a means to an end. It is at the further stage of land use planning, and projects in the area of natural resource management, that action gets taken. Land use planning does not only mean making plans; it covers their implementation and management, monitoring of progress, and revision. The most important scales are national level, for policy guidance and priorities, and district or project level, where developments are put into practice. Planning must be focused on the problems of land users, but these must be reconciled with other interests. The wide range of objectives, covering production, conservation, and the different sectors of land use, means that no standardized method is possible. The best that can be done is to provide a set of guidelines giving basic steps, with checklists of activities, together with decision support systems. Past experience and informed judgement will always be needed. Natural resource aspects should play a continuing role during the later stages of project planning, including during implementation and monitoring of progress.

The procedures of natural resource survey, land evaluation, and participatory methods, are carried out with the intention of improving the people's welfare. Throughout these stages, however, no one has yet taken any action to change, hopefully to improve, the present situation. The old land use systems are still being practised, soil is eroding, vegetation degrading, crop yields remain low, and the poor are still hungry – or whatever may be the problems of the region concerned. It is at the stage of land use planning, or rather, the implementation of the plan, that something actually gets done.

The end-products of soil survey and other types of natural resource survey are maps and quantitative data on natural resources and where they are to be found. Land evaluation takes the procedures several steps further: improved types of land use have been defined, the best places to practise them identified, and the consequences of one alternative versus others assessed. If anyone is going to benefit from all this work, three more stages are needed: deciding what to do, how to do it, and doing it. In more formal terms:

making a choice between the alternative land use options;
preparing a plan, consisting of practical measures which will promote or encourage the chosen uses;
implementing the plan (with monitoring and post-project evaluation).

LAND USE PLANNING

Land use planning is the systematic assessment of land potential, social and economic conditions, and alternative patterns of land use, for the purpose of adopting land-use options which are most beneficial to land users, without degrading natural resources; together with the selection of measures most likely to encourage such uses of land.

Planning may be carried out at international, national, district (project, catchment), or local (village) levels. It includes participation by land users, planners, and decision-makers, and covers technical, educational, economic, and legal measures. An element of natural resource management, for purposes of production, conservation, or, usually, both, is always involved.

Source: FAO (1993a), 86 (adapted).

In schematic terms, and perhaps from the point of view of the planner, putting the plan into practice appears to enter at a late stage (Figure 7). In terms of real time, the implementation of a development project takes at least as long as its preparation, perhaps 5 years for each in a typical case, and it is to be hoped that, under continuing local management, the benefits extend many more years into the future.

A critical point comes at the decision to implement, in aid-supported projects taken jointly by national government and funding agency. At that moment, the budget involved rises at least tenfold. The expense of survey, evaluation, and plan preparation is nothing compared with that of buying bulldozers, building dams, setting up marketing boards, training and paying an improved extension service, or whatever measures the plan requires. This is one reason why the benefit:cost ratio of good surveys and planning is so high; the entire preparation costs can be paid off by quite a small increase in efficiency during implementation. In practice, the gains from avoidance of real blunders, like growing a crop where it will perform badly, or incurring the opposition of a section of the local community, can be very much greater.

At the district level, there is no clear distinction between land use planning and project planning. Development projects that are in the sectors of agriculture, forestry, or conservation, and which involve an element of natural resource management, are essentially land use planning projects. This applies particularly where they involve more than one sector; indeed, some would say intersectoral elements are, by definition, essential to land use planning, although one can think of many purely agricultural or forestry projects which include the essential element of natural resource management.[1] Hence the land us planning cycle is a version of the well-established project cycle: identification – preparation – appraisal – implementation – monitoring and evaluation.[2]

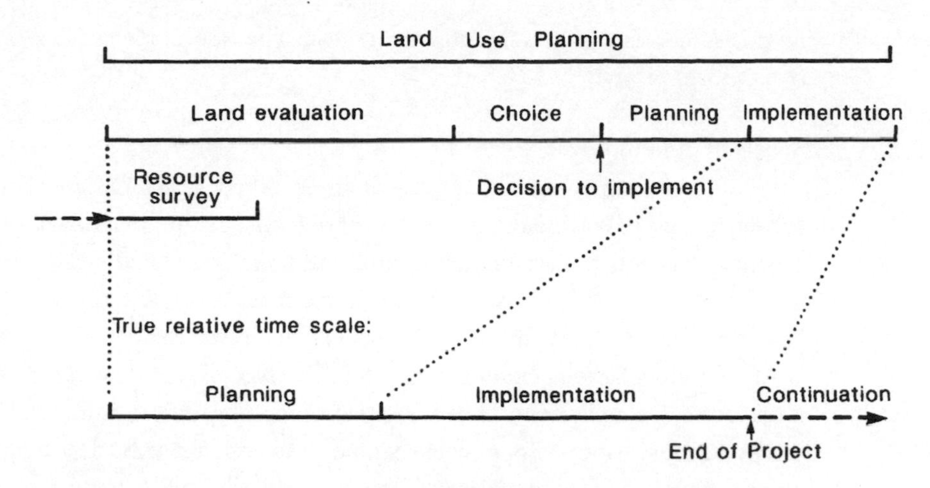

FIGURE 7 Resource survey, land evaluation and land use planning.

A basic change has come over land use planning during the second half of the twentieth century. In the period following the Second World War, it was sometimes a case of going into an area of empty land and allocating each part of it to different uses, clearing the forest or installing irrigation, for example. Under these circumstances, natural scientists had free play. Nowadays, nearly all projects are a matter of improving conditions in areas already being used, often quite intensively. In these circumstances, nothing can get done without working with the active co-operation of the local people, although the element of natural resource evaluation and management is no less.

Thanks, no doubt, to some effective lobbying by international agencies, these activities gained momentum from *Agenda* 21, the action programme arising out of the 1992 UN Conference on Environment and Development. They occupy the whole of chapter 10, 'Integrated approach to the planning and management of land resources'.[3] FAO is the Task Manager for the activities described in this chapter, and has issued a first appraisal of the issues.[4] *Agenda* 21 calls for the development of policies to encourage sustainable land use, the strengthening of planning and management systems, the promotion of awareness and public participation, and the improvment of data and information; it also asks for research that will lead to a better understanding of land resources and their management, for improvements in education, for regional co-operation, and for the strengthening of institutional capacity. Of course, it is all couched in international officialese ('Governments, in collaboration with appropriate . . . should strengthen . . . promote . . . support . . .') but these are real needs, and some progress is being made. The annual cost of implementing these measures is estimated at $50 million 'from the international community on . . . concessional terms'. There are likely to be difficulties in obtaining this sum, although it is no more than the cost on one single major World Bank development project.[5]

Scales and levels of planning

Land use planning is mainly applied at three broad scales or levels: national, district, and local. Beyond these lie two further levels: the international level, which is concerned with policy, priorities, and co-operation; and the farm level, the rational planning of land use and management on a single farm, in which the farmer may receive assistance from district extension services.

At the national level, planning is concerned with goals, policy, and the allocation of resources. Guidelines must be set between competing demands for land, such as for food crops, export crops, conservation, and urban needs. The demands of sectoral agencies, for agriculture, forestry, wildlife, etc., must be reconciled. Budgetary allocation is one means for governments to direct activity, for example by making possible, or otherwise, an adequate research service. Another is through environmental legislation. Governments can, and should, establish a national land use policy, and prepare something on the lines of a national master land use plan. Since the primary administrative divisions (states, provinces, etc.) of large countries are larger than many small countries, these may also have an intermediate level of planning.

The district level refers not only to administrative districts, but to any land area falling between national and local levels. The steps that can be taken at national level serve only to provide guidelines, give directives, and allocate budgets. It is at district level that practical action first gets taken: action to site irrigation schemes, for example, or to set up forest plantations, improve water supplies, build roads, provide better marketing facilities. It is also at district level where locally adapted systems of improved land management are designed, and applied through extension services. The needs of the district must be reconciled with national priorities. Watershed management projects are a type of district-level project centred on soil and water conservation.[6] This is the usual scale for development projects, and may also be called the project level. Many sector projects – for forestry planning, agriculture, establishment of national parks, etc. – operate at district level.

The unit for local-level planning is the village or group of villages. This is where the local people come to the fore, for it is they who plant crops and trees, apply irrigation water and fertilizer, or agree that no one shall clear and cultivate areas set aside for the village woodland. Local knowledge is put into action: for example, the project at district level may have proposed citrus fruit cultivation on some part of the village land, but farmers know from past experience that the soils there are too saline.

There are certainly links between these levels in a downwards direction. The national plan sets priorities which will help to decide the type and location of district-level development projects. These projects provide infrastructure (e.g. roads, supplies of fertilizer and seed), technical advice, and staffing, supported by a budget. Whether the aims of the project are successfully fulfilled depends to a high degree on how effectively they are applied at the local level.

There are also, at least in intention, links upward between the scales. The perceived needs of local communities, and their local knowledge, is transmitted to district level, possibly through district extension officers or local councils. The districts (in this case administrative) will then lobby at a national level, both through scientific communication and political means. This has unfortunately become known as 'bottom-up' planning, although multilevel planning is a better term. Multilevel planning requires a degree of decentralization of functions. The national government retains, for example, responsibility for flood control, mapping and statistics, fertilizer imports and export marketing, and some conservation areas, whilst decentralizing many functions in agriculture, livestock production, rural water supplies, and fisheries. Decentralization requires that skilled technical staff should be available at district level, and is more advanced in Asia than other developing regions.[7] The effectiveness of upward transfer between levels, and of multilevel planning in general, depends on how good are the mechanisms of communication, availability of skills, and efficiency of governance.

Examples of land use planning projects

The following list of examples are selected mainly to illustrate the variety of objectives to be found in land use planning and natural resource management projects. Whilst mainly from developing countries, three from the developed world are included for comparison. The projects are arranged in order of scale, national to district and local, with multisector projects preceding those with a focus on one sector.[8]

Ethiopia, national master land use plan National level, multisector. Initially, mapping and data-collection based on satellite imagery, followed by land evaluation for major kinds of use, the first time an overview of the country's resources had been made available. It was followed by a second phase of regional planning in selected areas, which then became the basis for siting of development projects.

Ghana, environmental resource management project National level, environmental sector. The project is to support implementation of the country's national environmental action plan (NEAP), by developing technical and institutional capacities for environmental monitoring. It will also develop methods to reduce land degradation through community farm management, and will demarcate and manage five coastal wetland sites.

China, red soils area development project District level but covering five provinces, multisector. The project is designed to check land degradation and improve productivity. Components include soil and water conservation, on-farm irrigation channels, biogas units, integrated development of swamp rice in valley floors and forestry on adjacent slopes, livestock, aquaculture, research, processing, and marketing. This is one of several very large-scale recent projects in China, based on 'loans' of over $100 million, but with incremental production intended to be 50–100 times this amount.

Brazil, Rondônia, ecological-economic zoning District (state) level, multisector. The state had received a vast influx of immigrants, with indiscriminate forest clearance and uncontrolled land use leading to degradation and land abandonment. It lacked a land use policy that was scientifically based and 'free of clientalism'.[9] Basic survey and suitability zoning was a first step, necessary but by no means sufficient, towards establishing some measure of rational policy and steps towards its implementation.

Malaysia, Jengka Triangle land settlement scheme District level, multisector, for new land settlement. In the 1960s, Malaysia was seeking land on which to settle landless people from the crowded river delta areas. Empty rain forest in the centre of the country allowed a 'clean sheet' approach to planning, and a series of schemes were implemented by a Federal Land Development Authority. A part of central Pahang State, the Jengka Triangle, was the largest of these. A soil survey was conducted within the forest. Land was zoned, for agriculture (oil palm or rubber), production forestry and protection forest. In accordance with agronomic requirements, the better soils were allocated to oil palm, the poorer to rubber. When the plan was implemented, the price of oil palm was higher so this was largely planted; a few years later there was a rise in the price of synthetic rubber (due to the oil price rise) and therefore of natural rubber, but with the long period that rubber trees take to come into bearing one cannot switch. The scheme was successful in giving a livelihood to former landless farmers, although the clearance of 2000 km^2 of primary rain forest would not receive international funding support nowadays.

Turkey, Eastern Anatolia watershed rehabilitation District level, multisector. The aim is to address jointly the problems of rural poverty and natural resource degradation. An increase in vegetative cover is a central concept, leading to reduc-

FIGURE 8 Land use planning, Malaysia: forest retained on the ridge crest, rubber on the hill slopes, oil palm on the richer soils.

tion in erosion and reservoir sedimentation. Methods include attempts to promote better range management, reforestation, greater production of fodder and fuelwood, with fruit tree cultivation and apiculture. Integrated watershed rehabilitation is being attempted in some 50 catchments over seven years.

Malawi, Lilongwe land development project District level, agricultural sector. Unlike many of the examples given, this programme covered one of the country's richest agricultural areas, with the most fertile soils. The problem was that these soils had been almost continuously farmed, with infrequent fallows and usually low inputs, for more than 50 years (as shown by air photography dating from 1947). Soil fertility and crop yields had declined, especially since 1960. The project established a network of local agricultural service centres which fulfilled two vital functions: they supplied improved seed and fertilizer (with credit) at the start of the growing season, and bought crops at the end of it. A system of soil conservation bunds was also set up, constructed by expensive mechanical means and subsequently not well maintained. In the initial years, crop yields were appreciably increased, but this improvement has not been maintained.

The Philippines, Luzon Central Cordillera agricultural programme District and local levels, agricultural focus, linked with forestry and conservation. Forest clearance and farming in a steepland area had led to severe degradation. With community participation, areas were identified which would be open to farming, whilst other areas, steeper or already degraded, would be reforested and allotted to catchment protection. At the same time, local level micro-projects were implemented, to give some direct and immediate assistance to farmers.

South Africa, Makhatini irrigation scheme District level, irrigated agriculture sector. This example is included as one among countless irrigation schemes, in which planning of land and water use is central to success. The basic water supply from dam and canal flow must be managed centrally, whilst use of the water, efficiently or otherwise, is in the hands of farmers. In this scheme, there are both commercial farms and smallholders, the latter with insecure property rights. Aims include water supply stabilization, security of private and communal land tenure, prevention of erosion through overgrazing, checking the spread of malaria, and employment through local processing of crops.

Zimbabwe, Binga District, wildlife management District level, wildlife management in relation to other land use interests. This is an example of a common situation where a wildlife reserve (national park) for conservation and tourist revenue had been established, but there are conflicts with local farmers and pastoralists. The reserve includes some good grazing land, not much of the tourist revenue reaches the local people, and elephants, etc. stray and cause damage to crops. Conflicts of this kind are difficult to resolve, but knowing how much land of different types exists, and how it is used, is at least a starting point.

Australia, Peel region (Western Australia), environmental management plan District level, intended to feed down to farm planning level; environment and agricultural sectors. Problems of eutrophication were being experienced in an estuary and coastal lagoon area. Farming, including horticulture and intensive pig production, was linked with this. The plan included a classification of land capability for effluent disposal. It was conducted at four scales, regional, local authority, catchment, and farm (the first three falling within what has here been called district level). Data and maps lead to strategies and controls; as may be expected, 'the degree of implementation is variable'.

Canada, Dauphin Lake management plan District and local levels, environmental sector, linked with land use. Livestock production in the lake basin over many years has led to erosion, increased river sediment loads, and water pollution. One technique is to restrict livestock access to riparian zones. The basis for action is to obtain partnership agreements between landowners and an advisory board.

India, watershed development District and village levels, multisector. Projects are operated in headwater catchment areas by the Ministry of Rural Development; funding is not international but from the central government, although assistance may come from external non-governmental organizations. The aims are jointly reversal of land degradation and poverty alleviation. There is a participatory approach, with watershed development committees, women's groups, village cluster councils, and the like. A shortage of skilled professional staff at this level can sometimes be compensated by sound use of local knowledge. The simple measure of agreement, adhered to, to keep livestock off gullied areas can have remarkable effects in only a few years. Some schemes of this type have been impressively successful, others not. They can be very small, for example in a watershed in Tamil Nad involving no more than 50 households, all of the harijan caste.[10]

Such examples could be multiplied indefinitely, not only in number but in variety of objectives. A few common features might emerge from a statistical analysis of such projects. They are all in areas with problems, of course. In the tropics, the most common environmental siting is in populated steepland areas which are cultivated or grazed, and have been deforested, leading to soil erosion, land abandonment and often fuelwood shortage: in short, in areas which have experienced the vicious circle of population increase, growth of rural poverty, and onset of land degradation. By its nature, this situation lends itself to a watershed planning approach, although watershed management projects tend to put too much emphasis on soil and water conservation at the expense of other benefits. Other common types are irrigation projects (large- or small-scale), sector projects for rural forestry and agroforestry, establishment or rationalization of a national park,[11] and, in semi-arid areas, improved food security with control of overgrazing and desertification. The wide range of scales, objectives, and methods of implementation is one reason why it is so hard to lay down rules for how to conduct land use planning.

Post-project evaluations show that, statistically, some types of planning are more difficult to implement successfully than others. One review found that the least likely to be successful are livestock projects, and plans for the integrated development of poor rural areas; another criterion for non-success was, regrettably, found to be projects 'which depend to a significant extent on co-operation between more than one government agency'.[12] Factors for success in projects include a stable political climate, support by central government for solving local problems, clear property rights, participation by local people, the existence of a single strong government institution to control the project, skilled technical staff, good scientific information, and good public relations.[13]

The context of planning land use in developed countries is very different.[14] In settled areas, land use is highly competitive and tightly controlled by legislation. A basic change in type of use generally requires planning permission; quality of land, often in the form of some grading from Class 1 to Class N, is one factor taken into account by planning authorities. There is more scope for constructive planning in regions with a proportion of forest and conservation reserves; it may be useful to set up a matrix of benefits, economic and environmental, as can be done with a computerized land use planning program, LUPLAN, developed in Australia.[15] In Canada, the national legal basis rests on a federal Planning Act; this deputes major responsibilities to the provinces, which set policy guidelines, and call for municipalities to prepare plans which observe them. In densely settled countries such as The Netherlands, planning is thoroughly researched and tightly controlled. Differences in government planning procedures between one country and another are a further reason for the difficulty of prescribing a standard method.[16]

People in planning

The people and the agencies with interests in land use planning, in modern jargon the 'stakeholders', include the following:

Land users Those who live in the planning area, and whose livelihood depends wholly or partly upon the land. Early accounts used to list these last, but, since development is for their benefit, and nothing will get done without their co-operation, they should head the list. The importance of participation by the local people is a recurrent theme, to which we return in the next chapter.

The planning team In developing countries this is often an external group of 'experts', from an international agency or consultant company, supposedly working in conjunction with local staff. There is a trend towards using local government staff, as in developed countries, which is much to be encouraged, even when they are technically less well qualified. The team's responsibility is to collect information, analyse it, and draft the plan. They come in the middle of a structure diagram, liaising with all other interested parties.

Sectoral agencies The government Departments of Agriculture, Forestry, Irrigation, Environment, etc. These supply local staff, particularly for extension services, make provisions for agricultural inputs, and organize marketing services. Their experience should also provide a bridge between land users and the planning team. Strengthening the capacity of these agencies, through training and technical support, is a common feature of projects.

The project implementation agency Once a plan is put into practice, the planning team becomes the implementation agency, hopefully with much continuity of staff. This agency controls the spending of the externally supplied budget, and so executes (often through contractors) the more expensive public works.

Donors or funding agencies Major projects in developing countries may be funded from international sources (UNDP, FAO, World Bank/IMF, etc.) or bilateral assistance (aid programmes of countries), all of which have their own criteria for approval of the competing demands for their resources. Once a country reaches an intermediate stage of development, funding will more often be from its own government. Non-governmental organizations (NGOs) bring their very limited funding along with their dedicated planning/implementation teams.

Decision-makers This is the jargon term for politicians who, from the national viewpoint, take the decision to go ahead with the plan. At local level this can be members of a council, but land use planning has such high political significance that the approval of major plans is unlikely to stop short of ministerial, and often presidential, level.

'You cannot please all of the people all of the time' is as true in planning as in politics. Clearly, interests of the land users should come first, since it is their needs that are the reason for starting projects of the problem-alleviation type. It is normal to give special attention to the poor, women, and minority ethnic groups. One can fairly balance local welfare against the interests of the national government, such as for export revenue or import substitution. Within a government, sectoral interests often conflict, as expressed in a 'Murphy's principle' of land use planning: 'For every serious land-use plan proposed by one part of a government, there will be another part of the same government which objects.' The planning team will still get paid if the project fails, although in practice their staff develop considerable personal commitment to its success.

Methods and guidelines

The efforts made to formulate guidelines for land use planning in developing countries came as a natural follow-up to work on the earlier stages of planning. By the early 1980s, standardized methods had been developed for soil survey and land evaluation, and had been widely applied. It seemed natural to suppose that the same could be done for the succeeding, and vital, stage of land use planning.

It did not work out that way, for many reasons. One was the immense variety of objectives in planning, another the different ways of doing things that have become established in each country. This last is determined in part by the socio-economic environment, but equally by modes of government, inherited legislation, and cultural propensities to approach problems in different ways – methodically, inspirationally, or otherwise. The range of objectives means that the requirements for information are equally varied. There are also the procedural steps, of planning and approval, required by different aid agencies. Some years ago a group of Dutch scientists decided to set out the disciplines required in regional planning and the activities; they assembled no less than 963 'identified activities'.[17]

The FAO's *Guidelines for land use planning*[18] was the outcome of many drafts and discussion meetings, delaying its appearance by several years. It is a landmark, although by no means the last word. It is of interest that, within two years, sales and distribution of this publication surpassed the distribution over 20 years of the *Framework for land evaluation*.[19]

The structure of the *Guidelines* is to set out ten basic steps in land use planning, for each of which checklists are given, stating who is responsible for the step and what activities are to be undertaken. Data sources and methods of collection are relegated to a separate, and summary, review. Because it has to cover all kinds of project, everything has to be written in general terms; most plans do not need to collect information on crop damage by elephants, but for a few it is vital. Since these guidelines mark a stage of progress in efforts, extending over more than 50 years, to devise methods for applying land resource information to practical development purposes, it is worth while setting out in outline the basic steps.

Steps 1–6 are, in essence, part of the procedures of land evaluation, whilst steps 7–10 cover what is done in preparing and executing a development plan. Funding is needed at two points: at the start, to prepare the plan, and, following the decision to go ahead with it, for implementation. The steps, stating who is responsible for each, are as follows:

1. *Establish goals and terms of reference* Decision-makers and planners jointly decide the goals of the plan, namely the problems it is designed to solve and the expected outcomes. Hopefully, this is done after consulting the local land users, and with their active approval. There may be intense lobbying to be resolved, for example over the spatial limits of the project. The goals are converted into project form as terms of reference. A rapid review of the available information is conducted, to see what further surveys will need to be done.
2. *Organize the work* The planning team, building on past experience of similar projects, lists the tasks to be done, and the staffing and other resources required, and draws up a time-chart of activities for the steps of plan

STEPS IN LAND USE PLANNING

1. Establish goals and terms of reference.
2. Organize the work.
3. Analyse the problems.
4. Identify opportunities for change.
5. Evaluate land suitability.
6. Appraise the alternatives.
7. Choose the best option.
8. Prepare the land use plan.
9. Implement the plan.
10. Monitor and revise the plan.

preparation, leading to its appraisal and, it is hoped, the final approval to implement it.

3 *Analyse the problems* The planning team now gets to work, touring the area, talking to the people and local officials, and looking at the landscape. This will often be done on an *ad hoc* common-sense basis, but if a methodology is required, diagnosis and design is as good as any.

4 *Identify opportunities for change* Based on the goals and review of problems, the planning team work out possibilities for improved systems of improved methods of land use, the 'land use types' in the terminology of land evaluation. Ways are sought to achieve increased production, reduce land degradation, improve conservation or whatever are the objectives, and to reconcile conflicting interests. Such proposals need to be referred to decision-makers, probably at ministerial level, before proceeding to investigate them further.

5 *Evaluate land suitability* At this point the planning team is in a position to go ahead with the central procedures of physical land evaluation, a comparison of the requirements of different kinds of use with the properties of areas of land.

6 *Appraise the alternatives* The planning team continue with the next stage of land evaluation, in which the alternative courses of action formulated in steps 4–5 are assessed in terms of their environmental, economic, and social consequences. Economists have been gathering their own kind of data throughout the last two steps, and an attempt is now made to link the results of economic analysis with the assessment of land suitability (with varying degrees of success).

7 *Choose the best option* This is the point where land use planning goes beyond the stage reached by land evaluation. Based on the results from step 6, a decision has to be reached on what is going to be done. The planning team may formulate proposals for this, but the decision-makers will have the final say.

A goals-achievement matrix may be constructed as an aid, but that is not the way the decisions will finally be taken.

8 *Prepare the land use plan* At this point, a report is prepared stating what it has been decided should be done, how it shall be done, what it will cost, and reasons for decisions taken. This forms the basis for requesting the further budget needed for implementation, usually higher by an order of magnitude than all that has gone before, a request either to aid agencies or for an allocation from the national budget. Good presentation is critical at this stage. If funding is secured, there may be a second, less interesting, phase of plan preparation, involving such matters as putting out tenders for borehole construction.

9 *Implement the plan* A step which is self-explanatory, easy to say, less easy to accomplish well. Planning does not simply mean making plans. It means seeing that staff are recruited, paid, and supervised; that equipment is purchased and maintained; that extension staff really do visit farmers; and that periodic meetings with local people are held to review progress. This work will be done by sectoral staff, usually co-ordinated through a project execution agency.

10 *Monitor and revise the plan* 'The best laid schemes o' mice an' men / Gang aft a-gley.'[20] A plan is not something fixed, to be implemented exactly as prepared. Changes in circumstances will invariably call for revision; even before the first harvest has been gathered, there may have been a change in prices. Quite early on, some aspects of the plan will start to go wrong, or at least work out differently from what is intended. Meetings must be held to discuss this, and perhaps methods changed and budget allocations revised. The older monitoring and evaluation ('M & E') methods were rather narrowly directed at matters such as whether money was being spent as scheduled, economic performance (farm incomes, benefit:cost ratios, etc.), with physical performance measured mainly by crop yields and production. Where reversal of forest clearance, soil erosion or other degradation is an objective, monitoring should cover these aspects.

The inclusion of implementation as a step in the planning process is a matter of principle. There is a tendency among planners, not only in developing countries, to think that their work has been completed with the writing of the plan. All such preparatory work is wasted without the managerial task of getting things done, involving forward planning and constant supervision to see that instructions get carried out. To give one example, suppose an element in the plan is the planting of improved hybrid rice seed; then, every year, the officer given responsibility has to order this seed, and do it several months in advance to overcome supply and transport problems.

A range of computerized techniques have been developed with a potential to serve the needs of planning.[21] Geographic information systems allow the overlay of climate, soils, and other environmental information with present and potential land use, a powerful means of analysing and integrating spatial information. Databases supply easily accessible sources of reference; for example, databases on climate and soils can be linked with the environmental requirements and growth responses of crops, such as are found in the FAO database, ECOCROP.[22] There are computer programs for multiple-goal analysis, for example LUPLAN, which assigns weightings to the various objectives of a project, calculates quantitative measures of their success, and assesses the degree to which alternative options will achieve this.[23] Such aids seem so powerful that when one is using them it may be hard to believe that they cannot solve all problems. The fundamental limitation is that, whilst databases on economic and social data are possible, modelling does not adequately represent people and their attitudes – family obligations, for example, or the hidden agendas of government officials.

Efforts are still being made to achieve the objective called for under *Agenda 21*, an integrated approach to the planning and management of land.[24] It is known that final decisions will be taken by people, and desirable that this should be done on the basis of informed judgement. The best means of promoting such a state of information is by means of a decision support system. The first draft of such a system is the same in essence as the steps above:[25]

Physical land evaluation: compare land use with land (databases on land resources, and on requirements of systems of land use), leading to an identification of land use options;

Socio-economic evaluation: economic analysis (database on costs and prices) and analysis of social factors;

Combine physical and socio-economic evaluation: carry out multiple goal optimization exercise to maximize achievement of desired objectives;

Select the best land use.

Details of how the 'multiple goal optimization exercise' is to be carried out are not specified. In practice, the circumstances of planning decisions are so varied that standard, possibly computerized, methods are likely to be confined to analysis of parts of the system, with the final decisions based on informed judgement and common sense.

There are a wide range of techniques which can assist in making decisions, but also severe limits to the value of the approach of databases, modelling, and goals achievement matrices. In establishing a new sugar plantation, for example, if forced to choose between a battery of data-analysis techniques and a consultant with past experience of sugar production, one would unhesitatingly choose the consultant (and still more so if he is an experienced local person). Combining the two approaches, computerized analysis and personal judgement, provides a mutual check, against false assumptions in the analysis, for example, or the overlooking of some indirect consequences by the consultant.[26]

Environmental legislation

Developed countries have systems of extensive and carefully formulated legislation on land use and the environment. There is usually a two-tier structure of national and local laws and regulations, and systems of planning permission, tribunals, and appeals. By and large, the laws are effectively enforced.

In developing countries, attitudes to legislation have gone through two swings of a pendulum. During periods of colonial rule in Africa and Asia, many developing countries acquired a set of laws or regulations intended to protect natural resources and regulate land use. These commonly included a forest act, watershed protection act, national parks act, and soil conservation legislation. Irrigation schemes were controlled by local regulations. Under this legal framework, government forests were protected by patrols of rangers, with fines for infringements. Legal enforcement of soil conservation legislation always presented problems. Cultivation was sometimes forbidden on slopes above a certain steepness, and close to river banks, but who was to enforce these laws? Agricultural extension staff were there to help farmers, not to prosecute them, and it was not a task considered important by the police.[27] When populations were smaller and pressures on land less, benevolent but autocratic governments were broadly successful in conserving a legacy of resources for the future.

A period followed when these structures, to the extent that they had been successful in the first place, very widely broke down. Laws remained on the statute books, but newly independent governments were reluctant to enforce what was regarded as carrying a taint of colonialism. However, there were more fundamental reasons than this. With growing pressures upon land, it simply became impossible to enforce environmental laws. There were not enough forest rangers to patrol reserves. If the only way to obtain food was to cultivate sloping land, it was socially and politically unacceptable to forbid this. The laws were not updated, leading to anomalies such as maximum fines which had been rendered meaningless by inflation.

The pendulum has partly swung back, and there is recognition of a role for environmental legislation.[28] In *Agenda 21*, the programme of action from the UN Conference on Environment and Development (UNCED), Principle 11 is that, 'States shall enact effective environmental legislation.' This has to be placed on the same basis as other kinds of law: that the community as a whole accept the need for controls, a high proportion of people abide by them voluntarily, infringements are first met by warnings, and the cumbersome process of prosecution is reserved for a small minority. Assistance with the framing of legislation now forms part of environmental and natural resource development projects. It should be included as an element in national land use plans.

The continuing role of natural resources

The role of natural resource analysis in the earlier stages of land use planning, up to the completion of physical land evaluation (step 6 above) is clear and well established. This is when the soil surveyors, hydrologists, forest inventory specialists, and other resource scientists are brought into project time schedules, and carry out their main surveys. To some extent the other activities must wait upon their results. Once that stage is completed, the view sometimes encountered is, 'Good, now that we have a soil map, etc., we can get on with the real work. Thank you, and good-bye.' The later stages of project planning, together with implementation, monitoring and revision, are dominated by engineering, agronomy, and, above all, economics.

This view rested on the older attitude that natural resources are a fixed quantity; once their distribution has been surveyed, that is the end of the matter. This misconception has now been replaced by a dynamic view of resources as ecosystems interacting with land management, but the consequences are not always appreciated. Just as farmers and foresters are permanently dependent on management of soils and plants, so there is a continuing role for natural resource aspects in the later stages of planning.

The stage of choosing between development options is based on analysis of the consequences of alternatives in three respects: environmental, economic, and social. the environmental analysis can be based on predictive modelling of the effects of different land management systems on resources, or the same estimate based on experience and judgement. Economic analysis is highly dependent on inputs of physical quantities, such as fertilizers needed and crop yields obtained. In social and policy analysis there are some inputs too, for example, are most of the poor on marginal lands? How dependent are they on collection of forest products, or hill grazing? At this and the subsequent stage of preparing the plan, the natural resource scientists, who, from having conducted surveys, know about the areas, should participate on an equal basis with technical experts and economists.

At the stages of implementation, monitoring, and revision, not all the resource survey and land evaluation staff will be needed. But at least some of the soils, water, forest, and range scientists should participate in implementation, co-ordinated by a post such as Chief Land Use Planner. The measurement of changes in forest cover, range condition, water flow and sediment load, and soil properties should be a major element in procedures of monitoring and evaluation, ranking alongside economic performance. The need to give equal status to environmental monitoring stems from the concept of sustainability, in which the longer-term conservation of the resource base for future generations is assessed alongside immediate production and economic performance. Indeed, if a true economic value is placed on natural resources, then their improvement or degradation will be reflected in economic analysis, as changes in natural capital. Monitoring of natural resources and their economic valuation will be considered in chapters 9 and 10.

Land use planning procedures cannot be systematized except on a principles-checklist basis, but a framework is needed, even if only for educational purposes. After that, much must be done by empirical judgement and good sense. Decision support systems, such as databases and modelling, are well named: they provide information which helps in the tasks of reconciling conflict interests and balancing the merits of alternative developments. Decisions on land use are a matter of high politics, so national land use planning agencies will always find themselves in the position of giving advice to ministers. This gives added importance to sound analysis and, equally, good presentation. Land resource considerations are fundamental at all stages of planning and implementation. They should not be allowed to drop out at the decision-making stage, nor in implementation, monitoring, and revision. Consultation with land users, through local councils, farmers' associations, and public meetings, should be an integral part of planning at all stages. In the final analysis, nothing will get done without their willing and active participation.

7 Land degradation

Land degradation lowers the productive potential of land resources, affecting soils, water, forests and grasslands. If unchecked, it can lead to irreversible loss of the natural resources on which production depends. The severity of two kinds of degradation, soil erosion and rangeland degradation (desertification), have sometimes been subject to exaggerated claims. Both are indeed widespread and serious, but satisfactory measurements of their effects have yet to be made. Soil fertility decline is more widespread than formerly realized, leading to reduced crop yields and lowered responses to fertilizers. About 5% of the agricultural land in developing countries has been lost by degradation, and productivity has been appreciably reduced on a further 25%. Some 10% of irrigated land is severely salinized. In the semi-arid zone where water is most needed, the limits to water availability have been reached. Over the past 10 years, forest cover in tropical regions has been lost at 0.8% a year, and there is no sign yet that the rate of clearance has been checked.

The direct causes of degradation are a combination of natural hazards with unsuitable management practices. Underlying these are economic and social reasons, fundamentally arising from poverty and land shortage. There is a causal link between population increase, land shortage, poverty and land degradation. Tentative economic analysis suggests that degradation is costing developing countries between 5% and 10% of their agricultural sector production. This affects the people through reduced food supplies, lower incomes, greater risk, and increased landlessness.

Land degradation is the temporary or permanent lowering of the productive capacity of land as a result of human actions. It covers the various forms of soil degradation, including erosion and fertility decline; adverse impacts on water resources; deforestation and forest degradation; and lowering of the productive capacity of rangelands. Two other kinds of adverse change to land resources, loss of biodiversity and human-induced climatic change, also have effects, direct and indirect, on productive potential. These are not normally included within land degradation, and are discussed as global issues in chapter 8.

It is critically important whether a type of degradation is reversible, and if so over how long and at what cost. The complete loss of soil through erosion is, for practical purposes, irreversible; the weathering needed to form a new soil from solid rock takes a period of the order of 10 000 years. Forest clearance for shifting cultivation is followed by recolonization with secondary forest, with a degree of soil

TYPES OF LAND DEGRADATION

Soil resources

Soil erosion: water erosion
wind erosion

Soil fertility decline: chemical degradation
physical degradation
biological degradation

Salinization

Waterlogging

Soil pollution

Water resources

Lowering of the water table

Adverse changes in river flow

Water quality deterioration

Forest resources

Forest clearance (deforestation)

Forest degradation

Grassland resources

Rangeland degradation (including desertification)

improvement, in 10–20 years, but restoration of the timber potential takes five times as long. Where soil fertility decline is caused by loss of nutrients, these can be replaced, at a cost, by balanced fertilizer inputs; where the cause is degradation of soil physical properties brought about by lowering of organic matter, restoration will take a number of years.

The costs of rehabilitating degraded areas is invariably greater, typically 10–50 times more costly, than that of preventing degradation in the first place. The projects for restoring salinized soils on the irrigation schemes of Pakistan demonstrate both the technical possibility of reclamation and its high cost.

Two concepts which underlie reversibility are those of resilience and thresholds. Resilience is the capacity of land for recuperation by natural processes. Thresholds are conditions of a resource below which it will not be restored by natural processes. The trees in savannas can recover from bush fires where these are neither too frequent nor intense; if fires occur annually or late in the dry season, and hence at high intensity, young trees are unable to survive. Moderate intensities of erosion will lower but not destroy the fertility of soil, although the actual loss of soil material is irreversible. With increasing intensity, however, a point is reached at which sheet or rill erosion gives place to gullying.

FIGURE 9 Soil erosion, Malawi: gullying has lowered the water table, destroying the former potential of this valley-floor grassland (*dambo*) for dry-season grazing.

Land degradation has both on-site and off-site effects. On-site effects are the lowering of productive capacity as shown either in reduced outputs (crop yields, livestock, or forest production) or in the need for increased inputs to compensate for the losses. Off-site effects refer particularly to changes in river flow, water quality, and sedimentation, as brought about by forest clearance and resulting erosion. Wind erosion has the off-site effect of sand deposition. On-site effects are the concern of individual farmers, whereas off-site consequences affect the wider community. Different institutional structures are needed to control each of these.

The central questions to ask about land degradation are: how widespread and serious is it? What are its effects upon productivity? And to what extent are the changes irreversible? From a policy standpoint, what matters most is not how much land has already been lost, but the current rates of degradation, and hence loss in the future. These questions cannot be answered unless degradation can be measured and there are indicators of change. Hence the problem arises of how reliable are methods for the survey of degradation and estimation of its impact.

Soil erosion

Soil erosion has been regarded as the most serious and widespread form of land degradation. At the same time, erosion illustrates the difficulties that are encountered over reliable measurement, and the consequent existence of widely differing views about the severity of its effects and the need for action. The term erosion covers the effects of both water and wind, but water erosion has received considerably more attention.

Water erosion

The most catastrophic form of erosion is gullying, found as frequently on pastures as on cultivated land. Measurement of its extent is no problem. Examples of severe gully erosion are found in Lesotho and Madagascar. The effects extend beyond the land directly destroyed, in that gullies lower the water table and much reduce the value of pastures. In Malawi, gullies are now found along some of the *dambos*, valley-floor pastures. These have developed within the past 30 years; none were to be seen during a soil survey of the country conducted between 1958 and 1962.

Sheet and rill erosion are more widespread than gullying, but it is here that measurement problems arise. Under experimental conditions, measurement appears to be straightforward: a plot on sloping ground is enclosed, conventionally 20×2 m, and the water and sediment which flows out of its lower end are caught in troughs and collected in tanks. This gives figures for water runoff, as a percentage of rainfall, and for soil as loss in t ha^{-1} (tonnes per hectare) per year. Very large numbers of such experiments have been conducted, with consistent results. Losses under dense natural vegetation are likely to be less than 1 t ha^{-1} per year, under well-managed crops or with conservation works 1–5 t, whilst, for crops such as maize or tobacco on moderate to steep slopes without conservation, rates of the order of 50 t ha^{-1} per year are recorded in the savanna zone and upwards of 100 t in the humid tropics.

Experimental plots, however, set up an artificial situation. Without the constraint of plot boundaries, soil removed from one part of a slope is carried downwards and partly redeposited. It is possible to measure erosion rates on farmers' fields, using a portable trap called a Gerlach trough, but this has a high local variability and has only occasionally been done.[1] It is more common to sample the sediment load of rivers and divide by the catchment area. This has problems of diversity of land use and of bed and bank erosion, but is a means of monitoring change. Comparison of measurements has shown that erosion rates measured on plots can be up to 10 times higher than in fields, a finding which casts doubt on the whole structure of erosion theory and practice.[2]

FIGURE 10 Soil erosion, Haiti: these once forested and soil-covered slopes will never again be productive.

Removal of soil by erosion and its renewal by rock weathering are natural geomorphological processes, so the question arises as to the rate of loss that is acceptable. The basis normally used is called the soil loss tolerance, defined as the maximum rate which 'will permit a high level of crop productivity to be sustained economically and indefinitely'. Tolerances were established for different soils of the USA, mostly in the range 5–12 t ha^{-1} per year; these values were decided by 'collective judgement . . . subjectively evaluated at regional workshops'.[3] In current work in the tropics, a value near the top end of this range, about 10 t ha^{-1} per year, is commonly taken as a guideline, because on cropland it is difficult to achieve much below this rate in practice. This is equivalent to losing a soil thickness of 0.8 mm per year, or 8 cm per century.

This estimate rests on dubious foundations, particularly as regards sustaining production 'indefinitely'. This word implies that it is the rate at which soil is renewed by rock weathering. The latter has not often been measured, but studies of rates of natural erosion show these to be more typically 1 t ha^{-1} per year, and it is reasonable to assume that weathering keeps pace with erosion.[4] It may be that 'tolerable' erosion rates will sustain production for one or several generations, but, since they imply loss of nearly one metre of soil in 1000 years, they are not fully sustainable.

Because direct measurement has proved difficult, the use of modelling has been widespread. This is founded on the universal soil loss equation (USLE), which states that the predicted rate of erosion, in tonnes per hectare, is equal to the product of five factors of erosion: rainfall energy, soil resistance, slope angle and length, crop cover, and conservation practices. For the USA, calibration of this equation is based on many thousands of plot-years of experimental data. It is employed as a field guide to conservation; having obtained the predicted erosion for a site without conservation, erosion is reduced to the tolerable level by conservation practices as necessary.

The universal soil loss equation was devised for this limited practical purpose, not to predict erosion over landscapes as a whole, although it has been widely used in this way. Other models have been devised for conditions in southern Africa and Australia, and process-based models of some complexity.[5] A high proportion of estimates and maps of rates of erosion are based upon modelling, often without any field measurement at all. An example is a report which states that on 40% of all cropland in the USA erosion exceeds the tolerable rates, and on 23% it is over twice these.[6] Earlier attempts to map erosion repeatedly declined to use field observations and fell back on mapping erosion hazard, derived from modelling.

It is not sufficient to know the rate of erosion in terms of soil loss. Before the considerable effort and expense of conservation works can be justified, it is necessary to know the effects on plant growth and crop production.

An order-of-magnitude calculation illustrates the effect of the loss of plant nutrients. There is clear experimental evidence that concentrations of nutrients in eroded soil are over twice those in the soil from which they are derived, owing to selective removal of fine particles. Assuming that a typical topsoil contains 0.2% nitrogen, erosion of 20 t of soil will remove 80 kg of nitrogen, together with other nutrients. This is equivalent to carrying several bags of fertilizer away from each field every year! A calculation of this effect for the whole of Zimbabwe showed that, expressed as the cost of fertilizer needed to replace lost nutrients, erosion from arable lands was costing the country $150 million a year, plus a larger amount from pastures.[7] There is a further effect from loss of organic matter by erosion, causing degradation of soil physical conditions.

Research is currently being directed at this question, without clear results to date. There have been cases in which the heavy erosion on control plots without conservation does not reduce crop yields over the first few years, matched by others in which there is almost total crop failure. This has now become a priority for research. Rather than work with micro-plots, it is preferable to measure yield trends over time on a field scale, comparing treatments with and without conservation.

How extensive and severe is erosion at the present day, and what have been its consequences? Three kinds of indicators have been employed to express this: land lost (taken out of productive use) as a result of erosion, rates of soil loss, and effects upon crop yield. Typical statements of this kind are:[8]

> It is estimated that accelerated soil erosion has irreversibly destroyed some 430 million hectares . . . about 30% of the present cultivated land. (1990)
>
> Worldwide soil erosion has caused farmers to abandon about 430 million ha of arable land during the last 40 years . . . about one-third of all present crop-land. (1994)
>
> Soil erosion rates are highest in Asia, Africa, and South America, averaging 30 to 40 tons ha^{-1} $year^{-1}$. (1995)
>
> 'In this region [the Machakos Hills, Kenya] erosion rates of 70–140 t ha^{-1} yr^{-1} are common . . . In Peru, the entire Andean region is severely affected . . . Erosion rates of 71.4 t ha^{-1} yr^{-1} are reported . . . In Indonesia, Suwardjo et al. (1985) reported that 22 million hectares of arable land are severely eroded. (1990)
>
> A recent source (The Times, 30/8/88:5) suggested soil erosion is costing Australia A$2000 M a year. (1991)

Such statements are not necessarily exaggerated, but from a scientific aspect they are open to criticism. An author may seek authority by quoting another in the academic style, but often the reference cited turns out to be a secondary source; the second quotation above cites the first, and the first probably stems from a Russian estimate of 1977. Given the problems of measurement, the exactitude of '71.4 t' is ludicrous. It is common to take rates of soil loss obtained on experimental plots and attribute them to the landscape as a whole. Estimates become somewhat more reliable for specific countries and regions. Notwithstanding these reservations, there can be no doubt that erosion in excess of 50 t ha^{-1} per year is common where steeply sloping land is farmed without conservation, and that erosion at such a rate has extremely serious consequences for the future.

There is a further indicator, qualitative but more reliable, the evidence of field observation. The world's most severe soil erosion is possibly found in Ethiopia, Lesotho, and Haiti. In densely populated parts of the Central Highlands of Ethiopia, entire hillsides have passed the threshold into the catastrophic stage of erosion and have been abandoned. In Lesotho, gullying has irreversibly destroyed substantial areas. In Haiti, whole landscapes which were soil-covered when under forest have been cultivated and then abandoned as stony wastelands. Highland areas of Sri Lanka have been abandoned following smallholder tea cultivation in which a soil cover was not maintained. Severe gullying can be seen in Madagascar, central Nigeria, and the Himalayan foothills of Pakistan. The Kenyan district of Machakos has been quoted as an example of good soil conservation, but there is an area, possibly former communal grazing land, that is an abandoned red-soil waste. An aspect not often emphasized in accounts of erosion is that pastures do not only suffer from wind erosion; they are as severely affected by water erosion (both sheet erosion and gullying) as is arable land. Evidence from personal experience is necessarily selective and qualitative, but serves to confirm published accounts in the most convincing away. Anyone who expresses doubt about the seriousness of the erosion hazard should visit some of these regions.

Wind erosion

Wind erosion is generally given less attention, by scientists as well as the community, although it was the dust bowl of the USA during the 1930s that first drew attention to erosion. It is not a process in which soil is carried from one place to another. The fine particles are removed as air-borne dust, leaving sandy residues behind. Wind erosion is largely a feature of the semi-arid zone, although sandy soils of subhumid and temperate regions can be affected. It occurs both on pastures and cultivated land.

Measurement is still more of a problem than for water erosion, and there is no standard technique. Some of the soil loss equations have versions devised for wind erosion, from which modelled estimates can be obtained. In areas of nomadic grazing, attempts have been made to estimate the amount of ground lowering by the height of earth pedestals on which bushes stand, but it is hard to know how much of this results from the opposite effect, the accumulation of blown soil around stems. The wind erosion which takes place during the dry season of semi-arid lands is closely linked to water erosion which follows it at the first, often the most intense, rains.

Soil fertility decline

Soil fertility decline is a summary term to cover changes in soil physical, chemical, and biological properties which lower its productive potential. It is made up of the following changes:

chemical degradation: loss of plant nutrients, and also acidification;

physical degradation: adverse changes in soil physical properties, including structure and water-holding capacity;

biological degradation: lowering of soil organic matter, with associated decline in the activity of soil fauna.

Although sometimes identified separately, these are closely interrelated. In particu-

FIGURE 11 Fertility decline, Niger.

lar, lowering of soil organic matter is the main cause of physical degradation and also affects nutrient supply. Degradation of soil physical structure has substantial affects on plant yield independently of chemical properties.[9] Maintenance of the soil organic matter content is a key feature of management, since this underlies many other properties: resistance to erosion, structure and therefore water-holding capacity, and ability to retain and progressively release nutrients. Recycling of organic material also helps to prevent the development of deficiencies in micronutrients.

Erosion is itself a cause of fertility decline, through removal of organic matter and nutrients. Even with no erosion, however, fertility decline can be brought about by other processes, notably:

- nutrient removal in harvest exceeding replacements, by natural processes and fertilizers;
- soil acidification, caused by continued use of some types of fertilizers;
- failure to replace, by organic residues, the soil organic matter that is continuously lost under cultivation.

Evidence is accumulating that fertility decline is extremely widespread, particularly in areas that have long been under annual cropping. Indeed, although it is a reversible form of degradation, the total consequences on lowering current agricultural production may be greater than those of erosion. In the Indian subcontinent, where fertilizers have been in use for 20 years or more since the green revolution, reports of nutrient deficiencies are becoming common. The explanation is that farmers first added nitrogen fertilizer, and obtained a good crop response; after some years, the augmented growth led to exhaustion of soil phosphorus reserves, and phosphate had to be added also; now, the same process is happening with respect to secondary and micronutrients, such as sulphur and zinc.

A result of fertility decline is that responses to added fertilizers are now less than formerly. In India, Pakistan, and Bangladesh, rates of increase in fertilizer use have not been matched by crop yields. There are also records from long-term experiments. A striking example is a 33-year experiment in Bihar, India; despite changes to improved varieties, wheat yields declined substantially with nitrogen, phosphorus, and potassium fertilization, whereas they rose with additional farmyard manure. In Bangladesh, soil physical deterioration is believed to be a cause of lowered fertilizer responses.[10]

For low-input farming systems in Africa, there are two indications of nutrient loss: nutrient balance studies and observed declines in crop yields. The soil nutrient balance can be estimated by comparing inputs (natural processes and fertilizer) with outputs (natural processes, accelerated processes and harvest). Such assessments have been made at country level for Africa, and also at district and farm levels for sample areas. Annual losses for Sub-Saharan Africa average 10–20 kg ha^{-1} per year for each of nitrogen, phosphorus, and potassium. The highest rates of nutrient depletion are found in Ethiopia and Kenya, followed by other East African states, Madagascar, and Nigeria.[11]

In 12 African countries, reported yields of cereal crops have declined over the past 10 years, in some cases (Angola, Liberia, Nigeria, Somalia) by more than 20%. Political strife accounts for some cases, but others result from continued cultivation with low inputs. In Malawi, during a 25-year period of stable government, yields of unfertilized local maize in four districts declined by 60%. For the country as a whole, fertilizer use has risen from almost zero to 50 kg ha^{-1} but maize yield has remained almost constant at 1.0–1.2 t ha^{-1}.[12]

Soil pollution can result from mining or industrial operations, and from agriculture. Mining wastes can lead to loss of land, for example the tin tailings of Malaysia and the 'red mud' produced by bauxite mining in Jamaica; both these examples are well controlled. Soil and groundwater pollution from agricultural activities has been up to now mainly a problem of temperate countries. As fertilizer and pesticide use increase in the developing world, it will become an increasing problem, calling for technical appraisal and legislative control.

FIGURE 12 Salinization, on irrigated land of the upper Indus plains, India.

Salinization and waterlogging

Salinization in its broad sense covers all types of degradation brought about by increase of salts in the soil. It thus includes both the build-up of free salts in the soil, salinization in its strict sense, and sodification, the replacement of cations in the clay complex by sodium. It is brought about through incorrect planning and management of canal-based irrigation schemes. Part of the water brought into the area is not used by crops but percolates down to groundwater. This leads to a progressive rise in the groundwater table, and, when this comes close to the surface, dissolved salts accumulate. Patches of salinized soil appear, as more or less circular areas of white, saline soil surrounded by a belt of stunted crop growth. A continued rise leads to waterlogging.

This process happened extensively on the Indus plains of Pakistan; the water table began to reach critical levels in the 1940s, and salinization has since become widespread. A sequence of costly reclamation schemes were necessary to check the rate of land abandonment, some 36 projects covering 3.7 Mha. Salinization is a feature of many early dam-and-canal irrigation projects. It can be prevented by construction of deep drains. Reclamation is a more complex process, involving tubewell construction, large-scale pumping to lower the groundwater table, followed by application of water much in excess of irrigation requirements in order to leach out salts, a wasteful and expensive procedure.

Because salinization is easy to identify, and also takes place on the 'managed' environments of irrigation schemes, estimates of its extent are somewhat less unreliable than those for other forms of degradation, meaning that their range of error is not much above plus or minus 100%. For Pakistan, a total of 4–5 Mha may be affected, about 15% of the irrigated land, of which at least one quarter is severely salinized. Most major irrigation schemes of India have also experienced salinization and waterlogging, and the problem is by no means confined to Asia.[13]

The global extent of soil degradation

The first attempt to estimate the severity and extent of soil degradation on a world basis was the Global Assessment of Soil Degradation (GLASOD).[14] As little or no quantitative data were available, the problems of mapping and comparative assessment of severity were estimated by the following method:

define soil mapping units, and types of soil degradation;

establish semi-quantitative definitions for degrees of severity of degradation;

ask national collaborators in each country to assess, for each mapping unit and each type of degradation, the percentage of the area affected at each degree of severity.

The original returns therefore give data of the following kind:

Nepal, Map Unit 14, area 450 km^2

Water erosion: moderate degradation 40%, strong degradation 15%.

Causes of degradation, and its current trend were also reported. The reports by national collaborators were moderated at a regional level.

A key feature of this study was that the degrees of severity were defined not in physical terms, as soil loss, nutrient decline, or the like, but on the basis of effects upon agricultural production. This was done partly because of the lack of systematic physical data, and also because it allowed comparison between different types of degradation. More importantly, it is the consequences for production which people wish to know. Four degrees of degradation were defined:

Light Somewhat reduced agricultural productivity; can be restored to full productivity by improvements to land management.

Moderate	Greatly reduced productivity, but still suitable for use; major improvements are required to restore productivity.
Strong	Non-reclaimable at farm level, biotic functions largely destroyed; major engineering works required for restoration.
Extreme	Unreclaimable.

The original returns give type, degree, and extent of degradation separately for each map unit. In presenting the results as tables, only the dominant form of degradation is given, resulting in a large loss of information; the original records can be obtained from a database. For mapping, an attempt was made to represent four different phenomena, the types of degradation, on one map, using a colour-shading system which combined degree and extent as 'severity' of degradation; the colours showed only the type of degradation that was most severe for each unit. This is cartographically unsatisfactory. Maps were subsequently produced showing each type of degradation separately.[15]

Out of a world land area of 13 000 Mha, 4300 Mha are deserts, mountains, rock outcrops, or ice-covered, leaving a balance of 8700 Mha of usable land, meaning land with potential for cultivation, grazing, or forestry. For developing countries (Table 3), about 1500 Mha or 25% of usable land are affected to some degree by degradation. The percentage degraded is highest in Africa and Asia, and lowest in South and Central America. About half the area of arable land and a quarter that of permanent pastures is degraded. Water erosion is given as the most widespread dominant type of degradation, with 836 Mha in developing countries, followed by wind erosion affecting 456 Mha, soil chemical and physical degradation 241 Mha, and salinization and waterlogging 81 Mha. These data are misleading, however, as they refer only to what was judged to be the dominant type of degradation; all land affected by erosion has necessarily also suffered loss of nutrients.

These results are based on personal judgement, there being no other way to obtain estimates at present. They are thus open to the accusation that they have been produced by 'interested parties', soil scientists whose institutional needs will be served if governments recognize the seriousness of soil degradation. There is no way of countering this view until objective methods of measurement are devised.

More reliance can be placed on the estimates of strong and extreme degradation. The definitions imply that these refer to land that is largely destroyed, and probably abandoned from agricultural use. Moreover, since they refer to gullies, hillsides stripped of soil, salinized patches, and the like, such degradation is relatively easy to recognize and assess in semi-quantitative terms. The total world area of strongly degraded land is 305 Mha, of which 224 Mha is due to water erosion and 21 Mha to salinization. About 95% of this is in developing countries, and amounts to 5% of

Table 3. *Results of the first Global Assessment of Soil Degradation (GLASOD)*

	Usable land	All degrees of degradation		Strong and extreme degradation	
	Mha	Mha	%	Mha	%
Africa	1 663	494	30	129	8
Asia	2 779	748	27	109	4
South/Central America	1 714	306	18	48	3
Developing countries	6 156	1 548	25	286	5
Developed countries	2 555	417	16	43	2
World	8 711	1 964	23	305	4

Sources: Oldeman (1994) and L. R. Oldeman (personal communication).

their usable land. If it is assumed that most of this loss has taken place over the last 60 years, probably at an accelerating rate, the current loss becomes at least 5 Mha per year.

The 21 Mha of severely salinized land, probably representing saline patches that have been abandoned, is also largely in the tropics. It amounts to over 10% of the irrigated area in developing countries.

These data mean that for the first time it is possible to make reasonable estimates of figures which in the past have often been quoted without evidence:

> *An area of about 1500 Mha, or 25% of usable land in developing countries, has been affected by soil degradation of some kind, to a degree which appreciably (10% of land) or greatly (15% of land) reduces its productivity.*
>
> *About 300 Mha, or 5% of usable land in developing countries, have been so severely degraded, mainly by erosion, that for practical purposes they can be regarded as lost.*
>
> *The current rate of loss is not less than 5 Mha per year, or 0.3% of usable land of developing countries.*

There have been some later modifications of the GLASOD results. In a review of land degradation in South Asia, it was noted that nutrient loss was reported to affect over 60% of Bangladesh and Sri Lanka, but was scarcely present at all in India or Pakistan. This was a clear case of observer bias, the respondents for these countries not being alerted to the phenomenon, and subsequently it was recognized.[16] In the light of widespread evidence, the estimates for fertility decline were raised to 26 Mha for India and 5 Mha for Pakistan, 16 and 20% of cropland respectively.[17]

The GLASOD study is currently being revised on a more detailed scale, beginning with South and South-East Asia; first results show that water erosion is considered less extensive than in the earlier survey, but wind erosion and fertility decline more widespread.

Water shortage and water resource degradation

Groundwater

Water resource degradation concerns groundwater and surface water (rivers and lakes), and covers changes in quantity and quality. This gives a primary division into lowering of the groundwater table, pollution of groundwater, adverse changes in river flow regimes, and river pollution including sedimentation.

Groundwater is replenished by downward seepage of the rainfall not used in the shorter element of the hydrological cycle, from soils via plants to the atmosphere. Under natural conditions, recharge is balanced by groundwater seepage into rivers. Abstraction from wells introduces a new element in the cycle. Where this is from shallow wells dug by hand, there is no substantial impact, but mechanical pumping from tubewells leads to much higher rates of use. Above a certain threshold rate, determined by local hydrological conditions, further pumping lowers the water table.

Overpumping has become a major problem in India, China, Iran, North Africa, the Middle East, Mexico, and the western United States.[18] Some countries of North Africa and the Middle East, Libya and Saudi Arabia for example, are knowingly treating water like oil, as a non-renewable resource to be mined. In areas of tubewell irrigation, such as the Punjab plains of India and Pakistan, the water table is falling at 0.5 m per year; the first to be affected are the poorer farmers with more shallow wells, whilst those with deeper wells can continue pumping.[19]

Water pollution by agricultural chemicals is a problem of developed countries with high inputs, such as The Netherlands. In the tropics, the main type of groundwater quality degradation is salinization, caused by leaching of salts in irrigated areas. Increasing salt levels render the water successively unfit for human drinking, for livestock watering, and, ultimately, too saline for use in irrigation. In the Indus plains of Pakistan and India, areas of non-saline ('sweet') water are found along lines of major rivers, maintained by downward seepage, whilst saline groundwater develops along the zones between rivers (*doabs*). For obvious reasons, the non-saline water is overpumped, lowering the water table; whereas there is little pumping of the saline groundwater, leading to rising water tables and soil salinization. Near coasts and in island countries, overpumping leads to intrusion of saline sea water.

Groundwater shows, in the clearest way, the problems of management of an open-access resource. When communities used hand-dug wells with low levels of abstraction, the water came as a free gift of nature. With the advent of power-driven pumping from tubewells, absence of controls led to overexploitation and inequitable distribution. In such circumstances, there is no alternative to a system of national controls; these must be designed by a combination of scientific survey with participatory planning, and then fairly and effectively enforced.

Surface water

Degradation of surface water refers to changes in river flow regimes and to drying up of lakes. The main adverse change to rivers is that flow becomes less regular. In particular, rivers which were once perennial cease to be so, drying up towards the end of the dry season when water is most needed; women have to walk further and further, sometimes taking several hours to fetch the day's water supply. The effect is upon water resources, but the main cause is deforestation of headwater catchment areas. Rainfall runs off the surface instead of sinking into the soil, giving flooding after storms and leaving less of the base flow from groundwater.

Where the deforested areas are also affected by soil erosion, sedimentation lowers the quality of the river water and causes reservoir sedimentation. Most Indian reservoirs will become silted up after less than half the lifetime that was initially predicted for them.

Many major rivers cross frontiers, and hence the countries downstream depend for water on flow from outside their boundaries. The most international river is the Danube, common to 12 European countries, but in the tropics the Niger, Nile, Zaire (Congo), Zambezi, Mekong, Indus and Ganges are all shared rivers, calling for international cooperation in planning for their use. The hard truth is that countries with upper sections can withhold water from those downstream, presenting a serious source of conflict, such as that which took place over the Indus waters following the partition of India and Pakistan in 1947. Egypt is almost entirely dependent on the Nile both for irrigation and all other uses of water, yet Sudan and Ethiopia have plans to abstract considerably more water upstream, a potential cause of future conflict.

A case in which human activity has surpassed the capacity of natural recharge is in the lowering of lake levels as a consequence of irrigation. The Aral Sea has been greatly reduced in area through abstraction of irrigation water from its feeder rivers, with loss of both natural ecosystems and productive capacity for fish. There is little practical possibility that this change can be reversed.

Water pollution and human health form a further element in water resource policy. In developed countries, the main concerns are river pollution by industrial wastes, a problem now generally under control, and pollution of groundwater by nitrates of fertilizer origin. In the less-developed world, much ill health is caused by taking drinking water from untreated river water, sometimes polluted by human wastes and carrying disease. Both for economic efficiency and on humanitarian grounds, the provision of safe drinking water for a high proportion of the world's poor is currently a United Nations development priority.

Water shortage: reaching the limits

The four elements of water resource degradation, quantity and quality of ground and surface water, come together in the problem of water shortage. A point is reached when the demands of growing populations exceed the supply capacity of the hydrological cycle. This has occurred in some 20 countries, and parts of many others. A working rule sometimes used is that a country experiences water shortage if supplies are less than 1000 m^3 per person per year. However, requirements are two or three times higher in countries dependent on irrigation. Hence it is countries with the greatest problems of water shortage that are the most dependent on water for food security.

Water requirements for food production are about 2.5 m^3 per person per day for plant food plus up to the same amount again if water for animal food (pastures, fodder crops) is added, a total of 1600–2000 m^3 per person per year for a balanced diet. In humid regions this can all come from rainfall, the so-called 'green' water. Semi-arid regions obtain about half from rainfall, and must find the other half from rivers and aquifers, or 'blue' water, whilst arid regions must find their whole supply from this source. As population grows, it becomes necessary to increase the percentage utilization rate of 'blue' water resources, and this becomes progressively more costly. Currently, some 1500 million people live in regions of water scarcity, mainly in North Africa, the Middle East, and South Asia. By 2025, aggregate water requirements will exceed accessible runoff, and half the world's population will lack the water needed for food self-sufficiency.[20] In the struggle to match population increase with food production, water is a major constraint.

Clearly, it is countries with the driest climates which first encounter water shortage, for example Saudi Arabia, Somalia, Yemen, Israel, and most countries of North Africa and the Middle East. Egypt 'imports', in the flow of the Nile, most of the water needed for its food. Egypt, Libya, Saudi Arabia, Mauritania, and some Central Asian countries are currently overdrawing their groundwater resources. Island states, such as Barbados, also experience chronic shortage. Some countries of the subhumid zone, Kenya and Malawi for example, have also reached the stage when there are permanent shortages in some regions. In Kenya's capital city, Nairobi, the daily 'queue' of 100 or more water containers, lined up by women beside a standpipe, is a common sight.

For water resource management in general, and large-scale irrigation in particular, there are two requirements over and above those of scientific assessment and planning; these are good management and good government. Individual farmers can conserve water and make efficient use of it, but can do nothing to control supplies. Those at the lower end of a canal distribution scheme are the first to experience shortage, as are owners of the shallowest wells.

As with all questions of resources there is a threefold basis for policy: scientific assessment, economics, and social needs. Water availability is surveyed, matched with demands, and requirements for sustainable use determined; costs of management alternatives are determined and different users consulted, with special regard to the poor. As a statement of principle, this would not be disputed. However, sustainable use then requires that the resulting management scheme should be applied with a reasonable degree of efficiency and absence of corruption, and this in turn calls for good government. Water management needs to be controlled at national level, by a set of reasoned policies, effectively and fairly administered.

Forest clearance and degradation

Forests are lost or reduced in value by clearance, degradation, and fragmentation. In clearance, forest or woodland is replaced by another land use, most commonly agriculture. Forest degradation means that the land remains under trees but their number or quality is reduced. This happens either because the best timber species are extracted by logging, or because cutting for firewood and domestic timber exceeds the rate of natural regrowth. Fragmentation greatly reduces the biological conservation value of forest, and areas affected are large; thus, for Brazil, whilst 23 Mha had been deforested by 1988, an additional 59 Mha had an edge effect, defined as being within 1 km of the forest margin.[21]

Whilst overcutting and degradation are locally severe, the main concern at national and international levels is forest clearance. This has been monitored through FAO world forest resource assessments conducted every 10 years. These rest on a definition of forest as land with a minimum crown cover of trees of 10% in the tropics or 20% in the temperate zone – a foresters' definition, since it includes areas which the layman would call savanna or thorn scrub. Fortunately the latter are identified as open forest, leaving the term 'closed forest' for formations with a crown cover of over 20% and trees more than 7 m high.

Forest cover can be measured by records of forestry departments, based on maps and ground survey, and by remote sensing. Since a forest cover is clearly identifiable on satellite imagery, it might be supposed that, by combining these methods, relatively accurate estimates would be available. This is far from being the case. Estimates of tropical closed forest during the 1980s ranged from 800 to 1400 Mha; the range of estimates for Brazil alone was 1.7–8.1 Mha. A preliminary FAO estimate for 1990 of 16.8 Mha was successively revised to 12.2, 10.7, and 13.0 Mha.

FIGURE 13 Forest clearance for cultivation, the Solomon Islands.

FIGURE 14 Forest clearance: after logging, this rain forest was cleared for the Jengka Triangle land settlement project, Pahang State, Malaysia.

The reason is that, 'a great majority of the tropical countries have insufficient institutional capacity to collect and analyse data on a continuous basis'. If data for a single year vary so widely, then confidence in estimates of the rate of forest clearance is even less. A survey of the three major forest regions of the tropics based on satellite imagery from the 1970s to the 1990s is in progress.[22]

With these reservations, the best information available is the FAO assessment for 1990 and the changes it shows since 1980 (Table 4).[23] Total tropical forest and woodland cover fell from 1900 to 1750 Mha, an annual loss of 0.8%. Rain forest fell from 760 to 720 Mha over the same period, and all moist forest (rain forest plus moist deciduous) from 1410 to 1300 Mha, the latter a loss of 10.7 Mha a year or 0.76%. The FAO survey voices the suspicion that, because it is based on official data, the true 1990 total could be lower – meaning that governments close their eyes to illegal clearance. Hence the total tropical moist forest for 1990 could well be 1100 Mha and the current loss 1% per year. By setting tropical deforestation against population increase, the relative changes become considerable. If current rates of clearance continue, the area of forest per person will fall from 0.6 ha in 1990 to 0.2 ha by 2020.

Clearance was most rapid in Asia and the Pacific at 1.2% per year. Bangladesh, Pakistan and the Philippines had clearance rates of over 3% a year, in the first two cases on a low base. The highest rate of loss is given as 7% for Jamaica. Forest plantations in the tropics are 31 Mha, meaning that the total reafforestation to date is equal to less than three years of current clearance.

The estimated loss of biomass between 1980 and 1990 is 2500 Mt per year. Loss of species is estimated at 2–4% over 10 years, although this figure is contentious.

In the temperate zone, the position is entirely different. Regional changes of plus or minus 0.1% a year balance out to give a stable total of just over 2000 Mha. This figure is relatively accurate.

The international community is strongly opposed to further forest clearance, not primarily out of concern for the needs of developing countries themselves, but because of effects on atmospheric carbon dioxide and loss of genetic resources. This is despite the fact that the forests which once covered humid parts of Europe and North America have long since been cleared. National governments in the tropics mostly support policy statements which seek to check clearance, although they have to set this aim against the quick revenue to be derived from logging. Governments should also be highly concerned about their fuelwood shortages, although this form of suffering by the people does not receive the attention given to food shortages.

Efforts to check clearance are co-ordinated through the Tropical Forestry Action Plan (TFAP), launched in 1985.[24] This is converted into country action programmes, in the first instance by holding a series of 'round table' meetings at which participa-

Table 4. *Results of the 1900 tropical forest resources assessment*

	Forest area (Mha)		Annual loss, 1980–1990	
	1980	1990	(M ha)	(% per year)
By region				
Africa	569	528	4.1	−0.7
Asia	350	311	3.9	−1.1
Latin America	992	918	7.4	−0.7
Total tropics	1910	1756	15.4	−0.8
By forest formation				
Rain forest		718	4.6	−0.6
Moist deciduous		587	6.1	−0.9
Dry deciduous		179	1.8	−0.9
Very dry zone		60	0.3	−0.5
Upland formations		204	2.5	−1.1
Other		8	0.1	−0.9

Sources: FAO (1993b, 1995a).

tion by all stakeholders is emphasized. Such meetings have no effect unless translated into action. The costs of the country proposals under the plan range from $15 M for small countries to $500 for a project to rehabilitate 18 watersheds in India where sedimentation of reservoirs was most serious. Attempts to implement the plan are currently in progress, although there are as yet no clear indications of success. An overall appraisal must await the results of the tropical forest inventory for the year 2000. A recent study by the World Wide Fund for Nature, based on satellite imagery, found continuing or, in some regions, accelerated recent rates of clearance; Brazil's deforestation 'continued out of control', with forest loss increasing by more than a third in the five years 1992–7.[25]

The main force which leads to deforestation is demand for agricultural land, and much of the clearance is illegal. Commercial logging can lead to forest degradation through the selective removal of the best species. This can be prevented in principle if a set of controls for cutting and rehabilitation are included in logging license agreements, but enforcement is frequently inadequate.

The forces which lead to deforestation, primarily demand for agricultural land, bear no causal relation to those which favour reafforestation. In temperate regions, the end of clearance more or less corresponded in time with the beginning of demands for forest plantations; in Britain, this turning-point in the forest cycle came after the First World War. In the very different circumstances of the tropics, there is no reason why these two trends should coincide in time.[26]

Fuelwood shortage and forest degradation

Pressures leading to forest degradation come from cutting for fuelwood and domestic timber. Fuelwood (including charcoal) accounts for 40% of world wood consumption and three quarters of the wood used in the developing countries. The fuelwood requirements per person in the tropics are 1.5–2.5 m^3 per year, larger in highland zones. Once a threshold is passed where the rate of cutting in a country exceeds the annual regrowth, there is an automatic acceleration in the rate of degradation.

Two situations can be defined, deficit and acute scarcity. The deficit situation occurs where needs are met only at the expense of overcutting (cutting in excess of the rate of regrowth), thus constituting a non-sustainable use of the resource. Acute scarcity means that the fuelwood deficit is so large that even overcutting does not provide the people with their minimum needs. It was estimated that, in 1980, 1300 million people were living in regions with a deficit situation, and 110 million under acute scarcity.[27]

Deficit and scarcity occur in two distinct situations: the semi-arid zone where forest is sparse and rates of regrowth low, and parts of the subhumid and humid zones with high population densities. Most countries in the sahel zone of Africa, and in North Africa and the Middle East, experience shortages. Large parts of the densely populated highlands of Ethiopia have been almost entirely cleared of trees, and fuelwood comes only from short-rotation eucalyptus plantations. In Asia, there is acute scarcity in much of the Himalayan zone of India, Nepal, and Afghanistan. Cattle dung is often substituted for fuelwood, making it no longer available as fertilizer. The human side of fuelwood shortage is seen in the long distances walked, often by women, carrying headloads of wood from increasingly distant sources of supply.

Although falling somewhat in relative terms, the absolute total of fuelwood used is increasing; set against a steadily falling forest area, this is inevitably leading towards more widespread shortages. Once a deficit situation has arisen, with continued population increase and declining forest resources it is extremely difficult to check its development into one of acute scarcity with widespread forest degradation. One method available is to establish communally managed plantations, often on hilly land more suited to trees than cultivation. Another is through agroforestry, in which fuelwood is grown on farms, using methods such as boundary planting.

FIGURE 15 Overgrazing, Kenya.

Rangeland degradation: 'the desertification story'

Rangeland degradation is a specific type of land degradation, brought about primarily by overgrazing, and most widespread in the semi-arid zone. It begins with changes to the species composition and density of the plant cover, usually a replacement of the former mixture of perennial and annual grasses with dominance by annuals. The reduced plant cover leads to a lowering of soil organic matter and hence its resistance to erosion. Wind erosion then occurs, selectively removing the fine particles and also dormant seeds by which the plant cover should recuperate. Gullying also occurs during the short rainy season, especially where there is cattle trampling along paths and around watering points. A stage is reached when the combined vegetation and soil degradation becomes irreversible: the land will not recover to its former state, even if pressures are reduced.

The distinctive, if complex, process of rangeland degradation has unfortunately become linked with the poorly defined concept of desertification. This word came into use during a series of drought years which affected the sahel region, the semi-arid countries south of the Sahara Desert. Drought years are to be expected in the semi-arid zone, sometimes for several years in succession. One such sequence affected the sahel region of West Africa in the 1970s; there was widespread loss of livestock, and much suffering by the people dependent upon them. This rightly attracted concern, and led to the first UN Conference on Desertification in 1977. It was not to be the only such meeting.

The story which followed was a regrettable one, aspects of it bringing discredit to estimates of land degradation in general. First, desertification came to be used in an emotive way, in the hands of the media, as 'the advance of the desert'. A '*World map of desertification*' was produced which in no way showed how much environmental change had occurred, but was a map of potential hazard: climatic zones, soil types, and human and animal densities.[28] Having attracted funding support for the study of desertification, the word was sometimes extended to other forms of land degradation, even to soil erosion in the humid tropics.

Finally, and based on such an ill-defined concept not surprisingly, exaggerated statements were made for the extent of desertification. It was claimed that 3700 Mha were affected, and 20 Mha of land lost from production annually. Once such figures are copied from one source to another, they acquire a specious authority.

In 1984, in a 'General Assessment of Progress', the UN attempted a survey by asking respondents in 91 countries to estimate the extent of desertification. With a lack of clear definitions, and the prospect of aid money if the problem was severe, there was a temptation to make exaggerated reports. Definitions of 'moderate, severe, and very severe' desertification were proposed, based on loss of livestock carrying capacity. As this value is hard to determine even for a small area, it is not surprising that estimates for large world regions are unreliable. The initial estimates of two independent consultants, for the total area of desertified rangelands, were 3072 and 1615 Mha respectively. This was later 'adjusted' to a figure of 3475 Mha.

The situation then improved. It was recognized that if the term was to retain any meaning it should be restricted to degradation in the semi-arid zone.[29] A new definition was adopted, that desertification is land degradation in arid, semi-arid and dry subhumid areas resulting mainly from adverse human impact.[30] This correctly implies adverse changes not only to plant cover but to soil and water resources. For an improved estimate, in 1991–2 the UN drew upon data from the Global Assessment of Soil Degradation (GLASOD), as it applied to dry regions. This showed that just over 1000 Mha were affected to some degree by desertification, less than one third the area of earlier claims.[31]

What is so unfortunate about the desertification story is that it led to scepticism

INFORMATION ON LAND DEGRADATION

Extent and reliability of mapping and statistical data for less developed countries

Comments refer to the data, not the severity of degradation.

Erosion

Poor. Mainly hazard mapping, based on modelling. Very few measurements of erosion under field conditions, or its effects on productivity.

Soil fertility decline

Poor. Systematic soil monitoring under field conditions has rarely been attempted, although methods are available.

Water resource degradation

Moderate. Methods available, but insufficient long-term monitoring of river flow regimes.

Forest clearance and degradation

Moderate. Ten-year world forest resource assessments provide benchmarks. National forest monitoring often inadequate. Official data probably overstate extent of cover.

Rangeland degradation, including desertification

Very poor. Estimates of 'desertification' probably exaggerated, but true extent not known. Methods for monitoring, from remote sensing and field observation, available but very inadequately applied.

Loss of biodiversity

Good. Monitoring maintained by conservation organizations.

Climatic change

Good. Detection of non-random change difficult, but much data collection and analysis are being directed at this.

about what is without doubt a widespread and serious problem. Rangelands (other than enclosed ranches) are a communal rather than an open-access resource; that is, some degree of constraint over where and when to graze is exercised by communities. But the linked increases in human population and livestock numbers have led to pressures which the fragile environment of the semi-arid zone cannot bear. Development projects aimed at improved livestock production have a poor record of success, particularly in Africa. Schemes for group ranching encounter difficulties. The situation is often compounded by the weak political position of pastoral peoples.

Hence, for rangeland degradation, as was the case with soil erosion, we are left with two apparently contradictory statements. On the one hand, there is a serious weakness in evidence for its extent, severity, and changes over time. But, on the other, observation and experience show that it is certainly widespread and in places severe.[32] Techniques are available for measuring and monitoring dryland degradation or desertification.[33] What is now needed is a co-ordinated and sustained effort to apply these methods.

The contrarian view

Given the paucity of quantitative information about land degradation, it is not surprising that an opposing view has been put forward: that things are not as bad as they are claimed to be. There are two elements to what will be called (from the term used by investment analysts) the contrarian view: that the degradation is less serious than it is claimed to be, and that there is no necessary connection with population increase.

There have been instances where soil erosion or rangeland degradation were reported, but found to be less serious than was supposed. In each case there was a perceived problem of degradation, the cause was attributed to the behaviour of the land users, and it was claimed that, if unchecked, there would be irreversible degradation. On visiting the sites, the evidence did not support these statements. Lessons drawn were that one should obtain evidence which is more than opinion, always consult the land users, and avoid extrapolating locally severe damage to entire areas.[34] There are far more cases where degradation was not formerly known about but only discovered on inspection, but these do not get described in the same terms. Better methods of monitoring would resolve such situations.

The other element is that, contrary to the usual opinion, population increase does not necessarily lead to land degradation. In the 1930s, the Machakos District of Kenya was described as a seriously degraded area, and photographs taken of eroded grazing land. Since that time, population has greatly increased, but output per head and per hectare have risen steadily. The degraded areas shown on the old photographs are now protected by conservation banks. There are, however, today some areas of grazing land which are heavily eroded. A theory which has received considerable attention has been built upon this case study.[35] It is suggested that higher population densities force people to farm more intensively and therefore to conserve their land. This is an extension, as applied to land degradation, of a much earlier theory known as the Boserup hypothesis, which stated that people will only change their agricultural practices if and when circumstances force them to do so.

Favourable effects of intensification on soil fertility, through greater inputs of material and labour, including recycling through animals, have been described for two parts of Nigeria, the Kano region and the Jos Plateau.[36]

Machakos District lies close to the markets of Nairobi, and has been highly favoured by aid projects. In particular, it was part of a sustained soil conservation project, funded by Sweden and supported at the highest level of government. For the example from Nigeria, there is the existence of a high fertilizer subsidy, funded by oil revenue. These special circumstances do not necessarily refute the general case. In Nepal, the Philippines and Java, for example, dense populations subsist on steeply sloping land by systems of terraces, which can only be maintained by high labour inputs.

To the extent that there are examples where land degradation has been reversed through local initiative and effort, not dependent on development projects, this is encouraging. However, it should not be allowed to obscure the very widespread existence of erosion and overgrazing. The lesson to be learnt from the contrarian view is the basic principle of geographical study and common sense: before taking action over a reported problem of land degradation, study the area in the field, and talk to the people.

Causes of land degradation

The causes of land degradation are made up of natural hazards, direct causes, and underlying causes. Taking soil erosion by water as an example, the natural hazards include steep slopes, impermeable or poorly structured soils, and high intensities of rainfall. The direct causes are unsuitable management practices, such as cultivation without conservation measures, or overgrazing. The underlying causes are the reasons why such practices are adopted, such as the cultivation of slopes because the landless poor need food, and non-adoption of conservation measures because farmers lack security of tenure.

For soil degradation, respondents in the GLASOD assessment were asked to name the principal direct causes. The results are indicative although not quantitatively significant. Water erosion was attributed more or less equally to deforestation, agricultural activities – the cultivation of land naturally at risk without conservation measures – and overgrazing. Wind erosion is primarily due to overgrazing and, to a lesser degree, overcutting of vegetation. Soil chemical and physical degradation result primarily from agricultural practices. All of this is much as would be expected. In more detail, the direct, physical causes of degradation are well understood, for example the deterioration of soil physical properties which occurs when farmers try to maintain crop yields by fertilizer use alone, without measures to maintain organic matter.

The direct causes of other types of degradation have been noted above: salinization through mismanagement of irrigation schemes, lowering of groundwater through extraction in excess of recharge, and adverse changes in river flow and sediment load as an off-site consequence of forest clearance and erosion. Deforestation is more often caused by clearance for agricultural use than by clear-felling for timber. Forest degradation may be caused either by overcutting for fuelwood, domestic timber, and fodder by local people, or by selective extraction of the best species in commercial logging. All of these direct causes, combinations of unsuitable management practices on land at risk, are well understood.

Economic and social causes

To view the cause of soil erosion as a failure by farmers to adopt conservation practices, or deforestation as illegal clearance within forest reserves, is only part of the picture; the root of the problem lies in economic and social circumstances. This is clearly seen by asking the question, why did soil conservation programmes in the past so often fail? The basic reason was that they attempted to impose conservation by legal compulsion, without asking why it was that sustainable management practices had not been adopted in the first place. Some observers viewed the situation from a socio-political stance, seeing a need for changes in the social structure of society if measures to combat erosion, and land degradation in general, were to be successful.[37]

Land tenure is rightly seen as a basic obstacle. Farmers are reluctant to invest in conservation measures if their future rights to use the land is not secure. Two kinds of property rights lead to this situation, insecure forms of tenancy and open access resources. Tenancy as such is not to blame, provided that there is legal security of tenure. In the 1980s, following a World Conference on Agrarian Reform and Rural Development (WCARRD), there was an impetus on reform of land tenure. Land reform programmes were attempted in many countries, with limited success owing to opposition by strong vested interests.[38]

Land shortage, brought about by population increase, has become a fundamental cause of degradation. Once farms are too small to support the children of a farming family, and all the good land is taken up, migration to sloping, semi-arid, or other areas with high natural hazards takes place. Frequently this will require clearance of forest or savanna woodland. The considerable growth in numbers of the rural landless add to this pressure.

Recognition of the fundamental economic and social causes of land degradation has led to a radical change in methods to check it. Soil conservation is normally applied on a participatory basis, through the approach of land husbandry. Forests which serve the needs of local people are more likely to be conserved if responsibility for their management is given to the village or community. Group ranching schemes are an attempt, not always successful, to apply the same approach to open rangelands. Degraded water catchments have been rehabilitated by mobilizing the effort of local communities.

Population, poverty, and land degradation

The direct and indirect causes of land degradation are linked by a chain of cause and effect, sometimes called a causal nexus. The driving force is an increase in rural population taking place on a base of limited land resources. This produces land shortage, leading to small farms, low production per person, increasing landlessness, and, in consequence, poverty. Land shortage and poverty together lead to non-sustainable land management practices, the direct causes of degradation. Poor or landless farmers are led to clear forest, cultivate steep slopes, overgraze rangelands, or make short-term, unbalanced fertilizer applications. These non-sustainable management practices lead to land degradation, causing lower productivity and lower responses to inputs. This has the effect of increasing the land shortage, thus completing the cycle.

Land degradation is by no means caused only by the poor. Irresponsible rich farmers sometimes exploit the land, but, by and large, farmers with secure tenure and capital are more likely to conserve natural resources. When natural disasters

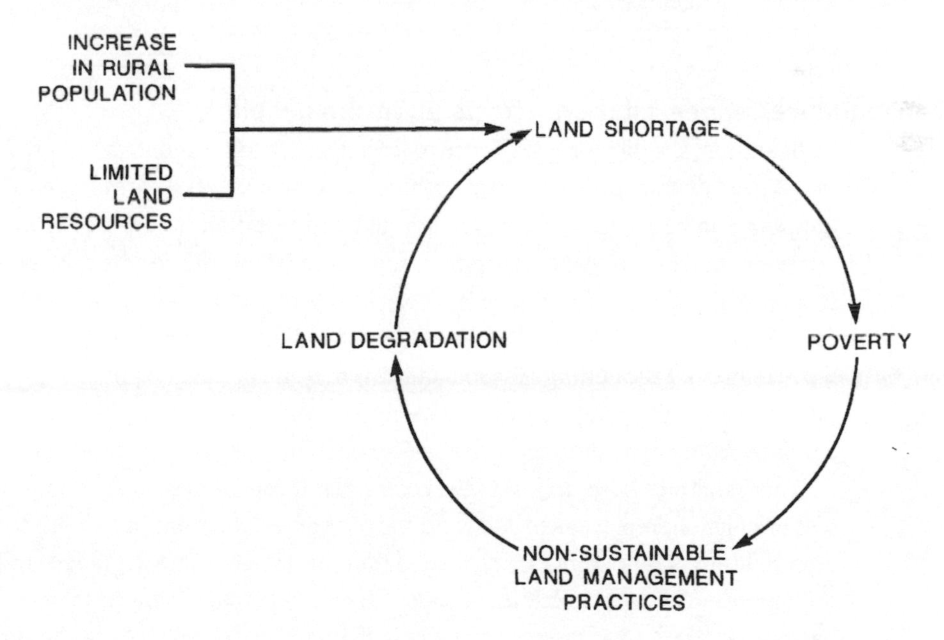

FIGURE 16 Population, poverty, and land degradation. Based on Young (1994a).

occur they can turn to alternative sources of income, or borrow and repay in better years. These alternatives are not open to the poor.

In the past, rural populations had access to adequate land to meet their needs. When a disaster occurred, whether of natural origin or war, there were spare resources to fall back upon. They could take new land into cultivation, kill livestock which fed upon natural pastures, or go into the forest and extract roots or hunt wildlife. Because of land shortage, these options are no longer available. Farmers are surrounded by other farmland, such common rangeland as exists is often degraded, and over large areas no forest remains. The options open are to work on the farms of others, non-agricultural occupations, enforced migration to the cities, or, ultimately, dependence on famine relief.

Through force of circumstances, it is the poor who take the major role in the causal nexus between land shortage, population increase, and land degradation. This link is now widely recognized. FAO writes, 'A lack of control over resources, population growth . . . and inequity are all contributing to the degradation of the region's [Asia's] resources. In turn, environmental degradation perpetuates poverty, as the poorest attempt to survive on a diminishing resource base'; and the World Bank states, 'Rapid population growth can exacerbate the mutually reinforcing effects of poverty and environmental damage . . . [of which] the poor are both victims and agents.'[39]

Consequences of degradation: effects upon the people

By definition, the direct consequence of land degradation is reduction in productivity. Where degradation is extreme, there is total loss of the resource. This is temporary in cases such as salinization and deforestation, and for practical purposes permanent in severe erosion. In soil degradation, the results may be lower crop yields, or a need for higher fertilizer inputs to maintain existing yield levels; soil physical degradation and micronutrient deficiencies cause lower responses to fertilizers. Once the cutting of woodlands for fuel and domestic timber passes their rate of growth, the potential of the resource is reduced, and can only be restored by a radical reduction in cutting, which is economically and socially unpractical.

There are three bases to assess the costs of land degradation in economic terms: lost production, replacement cost, and the cost of reclamation. In lost production, crop yields or other outputs are estimated for non-degraded and degraded land, and then priced. The two situations, with and without degradation, can then be compared. A weakness for the case of soils is the paucity of evidence for the physical reduction in output. A cleared forest has lost the capacity to produce timber for the 20, 30, or 50 years needed for its regrowth, besides which, continued use for agriculture will prevent any such restoration.

Replacement cost is based on estimating the costs of additional inputs, such as fertilizers, needed to maintain production at the same level as for non-degraded land. This is easier to assess than lost production, and also corresponds to what farmers seek to do. The higher pumping costs resulting from lowering of the water table is a similar example. The clearest example of reclamation costs comes from salinization. For Pakistan, the cost of reclaiming 3.3 Mha of salinized land has been estimated at $9 billion.[40]

For deforestation and erosion, off-site costs resulting from reduction of river base flows and sedimentation of reservoirs must be added. The presently assessed life of eight Indian reservoirs was compared with that anticipated at the time of their design; in four cases, this was 30–40%. In developed countries, off-site costs of erosion are often assessed as substantially higher than on-site loss of production, although in developing countries the opposite may be the case.

Full economic analysis requires complex assessments of future production or input changes, and this introduces questions of the discounting of costs and benefits over time, discussed in chapter 10. Provided the physical impacts can be quantified, however, it is possible to estimate the economic costs of degradation on an annual basis. This has been done for the South Asian region.[41] The GLASOD estimates were taken as a starting-point (modified by additional fertility decline, see above). Relative production loss for light, moderate, and strong degradation were assumed to be 5%, 20%, and 75% respectively, and these reductions applied to average cereal yields. Fertility loss was also assessed on a nutrient replacement basis. The cost to South Asia of land degradation was estimated as:

	Cost, $ billion per year
Water erosion (on-site costs only)	5.4
Wind erosion	1.8
Soil fertility decline	0.6–1.2
Salinization	1.5

Recent evidence would reduce the figure for erosion and considerably increase that for fertility decline. The total cost of land degradation to South Asia is about $10 billion a year, which is 7% of the agricultural production of the region. This estimate applies to soil-related forms of degradation, and refers only to on-site costs. The addition of water, forest, and rangeland degradation, together with off-site costs of erosion, would raise this figure substantially.

Similar estimates could be for the world as a whole, or for developing countries. Since the major objective of the GLASOD survey is to draw attention to the magnitude of the problem, and policy-makers respond most readily to results in economic terms, it would be valuable for the same organization to make a first estimate of economic costs. National land resource survey organizations should widen their horizons to do the same.

The results for South Asia may be taken as representative of the order of magnitude for developing countries as a whole. There is no reason to suppose that the South Asian subcontinent differs from Asia as a whole. The percentage of land affected by degradation is similar in Africa to Asia. It is just over half as extensive in tropical America, but for a smaller total area, but this difference is not significant at this level of guesswork. It is thus not unreasonable to say that the land degradation that has taken place up to the present is costing developing countries not less than 5%, and more probably nearer 10%, of their total agricultural sector production. Still more tentatively, this rate may be rising by 1% every 5–10 years. It is greatly to be desired that this highly speculative estimate should be replaced by a comprehensive study of the economic effects of land degradation.

The physical and economic consequences of land degradation are reflected in their effects upon the people. These include:

lower and less reliable food supplies;

lower incomes, resulting from loss of production or higher inputs;

greater risk: degraded land is less resilient, and higher inputs mean that poor farmers are risking more capital;

increased labour requirements, as when women walk large distances to collect fuelwood and water;

in the case of farm abandonment, increased landlessness.

In classical economic theory, 'land' was regarded as a fixed resource, to which factors of labour and capital were applied. With degradation occurring, it becomes a declining resource, and hence labour and capital become less efficient. If most farmers do not know about economic theory, they are well aware of it in practice. Land degradation means they must either accept lower production, or put in greater effort to maintain it at the same level.

Land improvement

Not all changes to land resources are in the direction of degradation. Its converse is land improvement, all relatively permanent increases in productive capacity brought about by human action. Irrigation is the most widespread kind of land improvement; besides its direct action in improvement of water resources, it often leads to an increase in soil organic matter and productivity. Swamp rice cultivation is a special case in which the natural soil profile is radically altered by the formation of a pan, an impermeable horizon which checks loss of water by downward seepage.

Other examples of land improvement are drainage of swamps, terracing, systems of water harvesting, and reclamation of land from the sea. Were it not for land improvement, the productive capacity of Egypt and Pakistan would be a small fraction of its present level. The best-known case of land reclamation from the sea is the polder system of The Netherlands, the only country which, in international statistics, regularly increases its total land area. The leading example in the tropics is Bangladesh, where land productivity and security have been raised by a series of flood control works. More generally, land management which reverses soil fertility decline, for example by raising soil organic matter levels, is a less spectacular but potentially widespread form of land improvement.

Land improvement also covers the reversal of degradation in its many forms, such as reclamation forestry on eroded hillsides, reclamation of salinized soils by leaching, and rehabilitation of degraded pastures. Watershed management projects have the objectives of land improvement through stabilizing river flow and reducing erosion and sedimentation, often by means of reforestation of degraded land. Controls on grazing frequency and intensity, best implemented by communal means, are the means towards pasture rehabilitation. In resource inventories, it is the balance between degradation and land improvement that gives the net change. Current monitoring of land resource changes is insufficient to be able to say precisely which countries are improving and which degrading their land resources, but the available evidence, not least from field observation, is that the balance is frequently negative. A leading objective of resource management should be to reverse this situation.

8 Global issues: climatic change and biodiversity

The atmosphere, oceans and biosphere are global commons, changes in which affect all human society. There is no doubt that human activity has caused global atmospheric changes, but the effects of these upon climate, and hence upon land productivity, are not known. What matters is not the climatic changes themselves, but preparedness by countries at risk to meet them. Conservation of biodiversity – plant and animal species, ecosystems, and genetic resources – is both a national and a global concern. Ecosystem services, such as regulation of the atmosphere, water, and nutrient cycles, and filtration of wastes, provide benefits to human society of the same order of magnitude as the whole of the world's gross national product. In developing countries, land shortage puts pressures on protected areas, and still more on ecosystem conservation outside these. Whilst international efforts help in the short run, the long-term solution can only come from recognition of the value of conservation by national governments and their people.

The atmosphere and oceans are natural resources which are shared world-wide. Human impact upon them results from the sum of actions by individual nations, and all have some degree of access to the resources they supply. These are commonly referred to as the 'global commons'.

For the case of the oceans, access is increasingly limited by legal constraints on fishing rights, enforced with difficulty and subject to frequent disputes. It appears that marine fisheries, other than under coastal aquaculture, have reached the limit of their sustainable yield. Effective controls to ensure sustainable use are an issue of much current importance, but one which lies outside the scope of this book.

A further factor of the environment is the natural biological habitats and species. Planning and management of these cover conservation of natural ecosystems, wildlife, biodiversity, and genetic resources. These certainly constitute an element in the land resources of individual countries, and their conservation is an element in national land use planning and resource management. At the same time they are regarded as part of the global heritage, and their conservation is a matter of concern to all mankind.

Policy and action on global issues can only be taken at international level; but international measures are necessarily the sum of the actions of individual countries, whether acting through agreed collaboration or not. These issues are therefore part of natural resource management at both international and national levels, and in the case of nature conservation also at district level.

Global atmospheric and climatic change

The issues involved in global atmospheric change have received wide discussion, both scientifically and among the general public, inevitably with some alarmist exaggerations. Leaving aside the pollution-related aspect of the loss of ozone, with respect to land resources there are essentially three stages to the question:

1. Are carbon dioxide and other greenhouse gases building up?
2. If so, to what extent is this affecting climate?
3. What will be the effects of climatic change on land resource potential?

Atmospheric monitoring was begun in Hawaii in 1957. Since that time, the level of atmospheric carbon dioxide (CO_2) has increased from 315 to 360 parts per million and the rate of increase shows no signs of reduction. From evidence of air bubbles trapped in ice sheets, we know that the major rise commenced with the Industrial Revolution, with a concentration around the year 1800 of 280 parts per million. The relative change over two centuries is 30%, half of this in the last 40 years. These rates are very much faster than any that occurred during the Pleistocene glacial period. Other gases which reduce outward radiation of heat from the earth, the 'greenhouse gases', are also increasing; the present-day atmospheric level of methane (CH_4) is approximately double its pre-industrial concentration, and that of nitrous oxide (N_2O) some 10% higher.

Changes in land use, primarily forest clearance and tillage for agriculture, are a substantial contributory factor to atmospheric changes, accounting for 30% of the rise in carbon dioxide and 25–40% of increase in other greenhouse gases.[1] The major cause, however, is fossil fuel consumption. An approximate doubling of the radiative effects of all greenhouse gases is predicted by the middle of the twenty-first century. It is extremely unlikely that this rise will be checked in the foreseeable future; any action taken by industrialized countries to reduce greenhouse gas emissions will be more than counteracted by industrialization in the developing world.

There is also no doubt that a small amount of global warming, an increase in the mean atmospheric temperature of 0.5 °C, has already taken place. This has been clear to scientists for some years, but only recently 'officially' recognized at a political level. In 1995, an Intergovernmental Panel on Climate Change agreed that, 'The balance of evidence suggests that there is a discernible human influence on global climate' – 'discernible' being accepted after lengthy debate on the alternatives of 'appreciable', 'measurable', and 'detectable'![2]

Returning to the scientific level, there is considerable uncertainty about the magnitude of further temperature change, with a possible rise of between 1.5 and 4.5 °C by the end of the twenty-first century. These values are small compared with the changes that have taken place over recent geological time. Effects upon global atmospheric circulation, and thus upon rainfall in specific regions, have been forecast by modelling. The complexity of global climate is such that there is still much uncertainty, but the two consequences which have been commonly forecast are higher temperatures in the north temperate zone and decreased rainfall, or a higher drought hazard, in the semi-arid regions of the tropics and subtropics.

Modelling has also been carried out to predict the effects on global food production, taking into account temperature, rainfall, and the direct physiological influence of higher atmospheric carbon dioxide. The trends are towards higher production in the grain-producing lands of the temperate zone but lower yields in the tropics; these effects are small in relative terms, and extremely uncertain. A rising sea level could cause immense problems of flood protection for cities, although there will continue to be gains in land from natural sedimentation and reclamation. In areas where rainfall is reduced, river flow decreases by a higher factor.[3]

Thus the answer to questions posed above are first, that a large and geologically unprecedented rise in carbon dioxide and other atmospheric gases has certainly occurred and will undoubtedly continue over the next 50 or more years, whatever actions governments may take to reduce emissions. Secondly, a small amount of global warming has already taken place and further climatic changes will continue, but the magnitude and nature of these is highly uncertain. As a consequence, answers to the third question, the effects on agricultural production, are highly speculative. They will not be unfavourable everywhere. The state of knowledge is well summarized in the conclusion to an FAO consultation on global climatic change and agricultural production. 'No clear-cut conclusions could be reached...on the geographic distribution of the impacts . . . or on the agricultural sectors most likely to be affected, because the longer-term biophysical responses to elevated atmospheric CO_2 levels [etc.] . . . are still incompletely known. At the regional and national level, moreover, the projections on changes in temperature and rainfall/evapotranspiration ratios are vague and contradictory.[4]

Further consequences arise from the direct effects of changes in atmospheric composition upon ecosystems.[5] Higher levels of carbon dioxide increase the photosynthetic efficiency (the rate of growth) of many plants, and also their water and nutrient use efficiencies. Some plants respond more than others, and those with faster growth have lower nutrient contents, altering the amount which herbivores must consume. There is also a large build-up of the stable gas nitrous oxide, N_2O, the consequences of which are less clear. More atmospheric nitrogen is fixed annually by man than by natural processes. These changes will certainly alter the balance of ecosystems, both natural and managed.[6] However, the magnitude of the net changes, whether on balance they will be favourable or adverse, and where they will occur, cannot be reliably predicted. A review meeting by FAO concluded that 'It is evident that the relationship between climatic change and agriculture is still very much a matter of conjecture with many uncertainties.'[7]

It is right that these great changes to the earth's environment should have become a cause of world concern. At international level, two lines of action can be taken. First, research into the processes and their consequences should continue to be funded. The primary objective is that if and when early warning signs of adverse consequences appear, for example a rise in sea level caused by melting of Antarctic ice, these should be recognized. Secondly, the current measures to reduce rates of greenhouse gas emissions should be pursued, since, whatever the consequences of change may be, adaptation to them will be easier if their rate is lessened. Developing countries are reluctant to modify their policies and economies, on the grounds of their relatively small contribution to global problems, and they rightly point to the vastly greater emissions to date from the western world. Yet with growing industrialization, concerted world action can only be brought about if developing countries accept a measure of responsibility for the global environment.

It should be made clear there are many actually or potentially serious problems concerning pollution aspects of the environment, discussion of which lies outside the scope of this book. Currently the most serious of these problems is atmospheric pollution in cities, which has already reached levels which are injurious to health and seriously affect the quality of life. Such pollution reaches the most severe levels in some Third World cities, Bangkok being an extreme example. Acidification of rainfall, with its adverse effects on vegetation, is another problem, the seriousness of which not in question. The consequences of ozone depletion are less well established. It is possible that another environmental crisis, so serious that it will cast global warming, the 'ozone hole', and other current concerns into the shade, will be encountered in the mid-twenty-first century. This will occur if the male sperm count, which in some Western countries has halved over the past 50 years, continues to decline at the same rate. The cause may be widely dispersed synthetic chemicals which mimic and disrupt male and female hormones.[8] All of these problems constitute valid and powerful reasons to control burning of fossil fuels, vehicle emissions, and other sources of pollution of the global environment. It is pollution that constitutes the major argument for reducing industrial and vehicle emissions, and not their uncertain nor entirely adverse effects upon global climate.

At national level, global climatic change must be met as and where its effects occur, by means of better land management. The effects will be small on average, and irregular in their spatial incidence. One of the most likely, an increase in frequency and severity of droughts on the semi-arid belt, may already have occurred in the African sahel although not elsewhere. The consequence is a further reduction in food security of semi-arid countries; but this is already such a serious problem as a result of population increase and land degradation that any effect of climatic change will add little in relative terms to the problem. 'Whether or not climatic change takes place, improving the resilience of food production and minimizing risks against weather variability are essential if agriculture is to meet the challenges of ensuring food security . . . This can serve at the same time . . . to prevent or mitigate the negative impacts of climatic change'.[9]

Climatic change as such is not dangerous: people will adapt to it. What is hazardous is climatic change allied to vulnerability. People will go short of food if they are vulnerable, and if their governments are not prepared.

Conservation of biodiversity

Biodiversity covers the three major components of biological resources: plants and animals as living species, their genetic basis, and the ecosystems within which they function. Biodiversity is a basic factor of land resources. Its conservation is an essential element of land use planning and management, and its loss a form of land degradation. From a national and district level policy viewpoint, biological resources should be considered alongside soils, water and other factors of land, with respect to survey, evaluation, planning, and degradation. The additional feature is that the world community also has interests in biodiversity, regarded as an element of globally common resources.

There is a distinction between the ecosystems within protected areas and those in the landscape at large. Protected areas cover a wide spectrum of degrees of protection, from nature reserves and wildlife refuge areas with very limited access, through national parks and forest reserves, to areas under agricultural use but with some degree of legal protection. There is a recognized international classification system for protected areas, formerly with 10 (partly-overlapping) classes, now reduced to 7 (nominally 6), which forms a valuable means of converting national terms to standard meanings. They are defined according to their primary management objectives:[10]

- Ia *Strict nature reserve* Managed for scientific research.
- Ib *Wilderness area* Managed to preserve its natural character.
- II *National park* Managed jointly to protect ecosystems and species, and for tourism and recreation.
- III *Natural monument* Similar in management objectives to Class II, but areas

which contain specific natural or cultural features of outstanding value.

IV *Habitat/species management area* There is active management to protect specific species and their habitats, but there may be some controlled use for production.

V *Protected landscape/seascape* Tourism and recreation, dependent on conservation of ecological and aesthetic values, are the primary objective.

VI *Managed resource protected area* The only class for which sustainable production is the primary objective, coupled with maintenance of biological diversity.

Still more concisely, farmers will be found in Classes VI and possibly IV, tourists in II, III, and V, and scientists in Ia and doubtless II–VI. In principle, there ought to be nobody at all in Class Ib except plants, animals, and 'indigenous human communities living at low density' – but the scientists cannot be trusted to stay out. National designations are often different from these terms; British national parks, for example, such as Dartmoor and the Peak District, contain farms and villages, and fall into Classes V and VI.

Protected areas play a key role in protection of endangered species, but biological resources cannot be conserved through these. Many functions of ecosystems originate from agricultural land and other managed ecosystems. Multipurpose forest management has an important role, since it is possible to combine conservation with controlled production, but agro-ecosystems, including cropland, pastures, and forest plantations, have important functions.

A further aspect of conservation is the view, held either on ethical or religious grounds, that humans are not the only species to consider. Animals and plants have rights of existence, the present dominance by the one species *Homo sapiens* being geologically of very short duration. This is now called the 'hard green' attitude, although it was long predated by the respect of the Hindu religion for all living things. An extension is to regard the earth itself as an organism, named Gaia after the Greek earth goddess, which is deserving of protection.[11] These views deserve respect, although unless there is a radical change in attitudes, it is hard to see them playing much part in practical policy.

Ecosystem services

Biodiversity conservation is by no means concerned only with protection of natural species, but covers a wide-ranging area known as ecosystem services. The principal ecosystem services are:

regulation of environmental processes, including atmospheric gases (notably carbon dioxide assimilation), water supply and filtration, and nutrient cycling;

the assimilation and filtration of waste products;

production functions, such as collection of non-wood forest products: thatching, gums, resins, medicinal products, etc., often a vital element in the economy of the poor;

nutrient cycling in agriculture;

biological control of pests and diseases;

nature conservation: the conservation of biodiversity in its own right and for leisure and study purposes, including the protection of endangered species;

genetic conservation, as a resource for future agricultural, industrial, and pharmaceutical uses, on aesthetic and ethical grounds, and as a precautionary measure;

protection of the environment for the remaining indigenous peoples.

A single example will serve to illustrate the vital nature of ecosystem services. The water supplies of many countries originate mainly from highland catchment areas. Suppose that the cover of forest and soils were to be removed from such areas. This would cause rivers to alternate between conditions of no flow and catastrophic flooding, and greatly reduce infiltration to the groundwater table. The additional construction and operating costs needed to maintain water supplies would be colossal. An attempt has been made at an economic valuation of the total world ecosystem services, using a range of different methods.[12] This estimate came to $33 trillion ($\$33 \times 10^{12}$) a year, which compares with the world's gross natural product of $18 trillion; that is, ecosystems bring benefits to human society of something like twice the amount of all production and services which are conventionally valued in economics.

Biodiversity monitoring and conservation policy

As with other land resources, there is a need for monitoring and hence indicators, and in this respect conservation institutions are ahead of other resource scientists. Protected areas are monitored at international level by the World Conservation Union (until recently the International Union for Conservation of Nature, IUCN), UNESCO, and the World Conservation Monitoring Centre (WCMC); detailed data with maps and a database are maintained. There are well-established programmes for monitoring endangered species.[13]

Indicators for monitoring biodiversity should include the composition of ecosystems (species present), their structure or pattern of habitats, and their functions (ecosystem services). Monitoring can be carried out at world, national, and district levels, and is at more detailed scales for critical ecosystems and species. Ideally, biodiversity indicators should be:

widely applicable spatially;

capable of providing assessment over a wide range of stress;

sensitive, so as to give early warning of change;

relevant to significant phenomena – species, systems, or functions;

easy to observe, and to measure objectively.[14]

To the public, preservation of large mammals and birds attracts the greatest attention, followed by small mammals and colourful plants. But protection, from killing or collection, is only one element. More important is conservation of habitats, and thus of the number and diversity of ecosystems in which species exist.

The importance of genetic resource conservation is felt particularly, but by no means exclusively, with respect to rain forest, where species diversity is highest. There is a combination of reasons for the concern expressed. Potential practical values are as resources for agriculture, especially crop breeding, for pharmaceutical uses, and for industrial applications. Beyond this are ethical aspects of species extinction and a precautionary principle: the loss of genetic resources is irreversible, and we do not yet know all of their potential uses.[15] Current practical problems arise with seed companies seeking patents on new varieties, and there are complex legal questions regarding national rights to genetic resources.

In developed countries of today, there are strong forces for protection: conservation organizations, the 'green' political lobby, and, most importantly, recognition of the value of nature by the general public. Protected areas are likely to remain largely as they are at present. There is a mutual recognition between farmers and the public of the ecological services performed by agricultural land.

In the less developed world, conservation forces are weak. The new urban elite shows a marked lack of interest in nature. Land use changes are often made with a total disregard of environmental costs (as was done in the developed world in the nineteenth century). Environmental legislation is frequently not enforced.

In many less developed countries, protected areas are in danger. To the local population they may appear to be 'unused' land, which could provide additional areas for cultivation or grazing. At present, the main argument for their retention is their value for tourist revenue, but much of this does not reach the local people. A recent study estimated that if all Kenya's national parks and forest reserves were converted to agriculture they could produce a net revenue of $203 million a year (in economic terms, the opportunity cost), as compared with the present $42 million from tourism and forestry, although this estimate can be questioned on grounds of under-valuation of ecosystem services.[16] International subsidies have been suggested. If, for its own interest, the developed world wants Kenya to protect its rhinoceroses, or India its tigers, it should pay them to do so. Actually raising the money to do this is another matter. 'Debt for nature' swaps have been proposed, protect your wildlife and we will cancel your foreign debts, but these can only be one-off measures.

Outside national parks, forest clearance and degradation of wooded savanna reduce the effectiveness of ecosystem services. On agricultural land, in an earlier age, sustainable shifting cultivation led to compatibility between production and conservation. With the present dominance of more or less permanent cultivation, many of the ecosystem functions have been irretrievably lost. Highland Ethiopia was once forested, but now is practically without a tree other than short-rotation eucalyptus plantations. A family subsisting from one hectare cannot set aside land for nature.

An attempt to overcome the problem of 'people versus parks', or versus forest reserves, is the approach of integrated conservation and development programmes. The basis is a participatory approach, planning in discussion with the local people and taking action with their active co-operation. The aims are to conserve nature within the parks, and at the same time meet the needs which led to pressures upon these – collection of natural products, wildlife damage of crops, etc. Specifically, there are projects to reduce pressure on natural forest 'islands' through encouragement of agroforestry on the surrounding land. Integrated conservation and development programmes are attractive in principle, but difficult to design and implement unless there is an adequate level of government efficiency.

In developing countries, the argument based on tourist revenue argument, coupled with international pressures for conservation, can help to check loss of biodiversity in the immediate future. In the longer term, the solution is for people and policy-makers of the developing countries themselves to recognize the reasons for conservation, in particular the value of ecosystem services. This in turn calls for an adequate level of education and responsibility among governments and people.

9 Monitoring change: land resource indicators

If policy and action to reduce land degradation are to be placed on an adequate basis, there is a clear need for land resource indicators, comparable to the economic and social indicators already in use. Only for forest clearance and biodiversity do such indicators exist. Measurable criteria, with potential to make comparisons between areas and to monitor changes over time, are needed for soil erosion, fertility decline, water resources, and the condition of rangelands. Whilst international organizations can provide guidance, assistance, and co-ordination, reliable data can only be obtained from a foundation of measurements by individual countries. The monitoring of changes to land resources should become a basic task of national resource survey organizations.

A recurrent theme of chapter 7 can be condensed into the statements:

> land degradation is serious, it has already had adverse and sometimes irreversible effects on production, and it will become worse if action is not taken;
>
> but from lack of good measurements, we do not know just how serious it is.

This dilemma is recognized in a classic understatement in Principle 15 of *Agenda 21*: 'Where there are threats of serious or irreversible damage, *lack of full scientific certainty* shall not be used as a reason for postponing cost-effective measures to prevent environmental degradation.'

There can be no doubt, for example, that soil erosion is severe in Haiti, Lesotho, and highland Ethiopia, or that pasture degradation is widespread in the hills of northern Pakistan and many parts of the African sahel. But statements of that nature are made on the basis of personal observation and judgement. As such, they direct attention to priority areas for remedial action, but provide no basis for comparing the extent and severity of degradation in different regions, still less for monitoring the effects of action to control it. From any viewpoint, whether of science, policy-making, or investment, this is a highly unsatisfactory situation.

There is a clear need for land resource indicators, measurable values that describe the state or condition of land resources.[1] The word indicator is used in its normal sense, to mean a number or other descriptor which is representative of a set of conditions, and which conveys information about changes or trends in these conditions. Examples of indicators are the use of body temperature as an indicator of human health, or visual symptoms such as the colour patterns on crop leaves as an indicator of nutrient deficiencies. Indicators can also represent in summarized form the total effect of many variables, as in the use of crop yield as an indicator of soil fertility.

In the area of economic and social data, indicators are already in regular use. National economic reports contain numerous indicators, such as gross and net national product, income per capita, wage rates, and price indices; these are used to observe changes in the economy, in particular to monitor the effects of policies. Sets of indicators of social conditions have been developed, for example life expectancy, health, literacy, school enrolment rates, the education of women; the World Bank compilation, *Social indicators of development* gives, by country, some 80 indicators of health, poverty, education, etc.[2]

By contrast, few indicators are available to assess, monitor, and evaluate changes in the condition of land resources. Such indicators are needed to guide policy changes and land management decisions, for example to monitor the effects of agricultural policies on soil fertility, or of conservation projects on erosion. Where forest clearance and wood shortage is a land issue, indicators are required to describe and monitor changes in forest resources over time.

The central objective of land resource indicators is to measure changes in the condition of land, in particular, to monitor land degradation or its converse, land improvement. Indicators provide a means to monitor the performance of agricultural, forestry, and other natural resource management programmes, at national, district, and local levels, and to measure the impact of policies and management measures upon the environment. Without land resource indicators, there can be no proper foundation for policy formulation and decision-making on matters affecting land resources.

The pressure–state–response framework

A programme has been started, through joint efforts by the World Bank, FAO, UNEP, and other institutions, to develop a set of land quality indicators. This rests on a concept that has already been found useful for monitoring pollution aspects of the environment, the pressure–state–response framework (Figure 17). The purpose of this is to place land resources within a context of policies, management measures and the results of these.

Indicators of pressure show the pressures exerted upon land resources by human activities. An example is the present and estimated future demand for groundwater as a percentage of the estimated rate of recharge. Indicators of state show the conditions of land resources, and changes in this condition over time. For

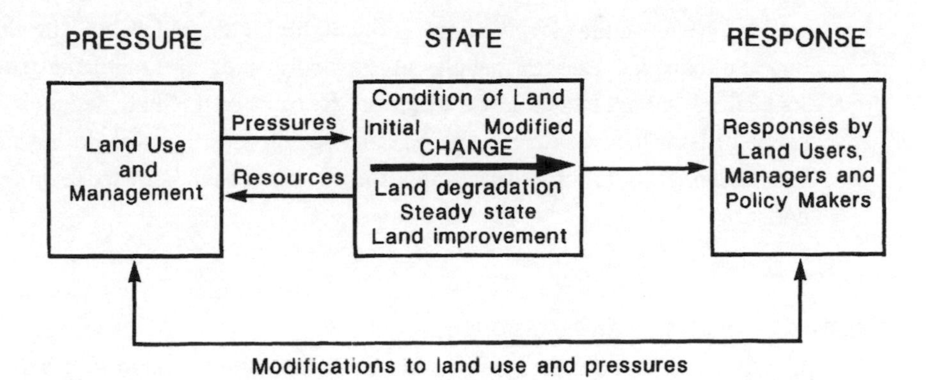

FIGURE 17 Land resource indicators: the pressure–state–response approach. Based on Pieri et al. (1995).

the same example, this could be lowering of the groundwater table, measured by an index which combines observations at a network of monitoring sites. Indicators of response show the action taken by society, as governments and individuals, as a result of pressures or changes of state. These might be the passing of laws to regulate water abstraction, indicators of the effectiveness of enforcement of such laws, or the adoption of management practices to increase water use efficiency.

Land resource indicators are needed at three levels of scale: national, district or project, and local. National indicators are a guide to national land use policy; in particular, by showing changes in the status of the country's soil, water, forest, and pasture resources they could show the need for action to check degradation. National indicators are also a guide to development priorities at international level showing, for example, areas of greatest fuelwood shortage, or most severe soil erosion (degradation 'hot spots'). The district level can refer to administrative units, but its main importance is that it is the usual scale for development projects. Indicators are needed first to guide project identification and planning, and later to monitor the effects of policy decisions and management measures. The local scale is the village or other community. This is the 'grass roots' level at which community action on land is taken, for example the management of village woodlands and communal pastures. It is also the level of community associations of farmers or other land users.

There is a further level, the farm scale. This is where decisions are taken on what to plant, how many animals to hold, and where the labour and inputs needed for farm management practices are ultimately employed. Resources will never be conserved unless individual land users choose to do so. However, farmers already employ indicators, based on a large store of local knowledge. Often these consist of plant indicator species, diagnostic of fertile, infertile, shallow, or saline soils. A well-known example is the colonization of highly degraded land by the grass, *Imperata cylindrica*. Pastoral peoples are skilled at judging rangeland quality, and make livestock management decisions on the basis of this knowledge. Within the scope of their local environment and current practices, farmers do not need help from scientists, although there is opportunity to put such knowledge on record for wider use.

Indicators of pressure and response

Identification of indicators of pressure on land resources, and response to attempts at change, do not present serious problems. Measurement of pressure essentially calls for a comparison between the requirement for a resource and its availability. Frequently it is not so much the absolute value of an indicator that is significant, but changes over time. For example, land shortage, and hence pressure on soil resources, is indicated by:

the ratio of cultivated land to land assessed as cultivable, expressed as a percentage;

mean or median farm size, and changes in farm size over time;

in systems of low-input farming, the ratio of cultivation to fallow;

the extent of cultivation of steeply sloping land, or other fragile environments;

a persistent negative soil nutrient balance, estimated by comparison between inputs and outputs;

land abandonment, indicative of severe degradation.

Similarly, pressure on water resources can be estimated by comparing abstraction with estimates of availability. Falling water tables, monitored at specific sites, indicate pressure and at the same time measure change of state. Fuelwood and domestic timber requirements can be estimated on a per capita basis, to give current and projected future requirements; these can be compared with the current rate of regrowth. The fuelwood requirements of Kenya have already exceeded replacement.[3] High or rising prices, for example of fuelwood and charcoal, indicate shortages and hence pressure on the remaining resources. Attempts to determine pressure on rangelands by comparing stocking levels with assessed carrying capacity have usually not been successful; fodder shortages, particularly in the dry season, and poor condition of livestock are alternative indicators.

Indicators of response are usually specific to countries or projects, permitting monitoring over time but not comparison between areas. The success of a soil conservation project, for example, might be monitored by measuring the extent of adoption of conservation practices, or the number of farmer associations active in conservation. The effectiveness of attempts to check forest clearance and degradation are measured most directly by a reduction in the rate of forest clearance; other indicators could be planting of community woodlands, or increased adoption of agroforestry practices. Indicators need not always indicate positive responses; thus abandonment of eroded or salinized land is an indicator of failure to check degradation.

Indicators of state and change

The primary need, however, is to establish indicators of state and change: to describe the present state or condition of land resources, and monitor changes in this condition. Without this basic information, measurements of pressures and responses are operating in a vacuum. Indicators are not simply data; they should convey the most significant aspects of the land factors they are intended to describe in a manner which is easily understandable, including by non-specialists. Ideally, indicators at national and district levels should be:

capable of being measured in quantitative terms;

representative of significant aspects of the resource described;

widely applicable spatially;

capable of providing assessment over a wide range of conditions or stress;

but at the same time sensitive, so as to provide early warning of change.[4]

Few indicators meet all these criteria. The examples discussed are expressed in terms of degradation, but also serve to monitor land improvement.

Soil erosion

For soil erosion by water, the indicator most widely employed is the rate of soil loss, as tonnes per hectare per year. This is widely measured on experimental plots but only rarely on farms, although the latter is possible by extensive sampling with sediment traps.[5] Because of the supposed difficulty of field measurement, there is a widespread practice of falling back upon modelling based on a range of models, the most widely used being the universal soil loss equation (USLE).[6] Rate of soil loss is a quantitative measure, universally applicable, but there are two problems with the way in which this indicator is presently employed. First, the models on which it is usually based have been calibrated under experimental conditions, but simply do not measure actual erosion in the field. Secondly, and in part as a consequence, the question which policy-makers ask is not the absolute rate of erosion but whether it is affecting crop yield, and there is no simple conversion from one to the other. Research is currently in progress over the effects of erosion on crop yield. One result which seems to be emerging is that, in the early stages of erosion, yield reductions may be quite small, under 20% or even not significantly measurable. A threshold is then reached when the well-structured topsoil is removed, at which point crop yields fall rapidly, by 50–80% in a few years, and the land is soon abandoned.

Crop yield loss itself is not a satisfactory indicator of erosion since it can result from many causes. Direct measurements of losses of soil nutrients would be satisfactory provided they could be made under field conditions, since these can be converted directly into costs of replacement by fertilizers.[7] Measurement of river sediment loads is not directly comparable between regions, but can be used to monitor changes in time over one catchment area. Guidelines exist for visible signs of erosion, but these need to be converted to semi-quantitative form and have rarely been systematically measured.

Through an excessive concentration on experimental plot measurements and modelling, soil conservation specialists have served well neither themselves nor the wider community. Observers have rightly been critical when regions predicted, by modelling, to have serious erosion show no signs of declining crop yields. Progress is currently being made on the relation of crop yields to rates of erosion, but, as already noted, crop yield is not a sensitive indicator of the early stages of erosion. What is now needed to complement this research is a frontal attack on the direct measurement of soil loss on farmers' fields.

Soil fertility decline

For the measurement of changes in soil fertility there are two approaches, indirect and direct. The indirect method is based on crop yields, in relation to inputs and management practices. Of course, yields are affected by many factors other than soil fertility, but some broad indicators are available. A highly generalized indicator is a divergence between increases in rates of fertilizer use and in crop yield. Potential yields for soils in good condition could be estimated, specific to input levels, and compared with observed yields. Surveys of indicator plants for soil degradation are applicable mainly at local and farm scales.

Direct measurement requires that soil monitoring should be regularly carried out as part of national soil survey programmes.[8] Countries need to establish indices of soil fertility, based on regular sampling and analysis from a network of sites, stratified by environment and land use. Initially, fertility trends would be established by national soil surveys for individual districts and countries. For the purpose of comparison, standards could be set up, giving acceptable ranges of soil proper-

INDICATORS OF STATE AND CHANGE OF LAND RESOURCES

Soil erosion

rate of soil loss (t ha^{-1} per year)
rate of loss of soil nutrients and organic matter
river sediment loads
extent and severity of visible signs of erosion (gullies, rills, soil slips, stony land, etc.).

Soil fertility decline

ratio between actual and potential crop yields under given inputs
divergence between change in fertilizer use and change in crop yields
changes in soil properties over time, measured by soil monitoring
occurrence of specific soil deficiencies, including micronutrients
presence of indicator plants for soil degradation.

Salinization

extent of patches of salinized soil, measured by soil monitoring
increasing salt content of soils, measured by soil monitoring
increasing salinity levels of groundwater, coupled with rising water tables
reports of crop failure and yield reduction through salinization.

Degradation of water resources

falling water tables, as monitored at specific sites
reports of tubewells drying up
longer periods of zero flow in former perennial rivers
higher flood peaks, measured by gauging.

Deforestation and forest degradation:

decrease in area of forest cover (absolute and relative)
fragmentation of former forest cover, measured by remote sensing
presence of degraded forests, measured by forest inventory
absence of natural regrowth.

Rangeland degradation:

offtake of livestock products, monitored over time
reduction in plant cover, as determined by remote sensing or ground transects
adverse changes in plant species composition, e.g. ratio of annual to perennial grasses, frequency of unpalatable species
extent of areas of trampled, crusted, or gullied land.

ties for agro-ecological zones and soil types. Soil salinization is a special case of specific regional applicability, which can be measured by the same approach of soil monitoring.

Crop yield is useful to diagnose the existence of a problem, and valuable in drawing attention to the consequences of soil degradation, but it is a highly generalized indicator. For the study of soil degradation to be placed on a sound scientific basis, including the supply of reliable data for economic analysis, there is no alternative to the establishment of systems of soil monitoring by national surveys.

Degradation of water resources

There are established methods for monitoring changes in the quality and quantity of both ground and surface water, although data for many countries are inadequate and unreliable. There are two stages toward the establishment of indicators. The first is to strengthen institutions responsible for river gauging, well records, and water quality analysis. This alone will show trends over time. The second stage is for countries to establish standards, as a basis for water policy, legislation, and management measures.

Deforestation and forest degradation

Deforestation is the only type of land degradation for which an indicator is already established and in regular use. As a result of the 10–year forest resource assessments co-ordinated by FAO, national data for changes in forest cover, by absolute areas and as percentages of remaining forest, are available for all countries. Some of the data are highly unreliable, but this is a matter of institutional competence and effort, not one of difficulty of measurement. Forest clearance as a percentage of remaining forest cover is the only indicator shown in the annual *World Bank atlas.*

Forest degradation can be monitored by well-established methods of forest inventory, repeated over time. There is no difficulty, in principle, for national forestry organizations to establish standards and monitor forest condition.

Degradation of rangeland resources including desertification

As discussed with respect to desertification, our knowledge of the condition of rangelands degradation is the poorest for any type of land resource. Despite the clear visible signs, we lack quantitative measures of its location, extent, severity, and trends over time. Pasture resource experts do not generally have a high opinion of regional surveys of rangeland condition, preferring to focus upon specific areas of pasture management projects. The success of such projects, however, can only be judged by scientific monitoring.

As with soil conditions, indirect and direct indicators are possible. The indirect approach is to employ offtake of livestock products, monitored as a trend over time. Direct measurement of rangeland condition at district level can be carried out by field transects, repeated over time. For larger areas, a remote sensing variable, the normalized difference vegetation index (NDVI), measures reflectance by living vegetation, which can be calibrated to give plant cover. Imagery is available on a regular time basis. There is no reason, other than institutional competence, funding, and effort, why regional and national standards for pasture resources should not be established and monitored over time. This would place knowledge of the extent and severity of desertification on a sound basis for the first time.

Loss of biodiversity

The situation with respect to indicators of biodiversity is radically different. Thanks to the existence of strong conservation interests, biodiversity is well monitored; the very title of the World Conservation Monitoring Centre shows its function. Where developing countries lack the desire or resources to do this themselves, international institutes step in. Comprehensive surveys of global biodiversity exist.[9] The results are given good publicity, the question of endangered species in particular attracting attention.

Biodiversity monitoring requires measurement of the species composition of ecosystems, their structure or pattern of habitats, and the functioning of ecological processes. Monitoring can be at the level of the landscape, ecosystem, populations of individual plant and animal species, or in terms of genetic resources. Ideally, indicators of biodiversity should be widely applicable spatially; capable of providing assessment over a wide range of stress; sensitive, to provide early warning of stress or change; and relevant to significant phenomena, but easy to measure.[10] Not all of these criteria are compatible, but indicators which fulfil different functions can be combined.

Examples of indicators of biodiversity conservation are:[11]

- plant and animal species numbers and populations;
- species threatened with extinction, for example at critical numbers of breeding;
- composite indicators of species at risk, combining a range of measured information;
- percentage of area dominated by wild species;
- percentage of area in strictly protected and partially protected status;

diversity of domesticated species, for example indigenous crop varieties;

rates of loss or specific habitats, for example wetlands.

An example of an indicator which is not in itself a major part of an ecosystem, but relevant, sensitive, and easy to measure, is the employment of populations of butterflies to monitor biodiversity in California.[12]

There are a range of comprehensive manuals on biodiversity assessment and monitoring.[13] In seeking indicators for land degradation, much could be learnt from comparison with the progress made in monitoring biodiversity.

Problems of data

The paucity and unreliability of data for developing countries has already been noted as regards land use statistics. The position is little better for sources of information on land resources. At national level, an accessible compilation is the two-yearly publication, *World resources*, which includes tables of land use, forest cover, agriculture and food, water and fisheries, and biodiversity. Most of the tables show changes over 10 years previous to the most recent data. Many of these data are based on FAO data; for agriculture and food, the *FAO production yearbook* is the basic source. For indicators of land degradation, however, there are major gaps. There are currently no data sources whatever which indicate national trends over time in soil erosion, fertility, water, or pasture resources. Forest cover is measured for the most part only at the ten-year intervals of the international forest resources assessments, since less than 30 out of 90 developing countries have ongoing monitoring programmes. For some countries, the most recent measurements of water use date from the 1970s.

There is a United Nations environmental monitoring programme, Earthwatch, co-ordinated by UNEP through its Division of Environment Information and Assessment. UNEP only has co-ordinating functions, and does not itself make measurements of environmental change. The greater part of these monitoring activities are concerned with the global commons, atmosphere and oceans, and with pollution aspects of the environment. Land resources are covered mainly through the Global Terrestrial Observing System (GTOS),[14] the mandate of which includes 'sustainability of managed ecosystems' and 'land use change and land degradation'. Earthwatch is supported by a global environmental information exchange network, Infoterra.[15] It is significant that out of 43 UN inter-agency contributors to this network, and 84 agencies, only 6 and 16 respectively are directed at land resources, inclusive of biodiversity. Truly, with respect to knowledge of the resources of our planet, we are 'eyeless in Gaia'.[16]

The current World Bank/FAO/UNEP programme on land resource indicators takes the initial step of making use of existing data sources, rather than going to the effort and expense of collecting new information. For example, in a recent proposal on land resource indicators, 'emphasis is placed on the potential for making better use of existing sources of data. Modelling may help to fill some gaps. There is a further possibility of using surrogate (substitute) indicators . . . These proposals are made with a view to economy of effort'.[17] Efforts are being made to provide guides to world data sources ('metadata banks').[18]

These efforts should not be allowed to obscure the fact that there are large gaps in data on the condition of land resources, and many problems over their reliability. The United Nations Conference on Environment and Development (UNCED) noted that, 'the gap in the availability, quality . . . and standardization of data between the developed and the developing world has been increasing, seriously impairing the capacities of countries to make informed decisions concerning environment and development'.[19]

It must be recognized that to obtain reliable, quantitative information on changes in the condition of land resources will require time, effort, and expenditure. This is no different from the situation for economic and social data, the collection and analysis of which requires much effort and skill. This work cannot be left to international bodies. It has to be undertaken by national institutions in developing countries, many of which are at present weakly staffed and poorly financed. International aid can provide guidelines, training, and temporary support, but real progress can only come from recognition by countries themselves of the need to monitor their land resources.

At the national level, the objective is to have tables, comparable with those found in economic and social publications, showing changes in the quantity and condition of national resources. These could be used to judge the success of policy measures directed at the achievement of sustainability. At district or project level, the objective is to monitor the success of policy and management in checking land degradation or achieving improvement, in the same way that economic performance is routinely monitored. The established procedures of project monitoring and evaluation should be widened to cover natural resource indicators.

At the top of the information pyramid, could a land resource index be developed? Gross national product per capita is widely employed as an index of the wealth of nations, and a human development index has been developed which combines indicators of health and education.[20] A corresponding land resource index would need to combine subindices of the status of soils, water, forests, rangelands, and biodiversity. Only one component, change in forest cover, is currently available. A land resource index is possible in principle, and represents a target for the future.

10 Costing the earth: the economic value of land resources

Despite growing recognition of the importance of environmental criteria, many investment decisions by development agencies are taken primarily on grounds of economics, and, specifically, on returns to investment. If natural resource considerations are to be allotted their rightful importance, there is no avoiding their conversion into money terms. Economic values are needed for analysis of soil and other land conservation projects, for estimating the loss to society caused by land degradation, and for national environmental accounting. Conventional economic methods undervalue natural resources; they may appear to come free, or to be priced at either their marginal or average values, ignoring the far higher price that would be paid if they became scarce. Above all, the practice of discounting, as employed in cost–benefit analysis, grossly underestimates future option values, the value of resources for use in the future. A consequence is that the economic losses caused by erosion, salinization, and other kinds of degradation are greatly undervalued. Economic methods as currently applied give equal weight to the needs of today's poor, but they steal resources, and thus welfare, from future generations. Assigning a value to land resources equal to their productive potential for at least 500 years, which virtually amounts to a sustainability constraint, would help to remedy this iniquitous situation.

Is it necessary, useful, or possible to assign an economic value to soils, water, forests, and pastures? There are certainly difficulties in doing so, and the first point to consider is whether it needs to be done at all.

There is an alternative procedure, which is to treat environmental considerations separately from economics. A 'sustainability constraint' is established, which requires that there should be no loss or degradation of certain, specified, natural resources – that the soils, water resources, etc. should be maintained or improved. Any development that does not meet this constraint is rejected on environmental grounds, and economic analysis of it is not carried out.[1]

To those who hold strong ecological views there are attractions in this approach, but for two reasons it is not a realistic way to proceed. The first is that developments which involve changes in land use often require an element of trade-off between gains and losses, both of natural and human capital. A reservoir built to improve water supply to a city results in loss of agricultural land, a forest cleared for agriculture necessitates a loss of forest resources. There need to be ways of comparing gains of one kind with losses of another, and whilst economic revenue is by no means the only way of doing this, it is the most commonly used.

The second reason is that we live in a world in which decisions are taken primarily on economic grounds, and, for all the added weight now assigned to environment, this situation is most unlikely to change. The cynical view, that this is because of the dominance of high-level government advisers and development agencies by economists, misses an essential point. All development involves investment, the allocation of scarce resources in alternative ways, and it is a universal practice to compare investments in terms of their economic returns.

So, if natural resources are to be assigned their proper place in decision-making, it is impossible to avoid considering them in economic terms. Some of the main applications of environmental economics to land resources are:

- project appraisal (pre-investment) and evaluation (post-implementation), to assess the value of changes in the capital stock (quantity and condition) of land resources;
- economic analysis of soil conservation projects;
- estimating the cost of land degradation;
- water pricing;
- environmental impact analysis of development proposals;
- national environmental accounting.

Environmental economics originated as a distinct branch of economics in the early 1970s, although certain aspects, such as the exploitation of mineral resources, had been treated earlier.[2] Much of the earlier work covered pollution aspects, treated as an externality. The pioneering work on land resources was on the cost of soil erosion, and hence benefits from conservation; for the arable lands of Zimbabwe, for example, it was estimated that the cost of replacing nutrients lost in eroded soil was $150, three times the cost of fertilizers actually applied.[3] Then from 1989 onwards came a series of studies demonstrating, at national level, that current land use was causing losses of the natural capital held in forests, soils, and coastal fisheries. In Java, the on-site costs only of erosion were causing a loss of $350 M a year, 3% of agricultural production. In Costa Rica, felling of forests, soil erosion, and depletion of fisheries, converted to economic terms, were equal to a loss of 5% of gross domestic product. In the Philippines, cutting of forests in excess of regrowth resulted in a loss of natural capital equal to 3% of gross domestic product.[4] It will be argued below that these were underestimates. Their implications, however, coupled with the rise in awareness of land degradation, led towards moves to include natural resources in national accounting procedures.

It is of critical importance that natural scientists should understand, and take and active part in, the economic evaluation of resources. There are two basic stages in such evaluation: understanding the physical changes and their consequences, and converting these to economic terms. The stages are linked through quantitative estimates of the effects of changes in land resources, particularly degradation, on inputs and production. If scientists do not supply these, economists are forced to estimate them and have sometimes, particularly for the case of erosion, made some very dubious assumptions. It is just as important to get the natural science right as the economics. Resource scientists should also appreciate how, by changing the assumptions or methods on which an economic analysis is based, radically different results can be obtained.

By now there is a large volume of theory on the economics of natural resources, initially concerned with conservation and more recently directed towards questions of sustainability. Whilst many questions remain in dispute, a number of concepts have been put forward about which there is some measure of agreement and which are relevant to the evaluation of land resources. This is a large and complex subject, of which only a brief account of some essential ideas can be given.[5]

Concepts and methods in the economics of natural resources

The natural environment as a free resource

Environmental goods come to us free. The classic illustration of this is in terms of marine fishing. The consumer pays for the boats and tackle, fisherman's wages, and transport to the market which together make up the price paid. But the fisherman does not actually pay to get the fish – they are there for the taking. Besides being free from any 'collection charge', marine fisheries are an open access resource, meaning that anyone can collect from it (apart from inter-governmental agreements to control catches). Many forests similarly offer open access with respect to the collection of firewood and minor forest products, a resource of special value to the poor. Sending livestock out to graze on the hills under village ownership means that they are fed without private cost; there may be communal controls on access, but the action of grazing is free.

The case of soils can be illustrated by the following hypothetical conversation between a land sales agent and a purchaser:

> 'Here, Sir, is the piece of agricultural land you have bought for $5000, exactly one hectare.'
>
> 'I see. Excellent climate, gently sloping, close to markets. But something looks wrong – there isn't any soil?'
>
> 'I regret, Sir, that the previous owner took the soil with him. Very concerned about natural resources is Mr Coke.'

'So to carry on my intended farming activities, I shall need to buy my own soil?'

'Anticipating your need, Sir, I have obtained quotations from local suppliers. The market rate is about $10 per cubic metre.'[6]

'So to cover one hectare to a depth of one metre I shall need to pay, let me see . . . another $100 000?'

'Regrettably so, yes, Sir.'

Considered as the cost of replacing lost soil, the value of this land has risen from $5000 to $105 000 a hectare, some 20 times its market price. This is due to the free nature of soil as a resource. Even the rates quoted by commercial suppliers of topsoil do not reflect the true cost since they refer to small quantities, possibly obtained from construction sites. Since taking soil from one site to another creates no new resource, we are left with the alternative of manufacturing new soil, by grinding up rock particles or dredging from the sea floor, then leaving it for 20–50 years for biological processes to convert to soil, as is done on industrial reclamation sites. The cost would be even more prohibitive.

The economically free nature of natural resources does not depend on them being subject to open access or communal tenure. For mineral resources, governments or landowners charge a fee for mining rights, and much capital and labour are required for extraction, but nothing is paid to the earth. The same applies to resources normally under private ownership such as agricultural land. The market price of agricultural land is paid to the owner for the right to take the produce from it, or trade it to others. The cost of agricultural production would be higher if a fee had to be paid to the earth goddess every time that the soil was used for crops. Where the resources which come free are not sufficient, we are willing to go to much effort and expense to augment them, supplementing rainfall by irrigation and natural soil nutrients by fertilizers.

Water pricing raises many points of difficulty. Water as rainfall comes free, and water from small-scale irrigation methods comes at low cost, that of the farmer's labour. On the other hand, on large-scale, dam and canal, irrigation schemes, the full cost of providing water is high. If a charge for amortization of the construction costs were to be included (as it is for dams constructed by private farmers), then the cost of irrigation water would become prohibitive. In practice, it is often subsidized, but this leads to the growing of low-value crops such as cereals, and to inefficient on-farm use of the water. A price needs to be set which encourages economy of use.

Suggestions have been made that a charge should be made for the use of natural resources, a form of an environmental tax. Governments would collect what the earth does not charge for. As noted above, this is done for mining rights, including the high fees charged by oil-rich states. The equivalent with respect to renewable resources would be for governments to charge for their use, as a soil use fee, forest

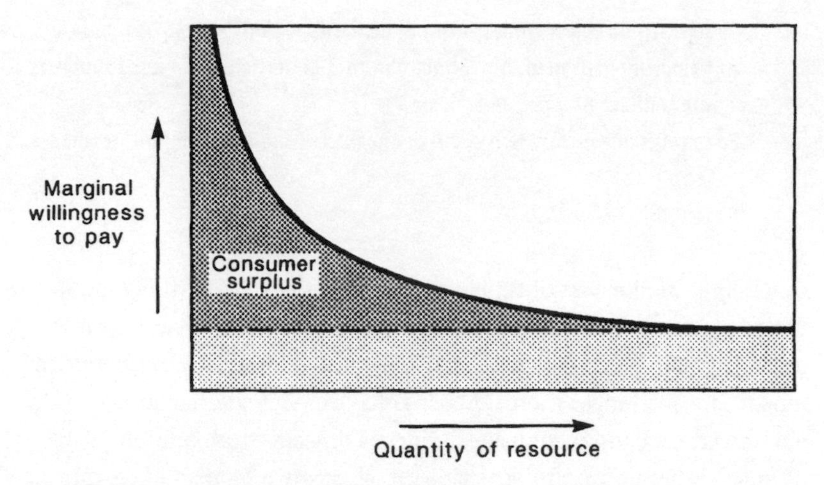

FIGURE 18 Marginal and total value.

harvest fee, a charge for extracting river or groundwater, or, in principle, even for the use of rainfall. Owners of fertile soils would pay more than those on poor-quality land. This is practised for forest harvest and water abstraction, but for soils it is an unrealistic proposition, as it would put production costs above those of countries which did not levy such charges.

The free nature of land resources means that they are undervalued, or even assigned a zero value. As a consequence, they can be degraded or destroyed without appearing as a negative item in accounts. Since land was not assigned a value in the initial assessment of capital resources, losses in its productivity are not reflected in economic analysis.

Marginal and total value

The market price of goods reflects marginal value, the willingness of consumers to pay for the last unit purchased, say, for one more litre of water. This is very much less than they would be willing to pay for the first few units of a scarce resource, as for drinking water in a desert. As supply became larger, consumers would be willing to pay a progressively lower marginal price for additional units. This is seen for soils by comparing areas with a similar standard of living but different population densities. The price of land, which is effectively a payment for the right to use its soil, is higher in densely populated countries or regions where the resource is in short supply.

Hence, if a country uses 5000 units of water, and the price which consumers pay is $1 per unit, the value of the water appears to be $5000. In reality, it is the sum of what consumers would pay for the first hundred units, the second, and so on, which is a very much higher value. In Figure 18, the value is the whole area under the curve of marginal willingness to pay, not that under the horizontal line which represents the market price. The difference is the consumer surplus; this is a further bounty from natural resources for which consumers do not currently pay, but which would be part of the value lost if the resource were degraded or destroyed.

This has important applications to resource depletion. In countries where fuelwood is abundant, its price may be zero (no marketable value) and its economic value low (representing the opportunity cost of collection). Where it is scarce, the market price may be $1 kg^{-1} or more, as in parts of India. If cutting exceeds regrowth, the standing biomass in a country becomes progressively depleted. Where, as a result of shortage, the unit price rises, then the value of the standing stock could appear to increase, if valued at the current market price, or marginal value, despite its depletion in physical terms![7]

Total economic value

The price of an item reflects willingness to pay by individual consumers. Economic value refers to the value to society as a whole. It is the sum of willingness to pay by all those with interests in the item, notably a combination of willingness by government (representing the community as a whole) and by private individuals.

Three further points are relevant to the discussion. First, both price and economic value are not properties which are intrinsic to a resource, but derive from the values that individuals and society place upon it. Secondly, value reflects the views and actions of present-day society, of people who are alive now; the interests of future generations are not directly represented. Thirdly, just as the price of goods fluctuates in response to changes in supply and demand, so also can economic values fluctuate considerably over time, reflecting changes in the desirability attached to them by society.

Thus the total economic value of land resources is the value as assessed by present-day society. This is said to be made up of four elements: direct use value, indirect use value, option value, and existence value.

Direct use value refers to the benefits obtained from land resources in the form of production of goods and services. It thus includes crops, livestock, timber, etc., and also services, such as recreation, tourism, nature conservation, or military training. These logically belong to direct use, since they reflect the willingness of society to set aside land (and thus indirectly to forgo the value of alternative use) for these purposes.[8] The function of land resources as a sink for waste products is also a form of direct use that is of great value to society.

Indirect use value covers off-site goods or services, those which are obtained outside the area of land covered by the resource. The function of forested catchments in regulating the flow of rivers is often cited, although, since the regular flow of water that is the desired product originates from within the catchment area, the distinction between direct and indirect use is blurred. More generally, indirect use covers ecosystem services, such as regulation of atmospheric, water, and nutrient cycles, pest control, and their function as a genetic resource bank.

Future option value[9] is the amount that society is willing to pay in order to retain a resource for use in the future. It refers to the summation of uses for an indefinite period in the future but – and this is a critical point – it is the value assigned by present-day society. It does not incorporate the willingness to pay of future generations.

A fourth element is sometimes included, the existence value, which represents what society is willing to pay simply so that a landscape, ecosystem, species, or the like should continue to exist. The underlying concept is that geological and biological processes have created such immensely complex and irreplaceable natural environments that humanity might be willing to pay simply so these should continue to exist, over and above their value as resources.

The direct and indirect use values are thus the current values of production of goods and services: agriculture and forestry production, provision of water, recreation, and tourism, and the values estimated for ecosystem services. The total economic value is the sum of use values for an indefinite period in the future. Since economic values normally refer to a time period of one year, in a literal sense future option values greatly exceed current use values. In the practical analysis of development projects, the 'present' is usually taken as something of the order of 20–30 years, known as the project life – equivalent to the interests of one generation of human life, and to the outcome of a long-term financial investment. The concept of project life thus reflects the present-day nature of economic value. Anything beyond the 'project life' can be considered as the future option value. In practice, this future period is treated in a generalized fashion, and is conventionally omitted from economic analysis altogether.

Investment decisions: welfare and self-interest

In taking decisions about the allocation of scarce investment funds, development agencies must take into account two sets of criteria, welfare and self-interest. On the one hand, the reason for their existence is to benefit the people of developing countries. On the other, the World Bank and the regional development banks are institutions of the highest financial standing, and need to maintain that status. The dilemma becomes greater because it is widely agreed that the welfare of the poor is the primary objective; yet, in the short term, this section of the community is least likely to be able to generate revenue to repay loans – this obligation falls to governments, acting in the longer-term interests of their peoples as a whole.

The dilemma between welfare and return to investors does not arise in the case of aid in the form of gifts, as in most projects of non-governmental organizations. In this circumstance, the welfare of the recipients is a self-justifying objective. Decisions are based on need, and what type of aid will give greatest benefit.

Investment agencies which provide loans must moderate their decisions on welfare grounds by considerations of ability to repay. A case in point is the costly World Bank agricultural development projects found in many parts of Nigeria during the 1980s. These generated few increases in agricultural exports; the benefits were almost entirely in the form of higher incomes and improved food supplies for farmers. Interest on loans was paid by government from oil revenues, the projects being essentially a method for transferring oil money to rural communities.

These two bases for investment decisions represent two branches of economics, positive and normative. Positive economics tries to describe and explain the observed behaviour, for example, what causes people or institutions to invest in agriculture rather than forestry. In particular, it observes how they take decisions which compare present with future consumption. Normative economics, of which welfare economics is a branch, is about saying what should be done, given that society has certain objectives, such as equality and sustainable development. Even though prediction is very far from perfect, positive economics rests on relatively firm ground, for it can always be referred back to observed economic behaviour. In development economics in which welfare, rather than return on investment, is the major criterion, the scope for altering the results of economic analysis by changing the assumptions on which it is based becomes far greater. Nowhere is this more the case than in social cost–benefit analysis.

Cost–benefit analysis and discounting

Outline of conventional practice

The greatest problems in valuation of natural resources arise in relation to their productive potential over time, and thus for sustainable use. This has frequently been based on procedures, namely cost–benefit analysis with discounting of future values, developed for analysis of investment in projects. Uncritical transfer of these procedures to the valuation of land resources has led to errors. The conventional procedures will be outlined first, followed by a critique of application.

Land development projects require expenditure in the early years in order to reap benefits in the future. This is most clearly seen in the case of irrigation schemes, but applies to all development of a project nature. In the early years, costs substantially exceed benefits. For a longer period in the future, whilst there continue to be costs, the benefits from increased production are greater. The 'without-project' costs and benefits, that is, those which would occur without the proposed investment, are subtracted from the year-by-year estimate of the costs and benefits under the proposed project, to give the net benefit.

Cost–benefit analysis is the most widely used method for assessing the economic value of investments for development. It originated in a quite different context, as a means of comparing alternative investments in business and industry. When used to compare development alternatives, it is applied in the form of social cost–benefit analysis, meaning that the benefits included are not just those to the investor but to society as a whole, for example better food supplies, or increased exports which benefit the national economy. Since the unit for comparison is the community as a whole, costs and prices are set at the full cost to the community after taking into account subsidies and taxes. For example, fertilizer prices include the cost to government of a subsidy; family labour costs may be set at zero if it would otherwise be unemployed (zero opportunity cost).

A way is needed of balancing the excess costs incurred in the early years of a project with the net benefits in the future. Since the decision to invest is being made now, all costs and benefits are converted to an equivalent value at the present time, called the present value. Discounting is a way of bringing future costs and benefits to present value, by reducing their value by a chosen percentage each year, the discount rate. The rationale is that, in business decisions, there is no point in taking risks if they yield a return less than could be obtained from safe investment, say in government bonds. Discounting can be regarded as the reverse of interest. Supposing that bonds yield 8% per annum, then the present value of a benefit from an investment obtained in one year's time is reduced by 8%, i.e. divided by 1.08, and a benefit obtained in two years' time divided by $(1.08)^2$. In the general case, with a discount rate of r, a cost incurred or benefit received of D currency units in n years' time has a present value PV of

$$PV = D\frac{1}{(1+r)^n}$$

To carry out a cost–benefit analysis, a project life is assumed and a discount rate chosen. The annual costs and benefits are assessed, and discounted to their present value. Three measures can then be employed to assess the value of the investment:

net present value: the present value of benefits minus costs, expressed as a money value, e.g. $1000;

benefit:cost ratio: the present value of benefits divided by costs, expressed as a ratio, e.g. 1.6:1, or simply 1.6;

internal rate of return: the compound return on funds invested in the project, expressed as a percentage, e.g. 15%; this is equal to the rate of discounting at which the investment has a net present value of zero.

The choice of discount rate greatly affects the outcome of economic analysis. In the field of commerce, the interest rate obtainable from savings sets a standard by which to determine the discount rate. In social cost–benefit analysis a 'social' rate of discounting is employed, the choice of which is more or less arbitrary: the investing agency simply says what it should be. Many developments (e.g. World Bank projects) take a rate of 10%, on the argument that investment in development should be expected to yield a high return. In order to give greater weight to the future benefits, the use of much lower rates, such as 6% or 3%, has been proposed.

The higher the discount rate, the less becomes the present value of a net benefit in the future. The consequences are considerable (Table 5). Even at a low rate of 5%, a benefit obtained after 15 years is reduced to a present value of half, whilst the situation 50 years hence has virtually no effect on the cost:benefit analysis. This is not due to the effects of inflation, calculations being made in terms of constant-value currency units.

Arguments advanced to justify the use of any discounting are:

1. Time preference: that people would rather have a benefit now than in the future.
2. Uncertainty: that the further into the future, the more uncertain is our knowledge of what is likely to happen.
3. Substitution: that, in the future, different investment opportunities will arise.
4. Marginal utility of consumption: that if the standard of living continues to rise, the same unit of benefit (e.g. 1 t of maize) will not mean so much in the future as now.

Table 5. *Examples of the effect of discounting*

	Present value of 100 currency units	
Years in the future	Discounted at 5%	Discounted at 10%
10	61	42
15	48	24
20	38	15
50	9	<1

Inappropriateness of discounting as applied to valuation of land resources

The process of discounting has a massive effect on the outcome of economic analysis. Quite simply, the choice of discount rate can have the effect of reversing the relative value of two proposed developments, or changing their absolute revenue from positive to negative. Resource scientists, agronomists, engineers, and sociologists rightly complain that proposed developments which have been assessed as desirable on all other criteria can be rejected by the apparently arbitrary assumptions of project life and discount rate.

Foresters were the first to object to the tyranny of discounting. The major expenses are land preparation, planting, and early care of the trees, whilst the main harvest comes after a period of 15–25 years for tropical softwood plantations, and 50–80 years for natural hardwoods. This makes it almost impossible for forestry to match the benefit:cost ratios of investments in agriculture, other than at discount rates of about 2% which are not normally accepted. Moreover, discounting favours fast-growing plantations compared with slower-growing natural forests. This is not because the average annual benefits, total revenue divided by number of years in the forest cycle, is any less, it is simply because of the timing of those benefits. 'The economic case against natural [forest] management would therefore be almost unanswerable but for one thing: it is almost entirely wrong . . . The high interest rates which act against natural management systems . . . and against forestry in general, have no more basis than somebody's opinion.'[10]

When applied to land resources, it is easy to show that conventional analysis gives results that conflict with common sense. As a highly simplified example, let us consider an irrigation scheme in which capital costs, all in Year 1, are $1 000 000, followed by irrigation maintenance costs of $20 000 a year. The land produces net benefits (revenue minus costs) of $20 000 a year without irrigation, and $200 000 with irrigation, a net benefit from irrigation of $180 000 a year. In this scheme, however, there is no drainage. The water table rises, the soil is salinized, and after 20 years the land must be abandoned. For the remainder of the 'project life', say at most 30 years, there is an annual loss of the $20 000 that would have been produced without the scheme.

At a discount rate of 10%, this scheme shows a net present value of $290 000 and a benefit:cost ratio of 1.27; the internal rate of return is 15%, well above the 10% threshold commonly used in project appraisal. On the basis of cost–benefit analysis, the scheme appears desirable, and would be selected in preference to investment alternatives showing a lower internal rate of return. Its merits would appear very different to the children of the present farmers, who would find themselves with worthless land, unreclaimable by their own efforts, and would join the ranks of the landless poor or urban unemployed. The damage is potentially reversible by a reclamation scheme, although the considerable expense of this pre-empts alternative investments at that future time.[11]

The position is similar, however, where damage to the land is irreversible, as in severe erosion. Table 6 represents a situation where land shortage has forced

Table 6. *Conventional cost–benefit analysis of soil conservation (all values are in US$)*

Year	Conservation cost	Annual revenue: With conservation	Annual revenue: Without conservation	Conservation benefits	Present value, discounted at 5%: Costs	Present value, discounted at 5%: Benefits
1	1 000	200	200	0	952	0
2	50	200	200	0	45	0
3	50	200	200	0	43	0
4	50	200	190	10	41	8
5	50	200	180	20	39	16
6	50	200	170	30	37	22
7	50	200	150	50	36	36
8	50	200	130	70	34	47
9	50	200	110	90	32	58
10	50	200	70	130	31	80
11	50	200	30	170	29	99
12–20:						
Annual	50	200	0	200	–	–
Sum	–	–	–	–	208	831
Present value					1 527	1 197

For project life = 20 years, discount rate 5%:

Net present value	−330
Benefit: cost ratio	0.78
Internal rate of return	2%

cultivation to spread onto steeply sloping land. For the purpose of this example, conventional conservation methods such as terracing, requiring substantial initial investment, are assumed. The capital cost of conservation is $1000, followed by annual maintenance of $50 per year. With conservation, land use is sustainable and the net profit from agriculture remains at $200 per year. Without conservation, crop yields remain stable for 3 years (as often found in short-term experiments), then begin to decline gradually. After some years, a threshold is reached when soil organic matter falls to a low level, reducing the infiltration capacity and structural stability. Erosion then accelerates rapidly, with corresponding fall in yields, until in Year 12 the land is abandoned.

Taking a 20–year 'project life', then even with a discount rate as low as 5%, investment in conservation shows a negative net present value, and an internal rate of return of only 2%. Extending the calculations to 30 years, the net present value of benefits just passes that of costs, i.e. the internal rate of return becomes 5%, still below the level normally accepted for investment. This is not an analysis from the farmer's point of view (with low capital and a short time horizon) but is supposed to be a calculation showing the welfare to society, the community as a whole. These conventional procedures produce the result that it is clearly not worth while investing in conservation measures which will prevent the land being abandoned, as a rocky wasteland, in 12 years time.

This example is somewhat stylized, not with respect to the reality of erosion and land abandonment but in assuming high-capital methods which are not the only means of conservation, but it serves to illustrate the problems in applying conventional economic analysis to land resources. Intuitively, such destruction of resources and future landlessness should be avoided. Since small farmers cannot borrow the capital needed to install conservation, government or aid agencies should support this investment for the long-term welfare of the people. Such action, however, does not meet the criteria normally taken in comparing alternative investments.

Turning the tables: analysis of a past investment decision

The artificiality and inequity of applying discounting to land resources can be further illustrated by a thought experiment. Suppose the projects outlined above were re-analysed not as of now, but from a future viewpoint. Instead of asking the question, 'Will this be a sound investment?' we take the standpoint of 20 years hence and ask, 'Considered retrospectively, was that a sound investment?'

The relative values that might be assigned to costs and benefits in different years will become radically different. The situation 'now' – i.e. in Year 20 of the above example – will be assigned its full value. Next in importance will be the immediate future, the former Years 21–25. The benefits received in the past are likely to be assigned lower values, on some such reasoning as, 'Well, things were good for a time, but that is past and gone'. The initial capital expenditure (irrigation dam or conservation works) will seem least important of all; it used up scarce investment funds at the time, but this has little significance now, when the debt has probably been written off. To make allowance for the assumed lesser importance of events in the past, one might even adopt a procedure of reverse discounting, that a cost incurred or benefit received *n* years *in the past* was discounted by, say, 5% per year (Table 7). On this basis, the same soil conservation project has a net present value of +760 and a benefit:cost ratio of 1.77.

Reasons for the failure of discounting

So what is wrong? Why do procedures of analysis which are legitimate when applied to choice between commercial investments give such intuitively unsatisfactory results when applied to land resources? Among the reasons are:[12]

Table 7. *Cost–benefit analysis of soil conservation from a future point of view*

Year	Years ago	Value in Year 20, inversely discounted at 5%	
		Costs	Benefits
1	20	377	0
2–11	19–10	249	319
12–20	1–9	355	1422
Present value		981	1741

Net present value +760
Benefit: cost ratio 1.77

The arbitrary nature of the choice of discount rate The existence of so much discussion as to what rates are appropriate is indicative of the lack of consensus.

Limited opportunities for reinvestment Commercial analysis is based on a limitless supply of alternative investments, both now and in the future. It assumes that, at all times, revenues can be capitalized and fully reinvested. For land resources in short supply, this cannot be done. Taking the example of irrigation, the reinvestment criterion requires that, having salinized one area, revenues could be reinvested in another project; but potential irrigation sites are already in diminishing supply. The action of pioneer coffee planters in Brazil in allowing soil degradation to occur was rational economic behaviour only so long as there was a supply of virgin land onto which to move.

The practice of discounting assumes that in changed future circumstances there will be different opportunities for investment. It amounts to saying, 'In 20 years time, alternative opportunities will be available to the farmers who hold the land which has become gullied/salinized'. For poor people dependent on the land, this is not the case.

Increase in the marginal utility of land with scarcity In the future, with population increase, there will be less land available per person; hence levels of inputs and productivity per unit area will be higher. One additional hectare of land conserved for use will be more valuable when it produces 5 tonnes of cereals per year than when it produced 3 tonnes. With the present apparently inexorable process of forest clearance, the real value of timber will rise. For all land resources, higher values should be assigned to their utility in the future.

This situation can also be brought about by overintensity of use, such as overfishing and overcutting of forests. The cause of this state of affairs is ecological, the passing of the threshold level for resource renewal. Such non-sustainable use causes increasing scarcity of the resource, and hence a rise in its price. The marginal utility per unit of the resource rises over time, although the quantity produced will fall drastically.

Failure to take account of the capital value of land resources

Human capital (buildings, machinery, etc.) is valued at the start and end of an accounting period, and allowance made for depreciation. Until recently, land resources, as natural capital, have not been included in the accounting process. Hence improvements to them do not appear as a gain in value, nor degradation as a loss. It is not sufficient to include the market value of a resource, for example the price of agricultural land, in the accounting because, as argued above, the apparently free nature of land resources means that their value to humanity is not reflected in their market price.

A time preference for present needs over those of the future

Arising out of its purpose of deciding between investments at the present day, conventional analysis intrinsically assigns a lower value to benefits to humanity in the future.

The last two points are related to the links between land resources and present and future production, and hence to sustainable land use. These aspects require further consideration.

Intergenerational equity: giving our children a fair deal

One further concept is as much a matter of ethics as of economics. Cost–benefit analysis is designed for the purpose of making investment decisions now. In the form of social cost–benefit analysis it gives consideration to all elements of society, including those who cannot pay; indeed, it is common to give additional weighting to benefits which accrue to the poor. However, it does not give equal weighting to the welfare of future generations.

Many people consider the needs of their children to be at least as important, often more so, than their own welfare. This applies as strongly among the poor as the rich; in Third World countries one finds people earning $50 per month who spend one third of their income on school fees. These children will, in turn, have the same level of concern for their own children, and so on. This is called the chain of obligation, and demonstrates that we have a personal interest in human welfare throughout future time. It may seem surprising that a theory is needed to justify why regard should be given to the needs of future generations, a fact which many would regard as self-evident.

Therefore, as well as being axiomatic on ethical grounds, the chain of obligation means that it is in the self-interest of today's society that the opportunities for welfare open to future generations should be at least equal to those which we have ourselves – in plain terms, that we should give a fair deal to our children and our children's children. That is intergenerational equity.

Reappraising the value of land

To recapitulate, one of the main objectives in placing an economic value on land resources is to determine the value of resources that are lost – eroded soils, lowered water tables, cleared forests, etc. – and thereby estimate the economic benefits from conservation. The assessment can be done at district level, as in natural resource management projects, or at country level through national environmental accounting. Progress has been made in recent years, but it has proved difficult to drag the analysis away from conventional approaches. The major obstacle is the effect of discounting, even at low percentage rates, in reducing future option value. This has the effect of making conventional analysis quite blatantly and selfishly favour the interests of present users as compared with those of future generations.

Explicit utility appraisal

An opportunity to reconsider the valuation of land resources is provided by the procedure of explicit utility appraisal. This rejects the practice of discounting all items, of whatever nature, at one standard rate. 'The only purpose for which conventional discounting can theoretically be recommended is for private, self-interested assessment of alternative investment strategies, with all benefits committed to reinvestment.'[13] Each cost and benefit is taken separately, its change in utility over time assessed, and the adjustment to present value made accordingly. Thus, instead of combining all items for each future year into one and discounting it as a whole, each type of future value is brought separately to an equivalent present value.

The pattern of change in value over time may be negative, zero, or positive. For example, if food supplies in a country are predicted to become scarcer, then one tonne of rice produced in *n* years time could be valued higher than the current market price. On the other hand, for a project having a known capital investment at some future date, say the dredging of silt from drainage channels, then discounting, not at the social rate but at the higher commercial rate, would be appropriate – indeed, the sum needed could actually be set aside as an investment.

Explicit utility appraisal therefore requires predictions of the trends in future real values. For land resources, there is an overall argument for higher future values owing to increase in population on a finite resource base, and hence increasing scarcity of land. This is illustrated by comparison with the converse situation of a loss of land. Consider a country which possesses 200 units of land. Let its population remain constant and the level of technology also stay the same. Now suppose that 100 units of land were destroyed by some means, perhaps a natural catastrophe. The value of the remaining units would substantially rise.

The almost inevitable doubling of the future population of developing countries presents the reverse of this situation. Since there can be no substantial increase in land resources, per capita availability of soil, water, pasture, and forest will decline. This will tend to raise both the market prices and the real value to society of crops, livestock, and forest products from these resources. These are arguments for increased marginal values in the future, and thus for negative discount rates – for a benefit of one tonne of maize in 20 years time to be assigned a higher value than its market price today.

Technological advances will mean that the same area of land gives more production. On the one hand, this has the effect of reducing land scarcity, but, on the other, it increases the amount of loss per unit of resource destroyed or degraded. With rising incomes, it could be argued that the products and services from land resources become relatively less important. This is all very well to those fortunate enough to have the rising incomes but not to the growing numbers of the very poor who, both as producers and consumers, are disproportionately dependent on land resources and their products.

There would be a reason for lower future values of land wherever substitution is possible, as in synthetic fibres for cotton and wool, or steel for constructional timber. Wood for fuel would be very much more costly but for the fact that Western societies have largely substituted mineral fuels, but this option is not open to the rural poor. Other types of substitution for land resources can be envisaged: hydroponics for soil in food production, desalination of seawater, even food from space. But to our present knowledge and vision, these are either technically or economically unrealistic.

The future option value of land resources

For assigning a future option value to natural resources, three elements are needed: the relevant discount rate, a decision on intergenerational equity, and a choice of time horizon. The value of the resource is then the sum of its production of goods and services to humanity over the period chosen.

For the discount rate, the balance of expectation is that so long as population continues to increase, which in developing countries means up to and beyond our present planning horizons, the real value to humanity of land resources is likely to increase. Granting that there are also arguments in the converse direction, there is an available assumption that is akin to a null hypothesis. It is that land resources, and the benefits they yield, should be assigned no less a value in future years than they have at present. In other words, the first component in the proposal is that natural resources should be discounted at a zero rate.

For the second element, intergenerational equity, the argument is based on ethical, not economic, grounds. To assign less importance to the welfare of future generations than to ourselves is a purely selfish decision. If it can be justified at all, it would be on grounds of uncertainty, the possibility of a nuclear or cosmic catastrophe, or some radical change in the world environment. The only ethically defensible course is to give future welfare equal importance.

The third element is therefore the time horizon, the 'project life' of humanity. This must be largely an arbitrary figure. If the argument is carried to its logical conclusion, there should be no time limit, but the resulting infinite value of resources would be rejected by many. It would not be unreasonable to take the millennium as a starting-point, to say that humanity has been dependent on land for more than 2000 years, and there is no prospect that this state will change for another such period. A more modest time horizon would be 500 years. The exact figure does not affect the orders of magnitude involved, so long as it is a long period, and held constant throughout the analysis.

Based on these elements, the value of a land resource, and therefore the cost of its irreversible destruction, is the sum of present direct and indirect use values, for both goods and services, plus the future option value. As a null hypothesis, the value in future years is assumed neither to increase nor decrease. In order not to discriminate against future peoples, the discount rate is taken as zero. Finally, as a working basis for analysis, an arbitrary time horizon is assumed. Since the value of the current year's production becomes insignificant, this leads to the result:

Total economic value = Future option value
= (Sum of present use values) × (Time horizon).

This valuation is applicable to land resources which, when properly managed, are renewable, but which can be degraded or destroyed. Thus, for a time horizon of 500 years, land which today produces a net profit of \$200 a year has a total economic value of \$200 × 500 = \$100 000.[14]

This valuation applies in its outline form to the total and irreversible destruction of a resource. Where the loss is either partial or reversible, it provides a basis for assessment of the appropriate cost.

Applications of land resource economics

The economics of soil conservation

Early attempts to estimate the cost of soil erosion were based on converting soil loss in tonnes per hectare into reduction in profile depth, and assuming a linear relation between present yield at present depth and zero yield with no soil remaining. This method led to unsatisfactory results. Rapid erosion, say at 50 t ha^{-1} per year, is equivalent to a loss of 4 mm of soil a year or 1 m in 250 years; the corresponding yield reductions were so small that conservation could rarely be justified.

Three methods have been employed: nutrient replacement cost, the value of lost production, and the cost of reclamation or restoration.[15] In the nutrient replacement method, estimated rates of soil loss, and tonnes per hectare, are multiplied by a measured or estimated percentage content of plant nutrients (usually nitrogen and phosphorus only). This is valued at the cost of replacing these nutrients annually in fertilizer. This method is attractive in that it lends itself to direct conversion to economic terms. It ignores erosion loss of soil organic matter and consequent degradation of physical properties, and also complete destruction of soil by severe erosion. The major source of error is not the nutrient content of eroded soil. It lies in the assumption that soil loss rates observed on experimental plots are representative of erosion in the field, when, in fact, the latter may be up to ten times lower.

Loss of crop production due to erosion is also easily translated into economic terms. The problem is that, despite recent research efforts, knowledge of the relation between soil loss and yield reduction remains poor. A number of experiments have shown no measurable yield reduction at all on control plots. This result, disturbing to conservationists, is due to the short-term nature of the trials; there is no difficulty in measuring yield loss where erosion has reached a more advanced stage.

The reclamation or restoration method is not applicable to severe erosion, since lost soil material cannot be replaced. Its application to fertility decline is noted below.

Difficulties in costing erosion stem as much from the inadequate physical basis as from discounting, time horizons, and other questions of economics. A review of 20 cost–benefit analyses of conservation projects concluded that, 'Physical quantification is a weak point . . . statistically insignificant results or ad-hoc assumptions play a distressingly large role.'[16] As a consequence, the margin of error in estimates may be unacceptably high; thus, in Mali, the current net farm income foregone due to soil erosion was 'estimated at US$4.6 to $18.7 million . . . 4% to 16% of agricultural GDP'.[17] In default of field data, economists often accept published estimates of erosion rates uncritically, or resort to the use of modelling, not fully aware of its limitations. Refined methods of economic analyses are carried out on a basis of highly suspect data.

This is a case where the scientists have failed to supply information in the form required for economic analysis. They need to leave the comfort of their erosion plots and go into the field, measuring crop production on fields which have and have not employed conservation methods.

Meanwhile, it would help if the capital cost of soil, considered as a farm asset, were added to analyses. Land producing $100 per hectare has a value to present and future generations, undiscounted and taking a 500-year time horizon, of $50 000 per hectare. Natural soil replacement by rock weathering is so slow that its effects can be ignored. Any permanent and irreversible loss of even 5% of the soil's productive potential lowers its value by $2500. Whatever the economic assumptions and methods, a loss of this order of magnitude will radically transform the economic value of conservation.

The cost of land degradation

Erosion illustrates to an extreme degree the difficulties of costing degradation. Other types of land degradation can be converted into economic terms with somewhat less difficulty. An example for South Asia, based on current production loss only, has been given above (p. 131).

Salinization can be valued by the restoration method, the cost of reclamation. It is noteworthy that Pakistan, losing its only available irrigable land, believed it to be worth while, given World Bank assistance, to carry out a series of extremely expensive projects for reclamation of salinized soils.

Soil fertility decline is a reversible form of degradation. The component of nutrient loss lends itself readily to costing in terms of replacement by fertilizers, although this may be difficult in practice where micronutrient deficiencies have developed. The component of loss of soil organic matter and physical properties can be costed either as the restoration cost of taking land out of production for, say, 10 years to restore these, or as direct loss of crop yield, the former a largely hypothetical course of action, the latter the normal situation. The basic requirement for economic analysis is an estimate of the annual sequence of indicators of soil condition (e.g. topsoil organic matter and nutrient content), their consequences for crop yields and responses to inputs, and the nature and duration of management measures needed for restoration. This is no small task, but one not beyond the capability of soil scientists and agronomists.

Lowering of the groundwater table is, in practice, irreversible, and some countries are knowingly mining fossil groundwater. The economic methods employed with respect to mineral extraction could, in principle, be applied. Changes to river flow regimes and water pollution present complex but not insuperable problems of economic evaluation. Where water is in short supply, domestic and industrial needs are likely to take precedence over agriculture; if abstraction exceeds freshwater recharge, this will potentially lead to a loss of agricultural production.

Logging in excess of the rate of forest regrowth has been found the most straightforward form of degradation to cost, by taking the market value of timber and multiplying this by the reduction in standing biomass; if the results are to be meaningful, year-to-year variations in the price of timber must be eliminated. Renewal rates by growth are relatively well known, although data on rates of clearance are suspect. In the case studies of Indonesia, Costa Rica, and the Philippines referred to above, decline in the standing stock of forest capital alone amounted to 3–5% of gross domestic product. In Zimbabwe, fuelwood harvest exceeded the mean annual increment of trees by an amount valued, at the current price of fuelwood, at $44 million. To the commercial price of timber, an appreciable sum should be added for the collection of minor forest products. For the intangible benefits of protection, nature conservation, and recreation, economic valuation techniques are available.

Desertification, and also pasture degradation outside dry zones, poses as many problems as soil erosion. Wind erosion is both a cause and a consequence, but any direct relation between erosion and pasture productivity is impossible to establish. Two sets of indicators are needed as the physical basis: the present productivity of pastures (adjusted for seasonal differences), and some estimate of the time taken for regeneration. These must then be translated into terms of livestock output, with allowance for seasonal movements, before the economic cost can be assessed; where ranching is practised, livestock offtake can be directly employed. The FAO *Guidelines: land evaluation for extensive grazing* gives a brief discussion of how economic suitability classes might be defined in terms of gross margin per hectare of land, which, in principle, might be applied to show declines as a consequence of pasture degradation.[18] Where the pasture degradation is reversible, the losses from desertification reduce revenue; where, through severe destruction of soil and biological resources, the capacity to regenerate is lost, they represent a loss of capital resources. The considerable practical difficulties of conversion are not a reason for supposing that the economic losses from pasture degradation are not large.

Whether degradation is reversible, and, if so, the time taken to restore the land, makes a large difference to assessing its cost. The irreversible destruction of a resource, as in gullying and severe sheet erosion, means a permanent capital loss, which, it is argued above, should be taken as not less than 500 times the annual production. Loss of a tropical hardwood forest, if regrowth is attempted, delays the next major harvest for some 70 years. To restore soil depleted of organic matter might require augmented inputs of biomass for 10 years.

Most attempts at economic valuation of land resources have been made by economists, taking such physical data as they can find, and sometimes making questionable assumptions. Natural scientists venturing into this area may be tempted to turn away on discovering the large differences that can arise from the same physical data by making different, and often arbitrary, economic assumptions. They should not be put off in this way. If investment to reduce land degradation is to be attracted, its consequences must be expressed in economic terms that command respect, and this will only come about if they rest on a sound scientific basis. This can only be brought about through a better mutual understanding between the scientists and economists. A practical step to achieve this is to include economists in natural resource institutes, and natural resource scientists in the staffing of agencies responsible for investment appraisal.

National environmental accounting

National accounts are intended to show the annual growth (or decline) in wealth of the country. Countries have their own systems, as part of their budgetary accounts, and these are adapted and compared internationally through a United Nations *System of national accounts*. It is from these that the widely used measures of gross and net national product are obtained. From the early 1990s, it became recognized that these systems were defective, and often misleading, with respect to environmental aspects. In particular, national accounts failed to take into consideration:

- resource depletion, for example mining and oil extraction;
- resource degradation, for example soil fertility decline;
- pollution of the environment, for example water and air pollution;
- consequences arising from scarcity of depleted resources, for example forests and groundwater.

In essence, a country could extract its oil reserves, clear its forests, degrade its soils, and pollute its water, and none of these changes would appear at all in the national accounts. In addition, work undertaken to combat degradation or pollution, called defensive expenditure, was valued as part of the national product. There was a disparity between the treatment of produced capital, such as buildings, roads, and machinery, which was valued with allowance made for depreciation, and the neglect of corresponding changes in natural capital. The effect of the treatment of environmental assets in conventional national accounting is that it is possible to generate income by exploiting a contracting resource base.

In the early 1990s, recognition of these defects led to moves towards environmental accounting, in formal terms, integrated environmental and economic accounting. Some industrialized countries, notably Norway and France, made pioneering attempts, the latter accounting for the *patrimoine national*, the national heritage of soil, inland water, flora and fauna, ecosystems, and mineral resources. It was shown that the apparent economic growth of some developing countries was reduced, or even made negative, if net change in natural resources was included. In Indonesia, for example, an annual growth in gross domestic product of 7% by conventional accounting was reduced to 4% if losses of natural capital by oil extraction, forest clearance, and soil erosion were taken into account.[19]

Once this problem was recognized, action at international level took place commendably quickly. It was decided that, rather than alter long-established national accounting methods, these should be supplemented by a 'satellite' system of *Integrated environmental and economic accounting*.[20] Moves are currently under way to encourage governments to prepare such accounts and, more importantly, to recognize their significance for present and future national welfare. Whilst the principles of environmental accounting are clear, their practical application is complex and presents many difficulties.[21]

Mineral resources, including oil, are the most straightforward, in principle, to incorporate in national accounts, since the physical stocks can be measured, and valuation made at market prices. A problem is the artificiality of apparent increases arising from new discoveries. Pollution and waste disposal aspects raise complex questions of valuation; these aspects interact with land resources to the extent that pollution is a cause of land degradation.

All the essential elements in environmental accounting are applicable to land resources. The key steps are, first, establishment of a physical asset account, and, secondly, determination of the net price per unit of the resource. The physical asset account consists in outline of:

1. Opening stocks: the soils (or agricultural land), water resources, forests, fisheries, and biological assets at the start of the accounting period.
2. Depletion: removal or destruction of resources.
3. Degradation: reductions in the productive value of resources, through land degradation.
4. Additions: increases in the productive value, through land improvement.
5. Closing stocks: at the end of the accounting period.

Items 1 and 5 are stores or quantities, the equivalent of the balance sheet in economic accounting; items 2–4 are flows of resources, equivalent to the income and expenditure account. The flows may be for productive purposes, such as harvest of forests, or non-productive, as in soil erosion. As in economic accounting, stores and flows should be made to balance. It is assumed that, for social welfare accounting, governments and not individuals own their natural resources.

In standard national accounting, gross domestic product (GDP) is converted to net domestic product (NDP) by subtracting net consumption of produced assets (or adding net formation). In integrated environmental and economic accounting, net domestic product is converted to environmentally adjusted net domestic product (EDP) by subtracting net consumption of environmental assets, or adding improvements to them:

NDP = GDP – consumption of fixed produced capital
EDP = NDP – reduction in natural capital.

In the present state of information, it is rarely the case that such data can be obtained on an annual basis. Many types of land resource estimates will be prepared at best only every 5 to 10 years; conversion to an annual basis, needed for linking with conventional national accounts, could be done by extrapolation of the most recent period. There are exceptions to this: a good forestry department should be monitoring clearance, planting, and growth on an annual basis.

The second step, placing a unit value on resources, raises the problems of valuation discussed above. For an asset which is at the same time the product consumed, such as forests, the usual method is market valuation. Reclamation cost is appropriate where this is physically possible, as in salinized soil. There is a body of economic theory on valuing biological resources and ecosystem services.

A fundamental problem arises in valuing renewable natural resources such as soils, which are potentially productive for an indefinite period in the future. The conventional means is the present value method, which takes the market value of net revenue flows (crops, etc.) obtained from the resource, discounted at some nominal rate and for its assumed 'life', both highly arbitrary assumptions. The present value method typically leads to the result that soil with a productive potential of, say, $100 per year is 'capitalized', i.e. reduced by discounting, to a present value which is typically 10, or at most 20, times the annual production, that is, $1000 to $2000.

A consequence is that, with methods of valuation now current, in countries with appreciable mineral resources, these dominate the environmental component of accounts. In addition, in the studies made to date, depletion of forests is assigned a considerably higher value than soil erosion. Thus, in the case study of Indonesia, net changes in the natural resource sectors over 1980–5 were valued for petroleum, forests, and soils respectively as −67%, −30%, and −3%. For the Philippines, resource depreciation in timber, soils, and coastal fisheries were valued respectively at −83%, −9%, and −8%.[22] Intuitively, these ratios are wrong. It should be added that the fault does not lie wholly with the economists, who were not been given physical resource data that they needed.

Adoption of the valuation approach outlined above, raising the future option value to society of soils to at least 500 times their annual production, would greatly alter these ratios. The irreversible destruction of 1% of the soil profile would then be costed at five times its annual productivity. A radical change in attitudes and accounting methods will be needed before this proposal, or something like it, is accepted.

It was a good decision to keep environmental accounts separate from conventional national accounts in the first instance, in order to see the nature and magnitude of differences between them. Three guidelines may be suggested for preparing environmental accounts. First, the contributions from different types of resource – mineral, land resources, ecosystems, and effects of pollution – and between the respective land resources should be clearly presented individually prior to amalgamation. Secondly, the assumptions made in unit valuation of each type of resource should be made explicit – indeed, until practices become established, they should be given prominence.

Lastly and most importantly, the assessments of the state of physical assets, as hectares of soil, forests or pastures, soil nitrogen content, pasture composition, and the like, should be presented as a separate output from the accounting procedure, before being assigned values. Even the starting-point for valuation, the present market value of a resource, may fluctuate widely from year to year through fluctuations in prices of inputs and products. The methods employed for valuation of off-site effects and environmental quality also greatly alter the analysis. As soon as assumptions with respect to time and future value are made, the results can be altered by a factor of many times. Taken together, the economic assumptions and methods can have a far larger effect on the results of economic accounting than changes in the resource stocks themselves. This consequence, distressing as it may be to scientists, is inevitable if the environment is to be costed at all. It is vital, therefore, to present the environmental changes as such, in the form of indicators, independently of their incorporation into integrated national accounts. This practice will also permit the taking of some decisions independently of economic valuation.

The physical data on land resources, degradation, and environmental change are at present far from being adequate for the needs of economic analysis. In this respect, the resource scientists and not the economists are to blame. Good environmental economic analysis calls for good indicators of environmental change.

11 Land management: caring for resources

The older approach to land management, based on the transfer of Western technologies, has been replaced by a new set of ideas. For management of the croplands, new approaches include the land husbandry basis for soil conservation, low-input sustainable agriculture, and small-scale irrigation. On open rangelands, reconciling the extreme complexity of land management needs with communal tenure raises problems which are almost insuperable. Multiple-purpose forest management has replaced the earlier focus on wood production. Agroforestry has helped to diversify farm production, and offered new means of soil management.

These new approaches have a number of ideas in common: understanding the processes in the soil, water, and plant ecosystem, as a basis for their modification; adapting management methods to the infinite variety of local conditions; and increasing production not by taking in more land nor with higher inputs, but by using soils, water, fertilizers, and plant resources with greater efficiency. Finally, it has invariably been found that best results come from a participatory approach, implementing changes through the joint efforts of resource scientists and the knowledge and skills of the local people.

The fundamental principle of land management is sustainability, the combination of production with conservation. Given the extent of poverty, the urgency of the food situation in the developing world, and the present low level of productivity of many farming systems, the priority must be to increase production. This has to be achieved in ways that do not degrade, and where possible improve, the land resource base on which production depends. The primary objective is, of course, the welfare of the people. Taking land resources as an alternative focus, however, provides a powerful means to integrate production with conservation, and so to lead towards sustainability.

A set of new ideas on land management has taken the place of the older approach, which was based on high levels of inputs and transfer of Western technologies to the developing world. The new approach has two common themes or principles. The first is to make use of external inputs, but at moderate levels and with higher efficiency. To achieve this objective means working in conjunction with processes of the natural ecosystems. The second principle is to implement changes through collaboration with the people, the approach of participatory development already discussed. Underlying these is a fundamental three-stage approach:

- to understand the functioning of the natural ecosystem, soils, water, plants, animals;
- taking this understanding as a basis, to construct a sustainable managed ecosystem, a land use system that will both be productive and conserve or improve the resource base;
- to reconcile the management needs for sustainable production with economic and social requirements and constraints.

Because the environmental conditions vary widely, standard recommendations, or extension 'packages', are not enough. Management methods have to be constantly adapted to the site conditions of climate, water, soil, and vegetation, not only to their variation in space but also their changes over time. Farmers have always made such adaptations, and extension staff should do so.

Much has been learnt about land management based on these principles. It is impossible to review the whole range of land management methods, so the focus will be on selected ideas which hold promise for the future. A framework is provided by the three major production systems – croplands, rangelands, and forests – together with the new science of agroforestry which overlaps these.

Managing the croplands

Land husbandry: the new approach to soil and water conservation

In the older, or conventional, approach to soil conservation, the objective was to reduce soil loss, measured as tonnes per hectare. This was achieved by means of earth structures: either terraces, or combinations of contour-aligned banks (often called bunds) with ditches. Water runoff was reduced either by causing it to sink in, as with terrace systems, or by diverting it into controlled waterways. Conservation of this type became a branch of civil engineering; manuals were published on how to build such structures for local conditions of rainfall, slope, and soil. Under the former system of land capability classification, only gentle slopes were classed as suitable for arable use; all steeper land was allocated to grazing, forestry, or conservation.[1] Agricultural extension work was based on the view that soil conservation should come first, as a prerequisite for agricultural improvements. It was commonly conducted on the basis of a prohibitive policy, either by forbidding cultivation of steeply sloping land or by legally enforced requirements for the construction of conservation works.

Devised initially for farming conditions in the USA, the conventional approach to conservation is technically successful in reducing runoff and erosion. As regards adoption it had some notable successes, in Zimbabwe for example, where some landscapes of bunds, waterways, etc. looked from the air like a conservation textbook. In Asia, some extensive terracing systems were constructed, for example in Taiwan.

But, in many cases, the older approach to conservation simply did not work. A clear example of failure is the case of Jamaica. Repeated attempts were made to introduce terrace systems to the hill lands, through a series of externally funded projects. These have not been maintained, and are largely abandoned. In many countries, land shortage has enforced widespread cultivation of sloping lands, and to prohibit this is both economically and socially unrealistic. In the Ethiopian highlands, whole communities have their land on steep hillsides. In Malawi in 1960, cultivation stopped at the foot of the hills; by the mid-1970s, cultivation had extended up the hills and onto the steeply dissected rift valley scarp areas. It is difficult to enforce legislative penalties. Farmers' co-operation could not be obtained unless they could see an immediate benefit in terms of higher crop yields, and, when conservation is carried out in isolation from other improvements, no such benefits occur.

Out of the failures of the former system a new approach to conservation arose, commonly called land husbandry.[2] Features of this approach are:

- The focus of attention is not upon soil loss as such, but on its effects on production; these arise principally through loss, in eroded soil, of organic matter and nutrients.
- More attention is given to biological methods of conservation, especially maintenance of a soil cover, including through agroforestry. Earth structures, whilst by no means excluded, receive less emphasis.
- In dry lands, there is greater integration between soil and water conservation. Farmers are able to see more immediate benefits from conserving water.
- It is recognized as politically and socially unacceptable to forbid the cultivation of sloping land. Ways have to be found to make such cultivation environmentally acceptable.
- In extension, it is recognized that conservation can only be achieved through the willing participation of farmers. For this to occur, they must be able to see benefits. It follows that conservation should not be a separate element, but an integral part of improved farming systems.

Thus there are two basic elements to land husbandry, technical and social. There is no doctrinaire reason to favour biological methods over earth structures if the latter are agreed to be the best solution, but the cost and labour involved in their construction, and more importantly maintenance, mitigate against their use. Taken to its extreme, their should be no soil conservation projects – only projects to improve sustainable production, in which conservation forms an element. In a project for the central hill lands of Jamaica, the primary, and explicit, objective is to improve production of perennial crops, coffee and cacao. More productive crop varieties, better managed, produce more leaf litter, and the only specific conservation-directed element of management is to ensure that this litter remains on the soil.

The current need is human and institutional. Existing staff of conservation departments will require retraining. Education in conservation has to be considerably broadened from its former, engineering, basis, to include skills in the use of biological methods, greater awareness of the wider problems of farming, and practical training in participatory extension. There is a need to consider how far soil conservation departments should remain separate. One solution is to abolish them, and incorporate technical conservation specialists within the general agricultural extension service. The improved opportunities which are offered by land husbandry can only achieve their potential through well-educated staff and effective institutions.

Low-input sustainable agriculture

The great improvements in crop yields in the period 1950–90 were achieved by scientific farming based on improved crop varieties. Soil constraints were overcome by fertilizers, and other environmental problems by inputs, including biocides and the expansion of irrigation. This high-input approach, based on Western technology, has encountered problems. Fertilizers are costly in terms of the energy resources needed to produce them; some modern farming systems are energy-negative, using more energy in inputs than they produce in food.[3] Continued high rates of fertilizer use lead to environmental problems. In many Asian countries, fertilizer use has consistently risen faster than crop yields, indicating a diminishing response. Further increase in the irrigated areas is now severely checked by limits to freshwater resources. An effective fertilizer distribution system requires a level of governance and management efficiency not yet attained in some countries. High inputs give best results on the more fertile soils, performing less well on the marginal lands which many farmers are forced to cultivate. Above all, large numbers of poor farmers simply cannot afford high levels of fertilizers and other inputs, nor do they have the capital to take on the risk which these involve.

These problems have led to a new approach to soil management called low-input sustainable agriculture, although it applies also to farming with medium levels of inputs.[4] Features of this approach are:

- to find ways of making farming on marginal lands sustainable, including by the development of plant varieties adapted to soil constraints;
- to maintain soil organic matter and biological activity, with benefits both for soil physical properties and balanced nutrient supplies;
- to augment nitrogen inputs through use of nitrogen-fixing plants;
- to use fertilizers and other inputs at moderate levels, improving nutrient cycling and nutrient use efficiency;
- to control pests and diseases by a combination of chemical inputs and biological methods;
- to use appropriate tillage, including minimum tillage;[5]
- to improve water use efficiency, both in rainfed agriculture (through dry farming methods) and under irrigation.

Two elements of this approach have their own technical terms. Making more efficient use of low to moderate fertilizer inputs, through biological nitrogen fixation, nutrient recycling, and maintenance of soil organic matter, is known as integrated plant nutrition systems (IPNS).[6] Pest management through combining judicious use of biocides with crop rotation, plant hygiene, and biological controls is termed integrated pest management (IPM). It helps to meet problems of the development of resistance to specific chemical controls.

To practise low- to medium-input agriculture is not easy. It requires greater management skill than simply adding inputs, and hence a more efficient system of extension services. There is a greater need to adapt methods to local climatic, slope, and soil conditions. This approach by no means aims at eliminating artificial inputs. Rather, it seeks to use limited levels of inputs to their maximum efficiency. If phosphorus is lacking from the weathering rock beneath the soil, then no amount of recycling can replace what is removed in the harvest. It is unfortunate that the phrase 'biological farming' has acquired the meaning of farming without chemical inputs at all. Biological methods, notably the management of soil organic matter, have their highest contribution to make where they are combined with fertilizers.

Low-input sustainable agriculture brings benefits that are both environmental and economic. It is intrinsically easier to limit pollution if inputs are at lower levels. Maintaining organic matter and soil biological activity brings a wide range of benefits to soil, including a better-balanced nutrient supply. Poorer farmers do not apply fertilizers at the maximum economic rate of return, because of lack of capital and the risk involved. Low- to medium-input methods are accessible to a wider range of farmers, including those on more marginal lands.

Improving irrigation management

Irrigation is the leading example of land improvement, changing the water supply so as to increase its overall productive potential.[7] Irrigated land is twice as productive as rainfed cropland: one sixth of the world's cropland is irrigated, producing over 35% of its food and over half the rice and wheat. In 1900, the world's irrigated area was about 48 Mha; it passed 150 Mha in 1964 and 250 Mha in 1994, 74% of this in developing countries. China, India, and Pakistan account for 45% of the world area and 60% of the total for developing countries. Egypt is the only country totally dependent on irrigation, but the percentage of cropland irrigated is over 75% in Pakistan and three of the central Asian republics of former USSR. North Africa and the Middle East depend on irrigation for their most productive lands, although they also have large areas under dry farming. Less widely known is the importance of irrigation in Central America, with about a quarter of cropland irrigated in Costa Rica, Cuba, and Mexico.[8]

The rate of growth of irrigated land is slackening, from a maximum of 2.3% a year in 1972–5 to less than 1% today. As this is well below population increase, the irrigated area per person has declined steadily from an all-time maximum of 0.048 ha in 1978–9 to under 0.045 ha by 1995. The underlying reason is the approach to the limits of available freshwater resources in many regions, leading to a rise in the cost of new developments. The average investment cost of new projects, as recorded in World Bank projects, is now over $5000 per hectare but this is highly variable. For projects 'rated as unsatisfactory' it is $18 000 ha^{-1}; for two projects, in Chad and Guyana, the unit cost was infinite, since the completed irrigated area at the time of evaluation was zero.[9] Counteracting developments which bring new land under irrigation is the loss of land to salinization.

There are three paradoxes to irrigated agriculture. First, it is a high-input, costly form of agriculture, best suited to high-return crops such as fruit and vegetables; yet it is often employed for production of relatively low-value food crops, in part as a consequence of subsidized water pricing. Secondly, the need for irrigation is greatest where it is least efficient. Whether measured in physical or economic terms, efficiency is highest for supplementary irrigation, improving water supply in areas which can still be farmed without it; yet it is in arid regions, like Egypt or much of Pakistan, that irrigation is most vital to the people. Lastly, it is in regions which are most dependent on irrigation that water shortage, and hence constraints to further expansion, are greatest. The volumes of water needed for irrigation are extremely large, 15 000 m^3 for one hectare of swamp rice. Consequently irrigation is, and will continue to be, the largest use for water resources, at 70% over twice that of world industrial and domestic uses combined.

The direct function of irrigation, to supply water to land, is only a means to an

FIGURE 19 Sustainable land management: swamp rice, Malaysia.

end. Its real objective is to increase the productivity of land by making good a deficiency in soil water, hence allowing the soil itself and inputs applied to it to be used to greater efficiency. It also serves to extend the growing season, and greatly to reduce risk. In a wider context, irrigation means improving production, employment, and incomes by using land more intensively than would otherwise be possible. The situation with respect to irrigated land reflects, in magnified form, that for cultivable land as a whole. Since irrigated agriculture is twice as productive as rainfed, expanding the proportion of irrigated land might seem an obviously desirable course of action. But opportunities to do so in the former manner, of bringing in more water, are constrained by the limits of freshwater supply of the hydrological cycle – indeed, demand has already exceeded sustainable supply in areas of falling groundwater tables.

There is no question of saying that large-scale forms of irrigation should be abolished; dam and canal systems are the only way to harness the water of major rivers. We can learn from the mistakes of the past, and well-managed irrigation schemes can produce good returns to investment. The focus of modern policy and practice is to use water with greater efficiency. Ways to do this are, first, to increase the amount of water which reaches farmers' fields; and, secondly, to improve farming practices on the irrigated land. In large-scale, dam and canal, schemes, water use efficiency, the ratio between water abstracted from rivers and that reaching farmers fields, averages 30–35%. There are some substantial unavoidable sources of loss, but efficiencies of 50% are attainable. The 'tail end' problem is well known, farmers at the lower end of the distribution system finding that the channels run dry at times when water is most needed. Methods employed to improve irrigation water management are:[10]

water user groups, representing the farmers;

a clearly defined division of responsibilities between the organization responsible for water distribution and the user groups;

a set of physical control structures which minimize losses and which can be operated under the above institutional arrangements;

good information collection and feedback systems;

effective agricultural services, advisory and for input supply and marketing.

It is relatively easy to establish such a management framework when starting from scratch on new schemes. To introduce one by reform of existing systems is harder, but makes a greater contribution to efficiency simply because these far outnumber new developments. Water user groups are a way to bring participation into large-scale schemes. Agricultural services should not be regarded as supplementary, but as an intrinsic part of an irrigation scheme.

Small-scale irrigation

Small-scale irrigation refers to methods of water collection and distribution which the farmers themselves, individually or collectively, can maintain; farmer-managed irrigation systems are an alternative name.[11] This has been a part of agriculture from the earliest times, with much ingenuity being displayed to achieve the basic operations, raising the level of water and transporting it to the fields. Among many remarkable examples are the underground channels, *qanats*, of Iran, and surface ducts which follow precipitous mountainsides in the arid mountains of northern Pakistan. Perhaps the most elegant method of all is the *anicut*, in which a distributary channel is cut directly from the river bank then taken downstream with a gradient less than that of the river, so that after some distance it is higher and can be employed to irrigate river terraces.

Small-scale earth dams, often constructed without machinery, are particularly widespread in the Tamil Nadu region of southern India, where they are known as tanks and often constructed as strings of reservoirs along river courses, and in Sri Lanka. These fulfil multiple functions, including cattle watering, and the growing of vegetables on seepage zones close to the dam. Hill irrigation schemes achieve the seemingly impossible in permitting swamp rice cultivation on slopes of 30% or more, through complex distribution systems serving the stairways of terraces constructed over centuries.

Because of the rising cost of new large developments, and the failure of some, there has been a revival of interest in small-scale irrigation. In many parts of Asia, the necessary skills have been handed down the generations, and there is an acceptance of the sheer hard work sometimes involved – in Vietnam one can still see water being raised from one channel to another by two people using a basket. Small-scale technologies are now being favoured in Africa, where a number of large projects have been notoriously unsuccessful. Relatively small rice schemes on alluvial lands have proved successful and popular, the rice finding a ready market in the towns. River flood plains and valley floors can become very much more productive in this way, although there is a loss of their former agricultural functions.

Tubewells, mechanically drilled boreholes operated by diesel or electric power, are a special case in which modern technology is applied to small-scale irrigation. It was through their development, initially through external assistance but later taken up spontaneously by farmers, that the upper Indus plains of the Punjab became one of the richest agricultural areas of India and Pakistan. They have great advantages: water can be raised from much greater depths than through hand-dug wells, the 'reservoir' is in groundwater, not subject to losses, and water can be raised exactly in the amounts and at the times wanted. The very success of this technology is partly responsible for its main problems. Uncontrolled abstraction of water has led to falling water tables. In this circumstance, sustainable use can only be reached through a system of controls limiting abstraction to the rate of recharge, and this in turn requires scientific planning, monitoring, and administration in a non-corrupt manner.

There is still much potential to increase production from irrigated agriculture, but the former solution of 'irrigate more land' will increasingly encounter natural limits set by supplies of fresh water. As with cultivated land as a whole, future gains must increasingly come from raising the productivity of presently irrigated land. This calls for a two-pronged approach: more efficient water distribution, and more productive methods on the irrigated land. Both in large and small-scale technologies, farmer participation is a key element, but government assistance can help to provide infrastructure and services. In larger schemes, with their costly infrastructure and complex management needs, sound and stable government is essential.

Making the best use of dambos

Competition between alternative uses for land on a local scale is illustrated by the case of valley-floor grasslands in the subhumid zone, recognized in Africa by local names such as *dambo*, *mbuga*, *fadama*, or *vlei*. They are a distinctive element in the extensive land resource zone formed by raised erosion surfaces, known to soil surveyors as the 'gently undulating plain', where dambos may occupy 10% of the total area. The essence of the vernacular names is not so much the valley floor as the tall grassland which grows there; 'meadow' would be the nearest translation.

In the traditional system of land use, the dambos form dendritic patterns of grassland amid the cultivated land on the interfluves. Their value was that the water table remained high, and the grass green, throughout the dry zone. Livestock could be grazed on hill pastures in the wet season, and taken into the dambos, for the dry season. The animals were brought into enclosures (*kholas*, *bomas*) at night and manure from them was applied to croplands, so transferring 'uphill' nutrients which the natural process of leaching had carried into the valleys. Marginal strips along the valley sides were put to growing vegetables (*dimba* gardens), making use of residual water left as the water table fell.

The productive capacity of some dambos has been lowered by a cycle of land degradation: overgrazing, gullying, lowering of the water table, and hence loss of the resource of dry-season grazing. There has been competition for arable use; comparison of air photographs of the Lilongwe Plain, central Malawi, taken 20 years apart, shows a pattern of land use which at first sight is similar, but which on being plotted shows that arable fields have crept down the slopes. Some governments legally forbid cultivation within so many metres of a watercourse, a highly unenforceable regulation.

In places, a new use has been introduced, rice cultivation. This can be swamp rice with the construction of bunds, but in Africa it is often wetland rice, varieties which do not require flooding but draw upon groundwater. This use is more productive per unit area than livestock, but it raises problems. The special role of dambos in the agro-ecosystem, to provide dry-season grazing, is lost. Moreover, such grazing was communal, regulated by the village custom, but ricefields are likely to be established by more wealthy and influential individuals. There could be a further use: agroforesters have noted the potential of dambo margins for wood production and fodder banks.

Grazing, vegetables, rice, trees, plus the need to protect this valuable component in the land resource pattern against degradation: these are the issues. Government advice and, still less, controls will have little effect. Land use in dambos can be managed only by the wisdom of local communities.[12]

Managing the rangelands

The problems of managing rangelands for livestock production are, if anything, still more complex and intractable than those of cultivation. Production is in two stages: the growth of plants, and its conversion to meat, milk, etc. by animals. The plant production is fixed in location, whereas the livestock can move seasonally.

FIGURE 20 Nomadic pastoralism, northern Nigeria: adapting traditional pastoralism to modern conditions has proved to be one of the most intractable problems of land management.

There must be access to water at all times. The biomass and composition of the pastures varies greatly across the climatic transition from desert margin to either the dry savannas or the humid Mediterranean zone; the livestock include cattle, sheep and goats ('shoats'), camels, or occasionally antelopes and other game animals. Some can only graze, others browse from shrubs. Year-to-year rainfall variability is the highest of all climatic zones. A further range of management needs arise from questions of animal health. This situation presents management problems which are difficult enough even in the most controlled of land tenure situations, a freehold ranch with full management control, including through fencing. It becomes more complex under the conditions of communal access which frequently apply to pastures; and still more so with the semi-nomadic system widespread in Africa and found also in dry plateau lands, some at high altitudes, of Central Asia. Migration across national frontiers may add to the problems.

Although much of the Third World's livestock production comes from managed pastures and stall-fed systems in more humid climates, a distinctive resource position is occupied by the rangelands of the semi-arid zone, the sahel of Africa, and similar climatic belts around deserts in other continents. This is an extensive zone, 12% of the area of all developing countries and 17% of Africa. It can be cultivated with difficulty or not at all, but has potential for livestock production.

There is a clear opportunity for meat production from the dry lands to complement grain production from humid regions, supplying a dietary need for protein especially to urban markets. A number of countries have sufficient environmental range for this exchange to take place internally, in others it can be achieved by trade across neighbouring frontiers. Yet systems of livestock production by extensive grazing have proved to be most difficult to adapt to the circumstances of the modern world. Notwithstanding the exaggerations of 'the desertification story', and the fact that outsiders may misunderstand the apparently devastated condition of rangeland at the end of the dry season, there is no doubt that degradation of pastures is extremely widespread. Attempts at improvement bring about new problems: establish boreholes to improve water supply and a trampled area results; improve marketing facilities, and the basis of subsistence production is disturbed. Of all types of land development projects, livestock schemes have the lowest rate of success.

Livestock grazing was also the basis for a theory published by Garrett Hardin in 1968, *The tragedy of the commons.*[13] It is ironic that this paper, one of the most widely quoted of all development publications, contained a fundamental flaw. Hardin based his argument on the common grazing lands of medieval Europe. Villagers each put out livestock to graze on these; when a bad year came (drought or disease) the same proportion of each herd died, so it was in the interest of each individual to have as large a herd as possible; this resulted in overgrazing, and everyone suffered. All would have benefited if each had exercised constraint, but it was not in the interests of any individual to do so. Hardin supposed this to apply to grazing lands and other communal land resources of the present-day tropics, and so to be a basic cause of land degradation. We now know that on commons, meaning communally managed land, social custom enforces controls over what may and may not be done, such as where and at what season animals may be grazed. It is not on commons but on open-access resources, those which belong to no one but are available to all, that the 'tragedy' of degradation through overexploitation occurs. Where there was formerly enough spare land for cattle to 'look after themselves', no rules for community control were set up; growth of livestock numbers and encroachment of cultivation onto grazing lands then led to overgrazing; India, Mexico and highland Ethiopia are among countless examples. Illegal extraction of wood from state forests, cultivation within them, and overfishing of lakes, illustrate the problem of open-access resources from other types of land use.

The standard approach to rangeland management, still widely applied, is based on an established set of scientific principles. The keystones are control of livestock numbers and rotational grazing. Stocking density is held at the livestock carrying capacity, defined as the maximum number of animals that an area of land can support, under a given management system, without non-reversible degradation of the vegetation. Different kinds of animal can be allowed for partly by reference to a standard tropical livestock unit, 250 kg weight. The management system can substantially alter capacity, through such practices as oversowing with forage legumes or supplementary feeding in the dry season. It is possible to obtain carrying capacity by measuring plant growth, obtaining nutrient content from leaf analysis, and comparing this with dietary requirements, but this is not often done. To ask an experienced ranch manager how many animals he can get by with in a poorer than average year may be as good a method as any other.

The second requirement, for rotational grazing, stems from the need for pastures to recuperate. Most zones contain a mixture of annual and perennial grasses, of which the perennials are the most nutritious. The perennials need a seasonal rest from grazing to build up sufficient leaf biomass to allow them to transfer stocks of carbohydrate to the roots (and below-ground storage organs), otherwise growth in the following rainy season will be poor. These requirements can be met on a fenced ranch by a system of paddocks, and less precisely on open rangeland through seasonal movements of livestock.

The concept of carrying capacity has always been viewed with suspicion, doubts being cast not only on its usefulness but also on its theoretical basis.[14] In a natural environment, there is a standing crop of plants and a standing number of animals, mutually limited by feed availability and grazing pressure. Omitting the complications of carnivores and human management, these reach an equilibrium, called the ecological carrying capacity. If it were not for year-to-year variation, the animals might be in poor condition; in practice quite the opposite happens, most wild animals appearing well fed, owing to the limitation of numbers in drought years. The ecological carrying capacity might be relevant for subsistence livestock production which depends on milk and blood. It is inappropriate for meat production, in which there must be an annual offtake of animals. By having a smaller number of animals than the ecological carrying capacity, and a correspondingly higher plant biomass, offtake is maximized, a theoretical situation known as the economic carrying capacity. The droughts which are inseparable from this environment further complicate the situation, reducing both animal populations and plant density. After single-year droughts, recovery of both is rapid, but sequences of drought years reduce the rootstock and seed availability, and recovery is slow. If long-term climatic change is added to variability, the situation becomes still more unstable.

What seems a formidably complex situation to outsiders is the basic circumstance of life for pastoral peoples. Migration is employed as a management tool: regular seasonal movements allow productive use of areas too dry to support livestock permanently, and also give seasonal relief from grazing in all parts of the circuit, whilst opportunistic movements can be made in response to local variations in rainfall. To reduce risk there is usually more than one kind of animal in the herds, for sheep and goats will survive from browse long after all cattle have perished. Traditional practices sometimes conflict with the supposed principles of range management, livestock numbers being increased until there is apparent overgrazing. But, as someone who had worked among the people of semi-arid Kenya put it, 'They know perfectly well that dry years will come, the cattle will die, and they will go hungry, but they increase livestock numbers to make the most of it in good years, that's their strategy.'[15] It is apparently a successful one: studies of five African open-range grazing systems have shown that their productivity (measured as kilogrammes of protein per hectare per year) is 50 to over 100% higher than that of commercial ranches in similar environments.[16] Many Westerners admire the tradition and dignity of the way of life of pastoral peoples, and would regret its replacement by commercial ranching.

In former times, some degree of ecological and social stability was reached among pastoral peoples. Why then, by combining scientific and traditional knowledge, are such difficulties encountered in attempting to develop these economies? The former equilibrium rested on circumstances no longer acceptable, such as absence of health services, and the practice of armed invasion of neighbours in times of need. Where cultivation encroaches into semi-arid regions, it takes the best grazing land. The welfare of pastoral peoples does not come high among the priorities of governments, and they become economically and politically marginalized. All of this adds to the formidable problems of land resource management.

A hard-line attitude would be to say that ranching with freehold tenure is the only long-term solution to the sustainable management of rangelands, but this is not likely to be acceptable in the near future. As with other forms of agriculture, the participatory approach is essential. This can be applied through adapting diagnosis and design to pastoral systems: analyse the constraints of the present system; bring scientific knowledge to bear on the design of possible improvements; then let these be tested by the pastoralists on a pilot scheme basis, and implemented more generally if they choose. Where possible, a dedicated range scientist should be on hand through this process – how much better it would be if this were someone drawn from the pastoralists themselves!

Multiple-use forest management

In recent years, forest management has seen a change in approaches if anything even greater than that found in agriculture. Again, there are three stages: a stimulus coming from failures of the older methods; a change in attitudes; leading to the development of new methods of management.

The older attitude to forestry, by no means absent today, was based on exclusion. Land currently under natural forest, or suited to plantations, was identified, and taken into what was called the 'forest estate'. This meant designated forest reserves, within which only the government forestry department and their employees could set foot. This provided the basis for a range of well-designed and, for a time, effective management systems. Forest plantations of selected species could be established, their growth monitored at sample sites, and the trees managed for good form (straight trunks without low branches); they could be thinned, and then clear-felled, operating a rotation around the estate. A difference is that, for softwood species such as pines, a rotation in the tropics lasts some 15–20 years, compared with the 50 years or more in temperate latitudes. The very much more extensive natural rain forests were managed for sustainable yield management (a justifiable use of the word 'sustainable' long before its recent vogue). Management methods, such as the Malayan uniform system, were developed, which envisaged rotations of 70–80 years, with selective harvesting rather than clear-felling, combined with methods to re-enrich the forest with commercially desirable species. This was a long-term system of planning and management, resting on a basis of 'keep the agriculturalists out'. Local people were allowed to collect minor forest products, such as gums, resins, and thatch, but cutting for fuelwood and of course cultivation must be resisted by legal controls, enforced by forest rangers.

During the 1980s, there was a serious decline in the effectiveness of the older approach, of which population increase was the underlying cause. The demands of agriculture led to clearance of forest for cultivation, sometimes officially sanctioned, more usually by illegal incursion. Cutting for fuelwood and domestic timber led to forest degradation. At the same time, political conditions were no longer such that it was practical to impose authoritarian control. This was not just a result of the replacement of colonial rule by independence. India, for example, has a strong policy of forest protection at both national and state government levels; but, where one forest ranger has to cover many thousands of hectares, control imposed from above is simply not practicable.

There are two foundations to the new approach: multiple-use forest management, and participation.[17] There is greater recognition of the many purposes of forestry, for production and conservation:

wood production: for sawn timber, domestic timber (roundwood), fuelwood, and charcoal;

the so-called minor forest products: gums, resins, thatching, rattan, food (fruit, roots), medicinal products, etc.;

sylvopastoral use, or forest grazing, especially in the subhumid and semi-arid zones;

FIGURE 21 Plantation forestry, on the slopes of Dedza Mountain, Malawi, making productive use of sloping land and bouldery soils.

protection, especially of headwater catchments, linked with maintenance of water supplies;

conservation, including plant genetic resources and wildlife conservation;

ecological tourism, as national parks.

Wood production is still of primary importance, not least because 80% of the wood in developing countries is used for fuelwood and charcoal, with demand projected to increase by 1.7% a year. Multiple-use management is based on recognition that these functions can be combined, especially that wood production can be compatible with controlled use for other purposes. The different productive and protective functions vary greatly with local conditions, and there are certainly cases which should be managed for one only: some plantations for pulpwood production, for example, or steep headwater areas which should be kept exclusively under protection forestry. But, in many areas, wood production on a sustainable basis can be made compatible with use by local people for hunting, gathering minor products, or forest grazing. Multiple-use management is more complex and difficult than the older methods; to be successful, it calls for both skilled design and effective local implementation.

The second keystone of the modern approach is to involve people in management of the forest. This means, first, ensuring that local populations have access to the benefits of forestry activities; and, secondly, to promote among them a sense of responsibility for the proper management of forest as a resource – the first is a precondition for the second. Access to the resources of forests and woodlands is particularly important to women and to the poorer sections of the community. Coming before all else is the need for recognition that it is in the best interests that certain areas of land should be kept under forest. In the 1980s, China transferred 20 million hectares of forest land to local households, for management as farm and village forests; India has also experimented with handing over state forests to local control. Forests will only survive if they are seen by the people involved to be more valuable under forest than converted to any other form of land use.

In competing for scarce development funds, forestry faces the further problem, already discussed in chapter 10 (p. 164). Conventional methods of discounting greatly reduce the 'present value' of forest products harvested 20 or more years hence. Forestry only gives high benefit:cost ratios if externalities (e.g. the value to river base flow) are taken into account, together with a low or even zero rate of discount. This is a clear illustration of the bias, in standard economic analysis, against the needs of future generations.

Multiple-use management is not a panacea, and problems have arisen as a result of increased pressures on forest land from population growth. It is effective at low to moderate intensities of use, but may break down and become unsustainable at high intensities of harvest. Intensively managed plantations retain an important role. Forest land use planning, in the form of zoning of forests with different objectives and systems of management, helps to achieve best use of land with varying resource potential.[18]

Foresters now accept that agricultural use will almost everywhere take priority on land of high potential. Forestry is promoted as the best way of making use of land with greater natural limitations. A better integration of forest land into the needs of whole economy, both at village and national levels, is achieved through multiple-use management, guided by research, and implemented by skilled staff working in conjunction with the local people.

Agroforestry

In 1977, a Canadian aid organization commissioned a study of what should be the priorities for research into forestry. The resulting report, *Trees, food and people*, recommended quite different directions from traditional forestry research. It said that more attention should be given to combinations of agriculture with forestry, and on helping farmers to grow trees, coining a new term, agroforestry. The report

FIGURE 22 Agroforestry: a home garden, Vietnam – highly productive and sustainable.

led to the establishment, initially on a very small scale, of the International Centre (originally Council) for Research in Agroforestry (ICRAF), and to the establishment of agroforestry as a new science. It was quickly seen that many, if not most, small farmers had always grown trees on their farms, not just by default, leaving selected species from the natural vegetation, but actively managing them and planting introduced species. Agroforestry became the new name for an old practice.[19]

By the late 1980s, ICRAF had achieved its first objective: the nature of agroforestry, and its potential to contribute to land resource development, was recognized. It was being applied in development projects, both by governments and non-governmental organizations, for which its low-cost, intrinsically participatory nature proved attractive. Indeed, for a time, enthusiasm overran research, and failures resulted from applying systems without sufficient knowledge. The necessary basis of research is now being built up.

Agroforestry is based on multipurpose trees, those which make more than one contribution to the land use systems in which they are grown. This is, indeed, a property of most trees, as witness the forestry term, 'minor forest products'. The intention is to distinguish the diverse roles which multipurpose trees play in farming systems, as distinct from wood production which is the dominant aim of large-scale forest plantations. It has been rightly said, echoing George Orwell, that, 'All trees are multipurpose, but some are more multipurpose than others.'[20] The

AGROFORESTRY

Agroforestry covers all land use systems in which trees or shrubs are grown in association with crops or pastures, in a spatial arrangement or a rotation, sometimes with livestock; there are usually both ecological and economic interactions between the trees and other components of the system.

Short definition:
'Growing trees on farms.'

Some common types of agroforestry:

- *Managed tree fallows* Selected species of fast-growing trees planted in rotation with crops, as in shifting cultivation but with the wood from the trees harvested.
- *Trees on cropland* Scattered trees amid crops, e.g. *Faidherbia albida*, which improves soil properties and crop yields beneath its canopy.
- *Multistrata systems, including home gardens* Densely planted, mixed stands of trees, shrubs, and crops, managed for wood, fruit, food and cash crops, sometimes with small livestock.
- *Hedgerow intercropping (alley cropping)* Hedges planted in rows across cropped fields, regularly pruned; the hedges improve soil organic matter, fix nitrogen, and improve nutrient cycling.
- *Contour hedgerows* A variant of hedgerow intercropping practised on sloping land, with the hedges aligned along contours; the hedges control runoff and erosion.
- *Biomass transfer (cut-and-carry mulching)* The trees are grown in separate blocks; leafy matter from the trees is transferred to the cropped fields, to improve fertility.
- *Trees on pastures (parkland systems)* Scattered trees amid pastures; the trees provide browse, shade for livestock, and improve soil fertility.
- *Fodder banks* Trees planted as separate blocks, harvested for high-protein fodder.
- *Farm and village forestry* Mixed woodlands on farms or communal village land, usually managed for multiple purposes, e.g. fuelwood and forest grazing.
- *Reclamation agroforestry* Trees are planted to restore fertility to degraded land, as in reclamation forestry; after a period, some of the trees are removed and the remainder managed for combined production and conservation.
- *Taungya* Farmers are brought into forest land, interplant timber trees with food crops for 2–3 years, after which the trees grow to maturity, the farmers moving to an area from which the trees have been harvested.
- *Aquaforestry* Combinations of trees with fisheries, e.g. fish ponds, mangrove.
- *Entomoforestry* Combinations of trees with production from insects, e.g. apiculture, mulberry with silkworm, shellac.

functions of multipurpose trees may be for production, of fuelwood, timber, fodder, fruit and nuts, thatching, oils, medical products, etc.; or for service functions, including shade and shelter, soil conservation and reclamation, and fertility improvement. Agroforestry is not a single land use system; there are about 20 agroforestry technologies, ways of arranging trees and other components in time and space, and of managing them for different purposes (see examples in the Box above).

Fuelwood was initially thought to be the major production objective of agroforestry, although diagnostic farm surveys showed that fodder shortage was even more widespread. Production of fruit, both for cash and subsistence, is also important, as is domestic timber, and poles for construction and agricultural purposes (e.g. tobacco curing, stakes for yams or pepper). More generally, having trees adds diversity of production to a farming system. They can supply capital when it is needed (e.g. for school fees). Through their capacity to withstand drought, trees can provide a fall-back in years of crop failure.

Agroforestry for soil and water conservation

Whilst the relative importance of the major production functions of agroforestry, for fuelwood, fodder, and fruit, varies between one region and another, there is no doubt that its greatest service role lies in soil management. Two primary functions have been identified and tested by research: soil and water conservation, and soil fertility improvement.

Trees and shrubs can be employed in conservation in supplementary or direct ways. In supplementary use, the trees are added to conventional soil conservation works: terraces, banks and ditches, grass strips. Trees help to stabilize the earth structures with their root systems; and they make productive use of the land which these occupy. In this way, there is more chance of the conservation works being acceptable, and becoming a permanent part of the farming system.

A greater contribution comes from two systems for direct use of trees in conservation: multistrata systems and contour hedgerows. Multistrata systems have for long been a means for combining production with conservation on steep slopes of the humid tropics. In the Philippines and Sri Lanka, rice is cultivated in the valley floors, whilst valley sides are occupied by perennial tree crops, such as coconut, oil palm, and rubber, sometimes as pure stands but often in mixtures with fruit trees such as jack fruit, mango, and durian. The 'forest gardens' of Sumatra form a sustainable method of using sloping land, contrasting with the degradation that has frequently resulted from annual cropping. Although the canopy in multistrata systems is dense, it is not this which effects the conservation. Selective removal of the canopy, understorey, and ground cover of litter under experimental conditions

FIGURE 23 Agroforestry: soil conservation by contour hedgerows, the Philippines.

has shown that it is the litter which checks erosion. Hence, even single-species stands of trees, such as plantations of rubber, coffee, or cacao, will control erosion provided that sufficient litter is retained to cover the ground surface.

In contrast to this long-established method, the contour hedgerow system is relatively new. It originated on Flores Island, Indonesia, in the 1970s, and was later taken up in extension work by a Baptist mission in the Philippines. Experimental confirmation of the effectiveness of the system followed. Demonstrations established at Machakos, Kenya, in 1983 are still producing steady yields. Hedges are planted parallel to contours, with about 4–8 m between the rows; they are pruned before planting crops and usually once during the growing season. The system is technically most efficient if the prunings are placed as mulch on the soil; but, as farmers are more likely to harvest these as fodder, it is fortunate that it has been found still to be effective with the hedgerows alone. How this happens is an interesting scientific story. Small terraces, about 50 cm high, form along each hedgerow, so it was first assumed that the hedges acted as barriers, exercising a sieve-like effect in holding back soil whilst the water flowed through them. This hypothesis was contradicted when studies with erosion plots showed that water runoff was also much reduced. The answer came from measuring rates of infiltration, which were found to be 3–8 times faster under the hedges than on the cropped land, probably due to the hedge root systems.[21]

Experimental data demonstrates a spectacular effectiveness in controlling erosion. On two occasions of heavy storms at the Machakos site, soil loss was reduced from 20–30 t ha^{-1} on control plots without conservation to about 0.2 t ha^{-1} on hedgerow plots, a reduction of 100–150 times; when an experiment is as effective as this, refined statistical analysis of results becomes superfluous. This studies have been confirmed by experiments in more than 12 countries, nearly all of which show reductions in erosion by a factor of at least 10 times, to well below the accepted 'tolerable' limit of 10 t ha^{-1} per year, with lesser but still substantial reductions in runoff. Experience with acceptance by farmers has been variable, but the cost and labour of establishment and maintenance are less than for conservation by earth structures. In areas of annual cropping, the contour hedgerow system now provides a viable alternative to conventional methods of soil conservation.[22]

Agroforestry for maintenance of soil fertility

It is clear from general observation that trees have the capacity to maintain or increase soil fertility. When soil is first cleared from forest, it is fertile. Shifting cultivators make use of natural tree fallows to restore fertility. Degraded soils have long been reclaimed through reclamation forestry, and, more recently, reclamation agroforestry, in which after an initial period of soil improvement an element of production is added, has been one of the recent success stories of agroforestry.

There are more than 20 processes by which trees improve soils. Among the most important are: increase in soil organic matter through litter and root residues, with a corresponding improvement in soil physical conditions; nitrogen fixation, by many leguminous and a few non-leguminous species; the uptake by tree root systems of nutrients released by rock weathering from deeper layers in the soil; and nutrient retrieval, the trapping of nutrients that would otherwise have been lost by leaching and their recycling through tree litter and prunings. Control of erosion also aids fertility, through reduction in losses of organic matter and nutrients in eroded soil. Recognition of these benefits led to the general soil-agroforestry hypothesis:[23]

Appropriate and well-managed agroforestry systems have the potential to:

> control runoff and erosion;
> maintain soil organic matter and physical properties;
> promote nutrient cycling and efficient nutrient use.

'Appropriate' means systems which are suited to local conditions, both environmental and socio-economic.

The capacity to maintain soil organic matter has been well demonstrated both experimentally and in the field. The potential to improve nutrient cycling is one of the biggest challenges of current research. It rests on the basis that, under natural forest ecosystems a high proportion of plant nutrients, more than 90%, are repeatedly recycled between trees and the soil, with only small natural inputs and losses. By contrast, annual cropping systems typically have about 40% rates of nutrient recycling, and consequently require high inputs of fertilizers to replace losses in the harvest. The proposition is that agroforestry systems can achieve a rate of nutrient recycling intermediate between these extremes. This is certainly true of dense multistrata systems. Finding ways of making recycling effective in association with annual cropping has proved more difficult; the early hopes placed on hedgerow intercropping (alley cropping)[24] have not been fully confirmed; there are problems of tree–crop competition for soil water, and farmer acceptance is poor. Systems of planted and managed tree fallows may turn out to be more effective, or alternatively biomass transfer, in which the trees are grown separately, where possible on hilly or other poorer land, and the litter carried to the cropped fields as organic fertilizer.

It is well worth pursuing research to find ways of improving nutrient cycling within productive and acceptable agroforestry systems. With low-input systems, the frequency and duration of fallow periods would be reduced. With the moderate-input systems which are now in favour, an increased rate of recycling would mean that limited supplies of fertilizer would be used with greater efficiency. The problems of designing systems are considerable, but progress is being made and, if this method is successful in improving nutrient use efficiency, the benefits would be immense.

Common threads

Land husbandry, integrated plant nutrition, small-scale irrigation, communal management of rangelands, multiple-use forest management, agroforestry – these are only a selection from the many new approaches to land resource management. Nothing has been said about aquaculture, water harvesting, dry farming, integrated pest management, or a host of advances which are being taken from the stage of research to development. The most substantial omission, in terms of its contribution to human welfare, is a discussion of the continuing improvements in rice farming, which have taken yields to levels unthinkable in earlier years; this advance called for basic research in the breeding of high-yielding varieties, sustained effort in improving fertilizer treatments and crop protection measures, and, not least, studies of how rice farming systems function in practice.

There are some common threads running through the ideas discussed. One is the need to understand the functioning of natural ecosystems – soils, water, and plants – as a basis for their sustainable management. This management must be jointly for production at the present and conservation to meet the needs of the future; and, as the natural environment is infinitely variable, the best methods of management will differ from place to place. Another is to meet requirements for increased production not by taking in new land, nor by adding higher levels of inputs, but by using resources with greater efficiency. A further theme is the active involvement of local people in land management. The participatory approach has a social value, helping to direct attention to the needs of the rural poor; it can equally well be justified on pragmatic grounds, in that no other approach to land resource management works so effectively. But participation alone is by no means sufficient; it will not solve problems of phosphorus deficiency, for example, or stem borer. These advances in land management, now all being widely applied, owed their origins to research.

12 Research and technology

The view that further research into land resource management is unnecessary, that all that is needed is more widespread application of existing knowledge, rests on a misunderstanding of the nature of science, in its pure and applied aspects. A spectrum of research is needed: fundamental, basic, applied and adaptive. Universities and international centres are best fitted to carry out fundamental and basic research, national institutions the applied and adaptive. The final stage of research is carried out by farmers, critically trying out new methods. The achievements of the 'green revolution' phase of research, based on improved crop varieties, led to threefold to fivefold increases in crop yields. Further advances will be achieved, but more slowly and with greater effort. The problems encountered with high-technology, high-input land use systems have led to a new approach to research, which stresses maintenance of soil biological activity and improved nutrient cycling, leading to more efficient use of limited inputs.

Whether assessed in economic terms or in its wider contribution to human welfare, research produces extremely high ratios between benefits and costs. There is a serious shortfall in funding. At the international level, donors should at least double the proportion of aid directed towards research; by doing so, they will bring longer-lasting benefits to farmers. Still more important is that governments of developing countries should recognize the need to strengthen their presently inadequate national research services. Farmers will not adopt improved methods without a basis of applied and adaptive research to ensure that these are convincingly effective.

There are two contrasting attitudes towards agricultural research. The first states that research brought about the achievements of agriculture in the past, and is the lifeline to the future. The productivity of modern agriculture could not be at anything like its current level, crop yields not one fifth as high in some countries, had it not been for the technologies resulting from research. Since the urgent need for increased production can no longer be met by taking more land into cultivation, it must be achieved through still higher yields which can only come from continuing research.

The second attitude is that the major need at present is not for further knowledge but for more widespread application of what is already known. Advocates of this point to the great disparities in average crop yields between different countries, average rice yields of 6 t ha^{-1} in Japan compared with 1.5 t ha^{-1} in Gambia, for example. If the countries at the lower end of this range were to close even part of this gap, far greater advances in production would be achieved than might come from further research. An adjunct to this attitude is the view sometimes encountered in developing countries that research is a luxury which may be left to scientists from Western countries, and institutes which they fund; agricultural sector budgets are so stretched by the day-to-day problems of field agricultural production that skilled personnel and funds cannot be spared for the supposedly non-essential activity of research.

Which of these attitudes prevails will determine the priority given to the financial support for research, both by international funding agencies and in the budgets of developing countries. This question is fundamental, since, without effective institutes, skilled scientists, and funding, no research will get done. It was the latter view which prevailed in FAO in 1995, when a major new programme to improve food security was based on 'dissemination of existing and proven agricultural technology' through a network of on-farm demonstrations.[1]

To the extent that the need for research is accepted, a basic question is: what kind of research should be carried out? Should the emphasis be on basic scientific advances, like the breeding of new crop varieties, or on solving the practical problems encountered by farmers? This is linked with intended methods of land management, whether priority should be given in farming by advanced technologies with high inputs, or to low external-input methods. A further aspect is the relative importance of scientific studies compared with research into farmers' needs, the constraints under which they work, their adoption of new technologies, and other socio-economic aspects.

The following discussion focuses on research and technology for crop production. In addition, research programmes are needed in the other sectors of rural land use, including livestock production and pasture management, forestry, agroforestry, irrigation management, and aquaculture.

Achievements

Scientific research into agriculture began in the nineteenth century with the discovery of crop requirements for the major plant nutrients, although fertilizers were not widely applied until the 1940s. In the inter-war period, substantial work was done in the tropics to improve the major export crops such as tea, coffee, and rubber, and such work continued into the 1950s and 60s. Older scientists deplore the way that results from this early work are often ignored.[2] The foundations of soil conservation research were laid in the United States from the mid-1930s onward.

The modern period of tropical agricultural research can be taken as beginning in 1960, the date of the founding of the International Rice Research Institute (IRRI), in the Philippines. This is the period known as the 'green revolution', centred upon the application of scientific methods to agricultural production. Among its many achievements were:

improved varieties of rice, wheat, maize, and other staple crops;

the control of pests and diseases, primarily through biocides;

the application of fertilizers, to achieve high yield in an economically efficient manner;

biological nitrogen fixation, applied by inclusion of nitrogen-fixing crops in rotations;

major advances in the understanding of soil constraints, for example the management of soil physical properties, low-activity clay soils, and effective cation exchange capacity, the wide-ranging functions of soil organic matter (humus);

the principles of pasture management, through control of livestock numbers and rotational grazing; pasture improvement through grass-legume mixtures;

advances in the improvement of livestock health and nutrition;

less spectacular but steady progress in improving the productivity of forest plantations;

a better understanding of farm systems: the constraints under which small farmers operate, their motivation, the role of women, the importance of minimizing risk; and the rationality of farmer adoption, or otherwise, of new practices.

The best known of the green revolution's achievements is the improved crop varieties. Formerly called high-yielding varieties, an early mistake in not including disease resistance and potential to grow on soils with constraints was remedied in later research. It was also said that the benefits largely reached richer farmers. Studies showed, however, that, although larger farms, with capital, inevitably gain the most from the new technologies, their benefits also reach smaller farms and the poor.[3] A more general aspect of research in this period was that it was orientated towards high inputs: fertilizers, biocides, and, in some areas, large-scale irrigation. This approach is still giving results. Wherever there has been a rise in crop yields, this has been closely linked with increase in fertilizer use.

The problems encountered with high-input technology have led to the new approach outlined in the previous chapter, low-input sustainable agriculture. This calls for a new focus in soils and agricultural research, which has been called the 'second paradigm':

> Rely more on biological processes by adapting germplasm to adverse soil conditions, enhancing soil biological activity, and optimizing nutrient cycling to minimize external inputs and maximize the efficiency of their use.[4]

One aspect, that of developing crop varieties adapted to soil constraints, requires continued, although reorientated, work on plant breeding. Research into integrated pest management is already directed at achieving control with reduced chemical inputs. Other parts of the low-input, high-efficiency approach call for a different emphasis to research. Management systems must be found which will make the use of marginal lands sustainable. A major role is played by the recycling of plant residues, both from above-ground litter and roots; this not only maintains soil organic matter but has both short- and long-term benefits for nutrient cycling. There is scope to make greater use of biological methods to improve soil fertility. Cheap but effective management systems are needed that will increase water use efficiency. In rainfed agriculture, these include soil water conservation, dry farming methods and water harvesting. Under irrigation, the onset of water supply shortages can be partly counteracted by making better use of available water.

This programme calls for a different type of research, more centred on management systems. It is wrong to call it a 'low-technology' basis, for the soil and plant processes which are harnessed are more complex than those involved in simply adding inputs. Nor is the study of low-input sustainable agriculture confined to trials of land use systems. It calls for the full spectrum of research, from fairly esoteric studies of the nature of soil organic matter, through the design and testing of systems, to the adaptive research which is vital if these methods are to function efficiently under the wide range of local environmental conditions.

Getting research done

Setting the research agenda

There is a spectrum in soils and agricultural research, from fundamental through basic, applied, and adaptive research. Fundamental research is the study of how specific processes act: how biological nitrogen fixation is accomplished, for example, or how nutrients move through the soil solution to plant roots. Basic, or component, research is the study of individual processes within systems: their nutrient budgets, water balances, methods of pest and disease control. Applied, or systems, research consists of trials of agricultural or agroforestry systems as a whole. Applied research answers the question, 'What happens?', basic research the question, 'Why does it happen?'. Adaptive research is directed at finding specific systems appropriate to local environmental conditions, appropriate crop varieties, rotations, and fertilizers, for example.

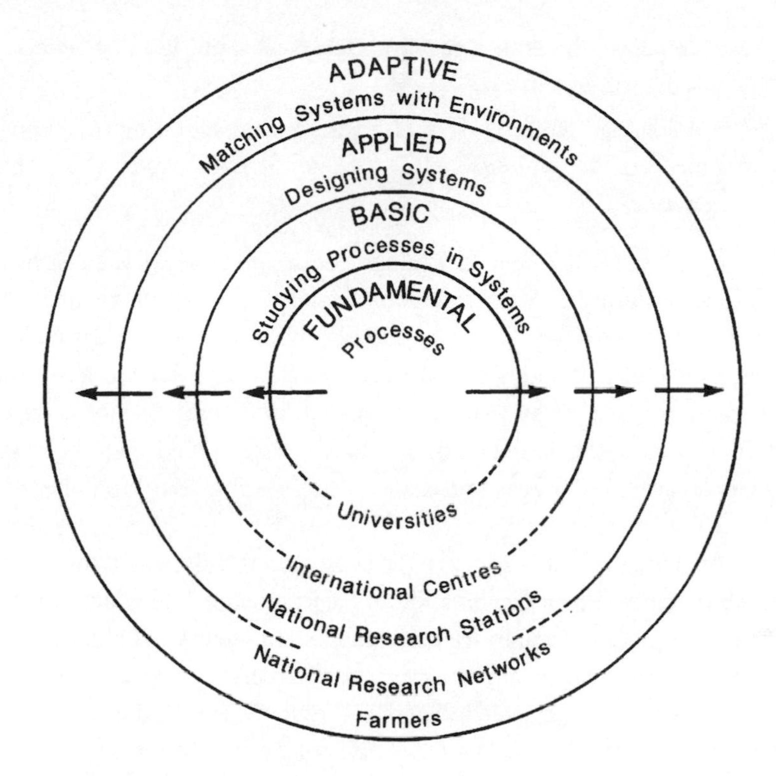

FIGURE 24 The research spectrum.

Attitudes to the relative importance assigned to basic and applied research vary. It is, of course, perfectly possible to do the right thing without knowing why it works, as farmers have always done.[5] System trials are what many laymen think of as agricultural research. The standard agronomic trial consists of taking a range of treatments, say crop varieties, fertilizer levels, and tillage methods, and running a statistically controlled trial to find which combination works best. The limitation to this 'try-it-and-see' approach is that it shows which works best on that particular soil, and under the weather conditions experienced during the trial. Without the knowledge of why these results have occurred, derived from research into processes, there is no basis for designing systems which will be robust and adaptable.

This was well illustrated in the new science of agroforestry. In the early years, donors to research sought quick results, and there was a concentration on multifactorial trials of agroforestry systems, varying the tree species, spacing, and management. What happened during this period may be called 'the practical systems fallacy':

> *True proposition* Agroforestry research should be directed towards finding practical systems, which farmers can use.

False deduction Therefore research should consist of trials of practical systems, which farmers can use.

Correct deduction Research should be directed at acquiring the knowledge of processes that will enable us to design practical systems, which farmers can use.

The fallacy can be illustrated by an analogy, the design of bridges. These have a clear function to perform, carrying transport across rivers. But, on visiting a research institute into bridge design, do you see rows of bridges, of different materials and shapes, with worried-looking men driving lorries filled with rocks across them? Of course not. You see stress tests being carried out on samples of materials, and computer-aided design based on the knowledge so acquired. Agro-ecosystems are too complex ever to acquire the precision of engineering design, but the principle is valid.[6]

This does not mean that system trials should be abandoned. A proper balance of research is achieved by a combination of basic, applied, and adaptive research: at the basic level, experimentally isolating plant components and studying individual soil-plant processes; at the applied level, system trials, testing a selection of the best estimates from present knowledge; and, at the adaptive level, finding modifications of systems suited to local climatic and soil conditions. In the course of the basic research, fundamental questions can be identified and referred to universities or specialist research institutions.

A basic dilemma faces research institutions of all kinds. As shown by cases like the discovery of penicillin, there is a large element of chance in scientific research; real advances are made erratically. Research grant administrators and donors to research institutions, on the other hand, like to know 'what will be found out, and how long it will take to do so', whilst the institutions themselves have to plan budgets for more than a year ahead. There is no easy way around this problem. As it is unacceptable to allot a 'contingencies' budget of more than about 5%, research directors must do their best to re-assign funds as scientific opportunities arise. A device would be to give scientists a personal budgetary and time allotment, say 5% of institutional budget, to spend on following up their own ideas irrespective of the approved institutional programme, although this is unlikely to be found acceptable by administrators.[7]

The research agenda can be set from two directions, each with limitations. One extreme is to let the scientists to do this, allowing them to work in areas that seem likely to produce results; often, this calls for fundamental research with only slender apparent connections to the problems to be solved, and taking many years before benefits result. Unchecked, this has an inevitable tendency to lead to high-technology solutions, developed without regard to their acceptability to farmers. The opposite extreme is the participatory approach to the design of research, as in the method of diagnosis and design (chapter 5, p. 75). This is based on identification of the problems of farmers, and analysis of the causes of these problems; an assessment of present knowledge is then made to see how far it can solve these, followed by the planning of research intended to fill gaps and improve the design of systems. The bias in this approach is towards systems research, seeking quick solutions.

It is a truism that both approaches are needed. The scientists need to be kept in touch with the realities of practical farming, whilst the farmers cannot know what new technical possibilities may be developed. There is one clear distinction, however, with respect to the agenda for different levels of research. The primary place for diagnosis and design is in adaptive research, taking existing knowledge and designing systems of meet local problems. Extension staff cannot be expected to do this, but neither is it an appropriate task (except on an occasional contact-with-reality basis) for research scientists. There is a need for a corps of specialist advisers to the extension service, the technicians or 'applied engineers' of agriculture.[8] For an extension district, there should ideally be specialists in soil and water management, conservation, crop production, pasture and livestock management, agroforestry, and other aspects related to local production. The inclusion of such specialists not only increases the effectiveness of an extension service, but raises its prestige.

In recent years, much attention has been given to research networks.[9] These are of different kinds. The most widespread are networks for information exchange, meaning that scientists working on the same topic meet and exchange information at regular intervals. Less common are networks in which the same varieties of crops, or species and provenances of trees, are grown across a range of climatic and soil conditions, a form of adaptive research. In a more recently developed form, scientists in different countries work on the same problem, not using identical methods but with a degree of local adaptation and initiative. This arrangement finds favour with research funding agencies. There are many ways in which networks of this kind can be organized, varying the ways in which staffing and activities are linked with the central organizing institution. One of the best arrangements has been developed for agroforestry research networks.[10] The research programme is planned as a whole; it is conducted at local centres, often national research stations, which are strengthened by stationing one or more internationally recruited staff members in them; and the central institution provides technical support, in the form of linked strategic research, laboratory facilities, technical advice and training.

Most research is directed at relatively short-term aims, those of year-by-year farm management. Attention should also be given to at least medium-term objectives, land resource conservation over periods of 20–30 years. Much of this work must be done by extrapolation, but there is an important function for long-term trials. The best known is the 'broadbalk field' at Rothamsted, UK, which has been cropped for more than 150 years.[11] A few long-term trials were established in the tropics during the 1950s, but funding for their continuation was hard to find and many were abandoned. In the light of sustainability, there has been a recent revival of interest, and a global inventory has been compiled.[12] Long-term trials provide a source of direct information on trends in soil properties and crop yields in response to known inputs and management. International research centres are best placed to conduct, or provide funding support for, such trials. Ideally, the major agricultural research stations of larger countries should set up small networks in their principal environments, but for them to secure continuity of funding is difficult.

Research and the advancement of knowledge

Research is the set of activities directed at the advancement of knowledge. It follows that what does not lead to improvements in knowledge is not research. As such, research is an extremely difficult activity, requiring skill, hard work, and an element of luck. The most visible activity of agricultural research, the conduct of field trials, is not sufficient. Trials need to be linked with a range of other activities: diagnosis of problems, reviews of the state of knowledge, laboratory analysis, formulation of hypotheses, modelling, and evaluation of results.

A key step in research is trying to find new ways in which a desired objective might be achieved, that is, generating new hypotheses. Many hypotheses turn out to give negative results, but it is the few that are successful that constitute real advances in knowledge. It is right to conduct statistically controlled trials at a late stage of research, but they are not the most important. The science of bio-statistics was only developed in the 1930s, so most of the great discoveries in the history of science were made without it.

In the case of agriculture, if a new method is to be adopted by farmers its superiority needs to be clearly apparent - not just significant in a statistical sense, but visibly better in terms of production. Good techniques also need to be robust, that is, to work when applied with less than optimum management. This is one area for on-farm research. Once a technique produces clearly visible results, and is robust enough still to function when applied only moderately well, then adoption will not be a problem. The spread of innovations through the 'informal extension service', that of farmers talking to each other, can be surprisingly rapid.

Crop yields: approaching the limits?

Average world yields of the three major cereals rose from close to 1 t ha^{-1} in 1950 to their current levels of 2.6 t for wheat, 3.6 t for rice, and 4.0 t ha^{-1} for maize. This

STATISTICS ON CROP PRODUCTION

Every year, figures are published for countries giving the area harvested, yield, and production of the major crops. These are widely used for analysis of trends and potential. Whilst more realistic than some other kinds of international statistics, too great a reliance should not be placed on their finer details. In developed countries, such data are often based on annual questionnaire returns by farmers. For most developing countries, the position is very different. Many do not measure either their areas harvested, yields, or production, other than at ten-year intervals by national agricultural sample censuses.

Production is always reported as exactly equal to area times yield. In the most favourable situation, local agricultural officers send in reports of the previous year's performance in their districts. Unless there has been a drought, few are likely to report a decline in yield. Nor do they wish it to be thought that sloping land is being cultivated, so the area under crops may be kept constant.

If there are no such reports, what probably happens is as follows. The government has some idea of the amount of a crop coming onto the market. It adds to this an estimate for subsistence consumption; as the population is rising, this figure becomes larger every year. It does not know the area harvested, so leaves it at last year's figure. As a result, the average yield appears to have risen, reflecting credit to farmers, government, and aid donors. No yield measurements have been made. Abrupt corrections to the sequences of data may be made when, normally every 10 years, there is a sample national agricultural census.

Occasionally the position may be worse. Suppose a government introduces a maize subsidy, to give farmers a good return whilst keeping the price low for urban consumers and the rural landless. Officials of marketing boards may purchase large amounts of non-existent maize from imaginary farmers and sell it to hypothetical consumers, pocketing the subsidy. There appears to be a sharp increase in yields and production, and policy advisers nod their heads sagely.

was primarily in response to the increased use of improved varieties and fertilizer. For developing countries, the values are only slightly less than the world figures for wheat and rice but lower, 2.6 t ha^{-1}, for maize (due mainly to the high production and yield in the United States). Taking rice as an example, the highest-yielding countries, Japan, North and South Korea, Egypt, and China, reach 6–8 t ha^{-1}, whilst, among countries which are major producers, the lower end of the range is 2–3 t ha^{-1}.[13]

In estimates of future food production, heavy reliance is placed on continued increases in yields. This is expected to come, first, from further advances in crop breeding, raising the genetic plant potential, and, secondly, by raising the performance of the lower-yielding countries. It is generally accepted that the quantum leap in yields brought about by the green revolution cannot be repeated. Certainly, in Europe, USA, and East Asia, a plateau has been reached in response to fertilizers, and rates of application have fallen.

Natural limits to plant genetic potential are set by photosynthetic efficiency, the proportion of radiation that can be converted to plant matter. There are three different photosynthetic pathways, biochemical means of achieving this conversion, on which all plants are dependent; the level of efficiency of these is intrinsically very difficult to raise. Limits are also probably being approached for improvements in the harvest index, the proportion of the total plant biomass that is consumed as grain, etc. The current biological limit obtained for hybrid rice, the highest yield obtained for small areas under ideal environments and management, is 10 t ha^{-1}.

Great advances were made when the older method of plant breeding, by selection of the best strains, was replaced by development of hybrids. Hopes are now being placed on the potential of genetic engineering, in which new varieties are developed by the artificial introduction of genes.[14] Initially, this method is being directed at resistance to pests and diseases, and also to tolerance of environmental stresses (e.g. drought, salinity) and improved storage properties. It will no doubt lead to further gains in biological yield potential, but at present it is impossible to assess the magnitude of any such gains. There are problems of safety and intellectual property rights, and also a substantial time delay, 10–20 years, before results are likely to reach farmers.

More reliance is placed on bringing the performance of lower-yielding countries closer to that of the best. This is sometimes referred to as 'closing the yield gap', a misleading phrase since it carries the implication that all countries can reach the averages of the best. The environmental conditions of climate, soil, and water, under which the higher yields are obtained, are not found everywhere. In the first place, yields under irrigation are some 50% higher than for rainfed agriculture, but water availability sets clear limits to further expansion in the irrigated area. Where new land is brought into cultivation, it will nearly always be less inherently fertile. An indication of environmental constraints is found in the low yields of sorghum and millet, 1.2 and 0.8 t ha^{-1} respectively for all developing countries; these are cereals normally grown on more marginal land. In many areas, land degradation has lowered both yield potential and response to inputs.

Cereal yields in developing countries from 1960 to the present have followed more or less linear increases. The most reasonable basis for estimating future change is by linear extrapolation of these trends. This would give average yields for developing countries in 2025 of about 3 t ha^{-1} for wheat, 5.5–6.0 t ha^{-1} for rice, and 4 t ha^{-1} for maize. Data are being watched closely for signs of a yield plateau, but this is not yet apparent on a world scale.[15]

Gains in agricultural productivity over the next 25 years will be much more difficult than in the past. A continued effort should certainly be made at the high-technology end of research, primarily plant breeding. As much or more benefit is likely to come, however, from research into soil management. In particular, it is a fallacy to assume that advances can be brought about merely by demonstrating improved methods. If farmers are to adopt a technology, it is critical that it should work, and work well; for this to be achieved, national networks of adaptive research are needed. Research needs to anticipate needs, since there is a lag of 10–20 years between the first scientific conception of a new advance and the development needed for its application.

For 50 years, research has bought time. With only the genetic potential of crop varieties available in 1950, food could not be produced for anything like the present populations of many countries. In some areas, primarily in Asia, production is also becoming dangerously dependent on large inputs of fertilizer, and the increases which are being called for in Africa and America will greatly increase fertilizer use, bringing problems of production, availability of energy, and pollution. There may just be the technical potential to meet food and other basic needs over the next 30 years. This will be very much more difficult than in the past, calling for a substantially increased, skilled, and well-coordinated research effort.

Getting it done: institutes and funding

There four groups of institutions involved in soils, agricultural, and forestry research: international centres, national agricultural research services, universities and specialized research institutions, and commodity research institutes. Their functions are not sharply separated, but each more appropriately performs certain types of research.

International centres

FAO and the World Bank are not organizations with mandate for research, but both carry out substantial work, mainly in areas related to their statistical, monitoring, and evaluation activities. Since 1971, international research has been conducted mainly by a network of International Agricultural Research Centres (IARCs), coordinated through the Consultative Group on International Agricultural Research (CGIAR), which exercises policy guidance and acts as a clearing house for funding. Founded in 1971 with 4 centres, the number has recently grown to 16.

The original centres had a focus on plant breeding, and retain mandates for work on breeding, management, and plant protection of specific crops. These include the centres best known to non-specialists, the 'miracle maize and rice' institutes, CIMMYT and IRRI. Some also have regional or environmental mandates, for the humid tropics (IITA), the semi-arid tropics (ICRISAT), and the Mediterranean zone (ICARDA).[16] The original mandates were extended to cover farm systems

INTERNATIONAL AGRICULTURAL RESEARCH CENTRES

With crop and regional mandates:

CIAT	Centro Internacional de Agricultura Tropical	Cali, Colombia
CIMMYT	Centro Internacional de Mejoramiento de Maiz y Trigo (maize and wheat)	Mexico City, Mexico
CIP	Centro Internacional de la Papa (potato)	Lima, Peru
ICARDA	International Centre for Agricultural Research in the Dry Areas	Aleppo, Syria
ICRISAT	International Crops Research Institute for the Semi-Arid Tropics	Hyderabad, India
IITA	International Institute of Tropical Agriculture	Ibadan, Nigeria
IRRI	International Rice Research Institute	Los Baños, the Philippines
WARDA	West African Rice Development Association	Bouaké, Ivory Coast

With mandates for irrigation, livestock, forestry, agroforestry, and freshwater fisheries:

IIMI	International Irrigation Management Institute	Colombo, Sri Lanka
ILRI	International Livestock Research Institute	Nairobi, Kenya
CIFOR	Centre for International Forestry Research	Bogor, Indonesia
ICRAF	International Centre for Research in Agroforestry	Nairobi, Kenya
ICLARM	International Centre for Living Aquatic Resources Management	Manila, the Philippines

Co-ordinating centres located in developed countries:

IFPRI	International Food Policy Institute	Washington DC, USA
IPGRI	International Plant Genetic Resources Institute	Rome, Italy
ISNAR	International Service for National Agricultural Research	The Hague, The Netherlands

studies and socio-economic research generally. The traditional separation between veterinary services and livestock management was for long perpetuated institutionally, but these have now been combined in a single centre for livestock research (ILRI). A wider range of research mandates has recently been introduced, with centres for irrigation, forestry, agroforestry, and freshwater fisheries. ISNAR is of a different nature to other centres, providing financial support and training to assist national research services.

The focal activities of the centres, plant genetic improvement and plant protection, have had an immense impact on tropical agriculture. It has been estimated that the economic value of the improved varieties released to farmers over 20 years is $50 billion; and that to feed the current populations of developing countries without these varieties would require at least 60% more land.[17] In economic terms, this achievement alone gives the centres an immensely high benefit:cost ratio.

More importantly, land pressures would be higher and food supplies vastly lower if it were not for their use. Corresponding benefits could be cited for the livestock sector. In the new science of agroforestry, ICRAF was founded in 1977, and its first task was to make governments aware of the potential of agroforestry to solve problems of land use. This objective was successfully achieved in the first 10 years, and it has now turned to experimental research, notably on the soil-improving potential of agroforestry. The international centres have clear advantages of scale, staffing, and equipment for basic research. With the aim of making sure their work is relevant to the needs of farmers, many also undertake applied and even adaptive research; the former is justifiable, the latter better left to national services.

National agricultural research systems

'National agricultural research systems (NARS) . . . are and will continue to be the cornerstone of the global agricultural research system.'[18] This fact may seem surprising in the light of the high profile of the international centres. It stems from two reasons: relevance to national needs, and funding. Advances in basic science must be followed up by a large quantity of applied and adaptive research before they can be applied to local agricultural and forestry management. National research systems have a much better knowledge of local land resource issues and management problems, and are better fitted to turn research into practice. The funding aspect is a matter of proportion. Despite a generally low level of government support, the sum of national research expenditures is an order of magnitude higher than the international total. Brazil and India each have agricultural research budgets approaching that of the international systems, and China's is considerably larger. Whilst it is true that, in research, quality is more important that quantity, it is implausible to suppose that the higher calibre of international staff outweighs this difference.

Most national research services in developing countries fight an uphill battle against inadequate funding, equipment, and staffing. Some of their most able staff are, regrettably if inevitably, lost to (and traded between) the international centres. Those who remain lack adequate operating budgets. There are exceptions; the Indian Council of Agricultural Research operates a network of 30 stations, staffed at PhD level, and with a total budget approaching (or in purchasing power parity terms exceeding) that of the CGIAR. China, Brazil, and Bangladesh are among countries which are expanding their research capabilities, although these remain small in relation to the rural populations served.

This unsatisfactory position stems from an absence of government recognition of the need for research, sometimes with the attitude that it is a luxury better left to developed countries. This view could not be more mistaken. The applied and adaptive research necessary if technological advances are to be properly applied in a country requires a cadre of research personnel covering each of the main branches, a minimum of 250 scientists.[19] Ways can be found for even small countries to have an adequate co-ordinated research service; Jamaica is an example.[20]

Universities and commodity research institutes

Universities in developing countries are severely limited in their potential for research by shortages of equipment and the low budgets of national research funding agencies. Some autocratic governments have even taken the view that their universities should confine themselves to teaching, leaving the expense of research to the Western world. These universities are the right place to tackle land management problems raised by the local environment.

Universities of developed countries are best fitted by staffing and equipment to take up questions of fundamental research, such as mechanisms of nitrogen fixation, or processes affecting low-activity clays; they also contribute to basic and applied research in the tropics. Formerly this was done by their scientists doing fieldwork in the tropics. Nowadays, a powerful synergism has developed between Western university scientists and graduates from the developing world, which in some cases is linked with the international research centres. This mode of co-operation has helped developing countries to produce research scientists of the very highest ability.

Commodity research institutes exist mainly for perennial export crops such as rubber, coffee, and tea. Their well-focused programmes provide a link between farmers and scientific advances, contribute to the selection and propagation of locally-suited plant varieties, and may provide advisory services. Some are funded by a levy on exports, an excellent arrangement since, at a very small unit cost (a few cents per tonne), there is a built-in link to growth in production and adjustment for inflation. National forestry research centres are fitted to perform similar tasks, but are dependent on government funding.

Research into land resources

The institutional arrangements for research into land resources are fragmented. None of the international centres of the CGIAR system has a specific responsibility for soils or agro-climatology, although at least six of them have developed strong programmes linked with their plant-based mandates.[21] International data on soils was formerly assembled, analysed and mapped by the Soil Resources, Management and Conservation Service of FAO, Rome; this function has now been partly taken over by the International Soil Reference and Information Centre (ISRIC), Wageningen, funded largely through the goodwill of The Netherlands government. There is an International Fertilizer Development Center (IFDC) based in Alabama, USA.

Arriving late on the international scene, the International Board for Soil Research and Management (IBSRAM) was founded only in 1985, with headquarters in Bangkok. It works through soil management networks of field experiments, based on regions and environments, for example networks for acid soils and for vertisols in Africa, one for sloping lands in Asia, and a network devoted to land management problems in Pacific island countries. Soil water and nutrient management and conservation are the principal fields covered. The experiments are conducted within, and by scientists belonging to, the participating countries. IBSRAM assists in planning the network programmes, provides advice, and arranges regular coordination meetings. It does not, however, undertake fundamental or basic research, and lacks headquarters laboratory facilities. An application for membership of the CGIAR system was rejected in 1990, on grounds that the governing body 'did not consider involvement in adaptive research and development activities of national programmes to be a desirable evolutionary trend' in the system.[22] IBSRAM's network arrangements were founded in the belief that what matters most is solving practical management problems of soils. It means, however, that there is no institution directed towards problems of strategic research, for example nutrient cycling, which provide the foundation for their management. This function is left to universities.

A success story comes from the Tropical Soil Biology and Fertility programme (TSBF), with headquarters in Nairobi. This has the objective of improving the fertility of tropical soils through knowledge of soil biological processes and its application in management. The research is conducted through a network of universities. TSBF was started by private initiative and ran for many years on a small budget. Recognition and associated increase in funding has stemmed from its success in achieving substantive research results; it would be hard to find an organization with a higher 'benefit:cost ratio', namely the ratio of advances in knowledge to cost of research. TSBF set out a series of hypotheses on soil organic matter, nutrient cycling, water management, and soil fauna, and has made advances in all these fields. It was, in particular, responsible for the 'synchrony hypothesis', that it is possible to improve nutrient use efficiency by synchronizing the release of nutrients from decomposition of plant residues with their uptake by crops, an hypothesis which found its major application in agroforestry.[23]

The foundations of research in tropical agro-climatology were laid by university research from the 1950s. When FAO came to assess the potential productivity of the world's land resources, it found that, on the scale on which an international assessment is conducted, climate was more important than soils. This led to the development of a system of agroclimatic (or agro-ecological) zones, onto which the soil types mapped in the *Soil map of the world* were superimposed. The resulting synthesis, now digitized as a database, forms the main basis for international exchange of agricultural technology.[24] Among the international centres, CIAT took an initiative on climatology. The CGIAR centres co-operate in seeking a common basis for agro-ecological characterization.[25] Research into ecology and biological conservation is well served, through the strong interest shown by western nations.

It is not necessary to have a strongly co-ordinated central organization of research; having a range of institutions operating in different ways offers more scope for scientific initiative. There is, however, an imbalance in the output of research in soils and related land resources. The international soil science journals are increasingly dominated by detailed studies of processes, mostly orientated towards temperate soils. It is very much more difficult to get support for research into problems of tropical land resource management. In this, as in other fields, it is a change in attitudes, in views on what is important, that is the precondition for action.

Paying for research

The international (CGIAR) centres are funded by donations from 20 countries and 8 foundations, headed by the World Bank. Their combined annual core budget is about $220 million, to which is added a further sum in project funding; this compares with an estimate of $10 000 million per year on research and extension in developing countries. Funding has expanded little in real terms since 1980, and there is no immediate prospect of an improvement. The budgets of individual centres, each responsible for research in a major field affecting the lives of millions of people in developing countries, are $10–25 million a year, equivalent to the running costs of no more than ten secondary schools in a Western country.

Donor countries are reluctant to support research, giving to it less than 2% of their total aid budget. Investment and development projects, considered to be of more direct benefit to farmers, take priority. This view is mistaken, and sets a bad example to developing countries; if Western countries, with their host of scientists and economists, do not consider research to be important, how can poorer countries be expected to do so? Studies have repeatedly shown that, in economic terms, the rates of return from research are extremely high,50–90% per year.[26] This is because, on the one hand, research is still an inexpensive activity in relative terms, as compared with projects for improvement of infrastructure and services; and, on the other, the multiplier effect of advances in knowledge is very large. A cogent plea has been made for donor countries to triple the percentage of aid directed towards research, to 6%.[27] This is unlikely to happen, but a change of attitude could lead to the proportion being raised to 3–4%.

National research expenditures can be measured against the number of farmers, area of agricultural land, or value of agricultural production. Developed countries spend about $200 a year per farmer, developing countries $4, whilst per hectare the contrast is smaller, $4 against $2. A better measure of a country's support is the agricultural research intensity ratio, its research spending as a percentage of gross agricultural production. In 1990, national agricultural research expenditures averaged 0.58% of the value of gross agricultural production for developing countries, compared with 2% for developed.[28] The consequence is visible in understaffed stations with poor facilities and a lack of working budget. For these countries where a productive agricultural sector is the foundation of the economy, the raising of research funding to a minium of 2% of production is a priority.[29]

This will only come about with a change in attitudes. Governments of developing countries deplore the fact that more of their farmers do not adopt improved methods. They will not do so unless there is sufficient national research, applied and adaptive, to ensure that the methods are convincingly effective.

13 Land, food, and people

Estimates of whether food supplies will be adequate over the next 30 years range from warnings to qualified reassurances. Food requirements, what is needed to avoid hunger, are not the same as economic demand. Food security requires more than simply meeting requirements on a world scale; food must reach every country and all sections of the community, with adequate provision for a bad harvest year. Projections into the future start from a position that is already in deficit in the 1990s: 800 million people undernourished, massive food imports by developing countries with an accompanying burden of debt, and widespread occurrence of land-degrading management practices. In developing countries between now and 2025, population change alone will require an increase in food supplies of 55%, and dietary changes will add to this demand.

Future growth in food production can only come from more land or higher yields. But the land still available for cultivation has been greatly overestimated; present cultivation is more extensive than shown in official statistics, and most remaining land is already under necessary alternative uses. Future growth in crop yields will be slower and harder to achieve than in the past; some regions are experiencing a yield ceiling in response to inputs. Even to prevent the situation from worsening, with continued land degradation and declining per capita food supplies, will require a greatly increased commitment to agriculture by governments of developing countries. But, unless accompanied by efforts to reduce the rate of population growth, even such an increased commitment may not be enough.

Given the inevitable growth of world population over the next 50 years, will there be sufficient food for all? In the area of land resources, this is the question most frequently asked. It is clearly a matter of concern for ourselves and our children. If there is a substantial shortfall in food supplies, then the political and social repercussions, and the sheer magnitude of human distress, will be very serious indeed. Even the 'if' in that statement shows that it comes from the Western world; those who are still more concerned are the 800 million people who are already undernourished.

It is well known that the debate was started by Thomas Malthus in 1798, in *An essay on the principle of population.*[1] He argued that population grew geometrically (exponentially) but the means of subsistence could increase only arithmetically (linearly), so there would inevitably come a time when the needs of population exceeded the productive capacity of resources. It would then be checked by 'misery' (starvation, war) or 'vice' (abortion, etc.). In a second edition of 1803, he added a further check of 'moral restraint' or late marriage, postponed until the family could be supported (and to which, had it been available, he would certainly have added contraception).

For 150 years, the Malthusian principle was made to seem irrelevant by the expansion in cultivated land. This was the solution not only at the international level, through migration from the old world to the new, but also for the village and the family. If farms were needed for your children, you took in more land from the forest; when this supply ran short, you could migrate to a less crowded region. Cropland rose from about 3 Mha in 1700 to 13 Mha by 1950. Then agricultural research came to the rescue in the form of the steady rise in crop yields, from a mean world cereal yield of 1 t ha^{-1} in 1950 to nearly 3 t ha^{-1} today.

Concern was revived by the appearance of two books, Ehrlich's *The population bomb* in 1968 and Meadow's *Limits to growth* four years later.[2] These drew attention to the period of rapid and exponential population growth which the world had entered upon. If continued at its present rate this would lead, in less than a 100 years, to numbers which could certainly not be supported; indeed, continuation of the present rate of exponential growth would hypothetically lead, in only a few centuries, literally to 'standing room only' on earth. Whatever might be achieved by further advances in agriculture, immediate steps to put a brake on population growth were a priority. These warnings attained wide circulation and led to further studies. The US President's Science Advisory Committee, in a report of nearly 1000 pages, concluded, 'Unless the rate of population increase can be sharply diminished, all the efforts to augment agricultural production will merely postpone the time of mass starvation, and increase its agony when it inevitably occurs.'[3]

Twenty-two years after *The population bomb*, Ehrlich followed this up with *The population explosion*.[4] This has been the most influential recent account. It ranges beyond food supply, to matters of energy and pollution, with the overall conclusion that the 1990 population of 5200 million was probably higher than the world's resources can sustainably support.

Highly misleading statements on land availability are sometimes made. 'Famines and food shortages occur mostly in sparsely populated subsistence economies such as Ethiopia, Tanzania, Uganda . . . [where] land is abundant.' 'As for water and land for cultivation, both are abundant . . . there is an intellectually respectable view that the Earth is capable of supporting any population that can realistically be envisaged.' 'The world could readily feed another billion people, right now, without stressing any fragile acres or putting on heavy doses of farm chemicals.'[5] Such statements would not be worthy of attention but for the fact that they are occasionally quoted as having authority. They rest at best on a misunderstanding of scientific findings, at worst on assertion, coupled with ignorance of conditions in the real world. They should not be confused with reasoned studies.

Past estimates

Many attempts have been made to estimate the food production and sustainable population supporting capacity of the earth, differing widely. Out of 65 such estimates, 35 lie between 4 000 and 16 000 million population, 7 below the present population.[6] A number of the most recent and thorough studies have been selected to show how widely the results differ, and also to illustrate the range of issues involved. Some cover developing countries only, others the whole world.[7] Target dates with respect to population and food requirements range from 2000 to 2050. The studies can be placed in three groups: reassurances, warnings, and what will be called the orthodox view, appraisals by the major international institutions.

Reassurances

Computation of the absolute maximum food production of the world A hypothetical estimate based on photosynthetic efficiency, water and soil limitations, and an inventory based on the FAO world soil map. There could be 3419 Mha of cultivated land of which 470 Mha is irrigated. If 65% of this land is used for cereals, and based on a world average yield of 13.4 t ha^{-1}, the 'absolute maximum production' could be 32 390 Mt, over 15 times the present cereal production.[8]

How many people can the earth feed? Based on a 20% increase in cultivated land, a 35% increase in average yields, and a 20% reduction of post-harvest losses and waste, the world could support 10–11 billion people by 2050. There is a lot of 'slack' arising from present inefficiencies, both in production and in use of food. 'Assurance of a globally adequate food supply . . . will have to contain a strong component of what I would label the rich world's self-serving altruism [including] greatly expanded transfers of efficient farming techniques abroad, as well as the modification of unsustainable diets at home.'[9]

How much land can ten billion people spare for nature? By doubling the arable land area to 2.8 Bha, reducing consumption in developed countries from 6000 to 3000 kcal per day, and raising yields, we can feed 10 B people and still spare some land for nature.[10]

Population pressure and food supply in the developing world By increasing cultivated land by 20% (an additional 170 Mha), cropping frequency by 20%, and crop yield by 100% (to 4.5 t ha^{-1}), food supply could be doubled between 1989 and 2050, sufficient to meet a 70% population increase plus an improvement in diets.[11]

Sustainable world food production and environment Total world productive areas could be 3.8 Bha crops plus 4 Bha grasslands. World food production, as grain equivalents, could be 31 Bt at low inputs or 72 Bt at high inputs. Assuming no limitations in the distribution and transport of food between regions, the attainable maximum production at low input levels could support 65 B people on a vegetarian diet, 35 B on an intermediate diet, and 20 B people on an affluent diet. 'While there is an upper limit to food production, global agriculture is still far away from it, and . . . for the world population in 2040 food security will be ensured.'[12]

The first and last studies listed are based on inventories of land resources, combined with assessment of the scope for increase in average crop yields. The other estimates are mainly based on scope, or potential for improvement. These studies also draw attention to the demand side of the question, the effects of possible changes in diet and reduction in waste. All assume perfect redistribution, that food produced anywhere in the world can reach population anywhere else. The overall message of this group is that there is still ample potential to increase food production sufficiently to meet the needs of world population at least up to 2050.

Nevertheless, studies of this kind can mislead through being quoted out of context. It will be said that 'scientists have proved that' there is ample capacity for further increase in food production – misunderstanding the nature of science, which is concerned not just with radiation, soil and water but with people and the circumstances of the real world. There is even a school of super-optimists, which holds that the earth can support 40 billion people or more; free market mechanisms will solve all problems as they arise. Presumably this includes the assumption that high prices of commodities will stimulate private financing of the necessary research.[13]

Sustainable world food production and environment is an extremely detailed computerized calculation; it is based on a world grid of 15 000 cells, follows about 120 steps of calculation and has 21 technical appendices. Within its lights, it is carried out with the greatest scientific rigour. The estimates of maximum production are heavily qualified by reservations, to the effect that this is a theoretical exercise which omits some of the constraints of the real world, and is not an attempt at prediction. This estimate of a supposed theoretical maximum production makes the following unrealistic assumptions, about half of which are explicitly recognized in the original:

knowledge by farmers, and their application of management methods, is optimal;

there are no limitations on distribution: all food produced reaches those who need it;

all land suited to arable use is under such use, only what is left being grassland;

rain forests will be cleared from all productive land;

no land suited to production is assigned to purposes other than food production: non-food agriculture, cities, nature;

the effects of land degradation on lowering productive potential are ignored;

the irrigation requirements that are assumed vastly exceed available resources of fresh water;

there are no constraints (of energy, economics, etc.) to optimal inputs of fertilizers;[14]

food consumed by pets and draft animals is ignored;

no allowance is made for war, civil strife, and misgovernment.

It is hardly surprising that the study concludes:

'8.7.3 (Un)certainty of the results

There are so many unquantified aspects that contribute to the (un)certainty in the final results, which hinders giving a quantitative estimate of this uncertainty.'[15]

Warnings

Food security, population and environment It is doubtful . . . whether food security could be achieved indefinitely for a global population of 10 or 12 billion people. Rather, it seems likely that a sustainable population, one comfortably below the Earth's carrying capacity, will number far fewer than today's 5.5 billion people.' How many fewer depends on the extent to which current degradation can be checked.[16]

Constraints on the expansion of the global food supply The world's arable land could be expanded at most by 500 Mha. The discussion is based on three scenarios. On the 'business-as-usual' scenario (continuation of present trends), 'the world is unlikely to see food production keep pace with population growth . . . [It] will experience a declining per capita food production . . . spreading malnutrition and increased pressure on agricultural, range, and forest resources.' On the pessimistic scenario, per capita grain production will be 40% lower by 2050. On the optimistic scenario, 'If rapid population growth stabilization can be effected, leading to a world population of 7.8 billion instead of 13 billion by the year 2050, then grain production adequate for the population might be achievable.'[17]

Full house: reassessing the earth's population carrying capacity The limits for sustainable production from the seas and from rangelands have already been reached. The limit for further responses to fertilizers are being approached in high-use regions such as Asia. If present trends continue, the world will face grain import requirements that exceed exportable supplies by 526 M t. If people living in developing countries want to consume more livestock products, then 'the house is already full'. If population increases by 3.6 billion, 'the only real question is whether the average global grain consumption per person in 2030 will be closer to China's current 300 kilograms a year or India's . . . 200' (this compares with the USA's current annual consumption of 800 kg per person). 'It is difficult to see how an acceptable balance between food and people can be achieved without a broad-based reduction in population growth.'[18]

These three reviews are based primarily on reviewing what is happening at the present day, and projecting recent trends into the future. Besides land availability and its productive potential, they take into account problems in the redistribution of food supplies from surplus to deficit regions; and emphasize land degradation and limits to the responses to agricultural inputs. *Constraints on the expansion of the global food supply* adopts the technique of alternative scenarios, of assessing what will happen if actions and trends continue as at present, or get worse or better. The overall message of this group is that the sustainable carrying capacity of the earth has already been exceeded or will shortly be reached. If present trends continue, the consequences of famine and malnutrition will be extremely serious.

The key element of these assessments is that action within the area of land resources and agriculture alone will not be enough. A greatly increased level of investment and effort in research and better land management would reduce the magnitude of suffering, but there can be no acceptable solution without action to check population growth.

The orthodox view

By FAO:

Agriculture: toward 2000 If present trends continue, by 2000, 'there would be a horrifying increase in numbers of seriously undernourished, to some 600–650 millions.' On a 'modestly ambitious' Scenario A, 'production rises 12% above present trend rates; the deterioration in cereal self-sufficiency [in developing countries] is arrested . . . [but] numbers of seriously undernourished decline only marginally'. On an optimistic, but desirable, Scenario B, agricultural production in developing countries rises by 3.7% a year between 1980 and 2000. 'Food consumption improves appreciably everywhere, and *although serious undernourishment is not eliminated - no remotely feasible rate of economic growth could by itself bring about that result within two decades* - the numbers of nutritionally deprived people are sharply reduced.' On Scenario B, arable land in 90 developing countries (excluding China) is forecast to increase from 728 Mha in 1975 to 823 Mha in 1990 and 882 Mha in 2000.[19]

Land, food and people An assessment based on three alternative levels of inputs, low, intermediate and high. At the low input level, by 1975, 38% of the area of developing countries, mainly in the semi-arid zone, was already carrying more people than it could sustain; 64 countries would be below their supporting capacity by 2000. Of these, 36 countries would remain critical at intermediate input levels, and 19 at high input levels. Omitting countries with substantial oil or manufacturing exports, the most critical countries by 2000 include Rwanda, Yemen, Barbados, Haiti, Burundi, Lesotho and Mauritania.[20]

World agriculture: towards 2010 There is still substantial cultivable but not cultivated land; but 'there will be little land left for further expansion beyond the year 2010'. 'The growth rate of world agricultural production at 1.8 percent p.a. will be lower in the period to 2010 compared with that of the past.' For major crops, growth rates in yields 1990–2010 'can be expected to be much below those of the last 20 years'. 'The slowdown in world agricultural growth is also due to the fact that people who would [=wish to] consume more do not have sufficient incomes to demand more food and cause it to be produced.' Net cereal imports of developing countries will grow from 90 Mt to 160 Mt in 2010. The combined agricultural trade balance (cash crops plus food crops) of developing countries will switch from net surplus to deficit. 'It is . . . reasonable to foresee a continued role for food aid for a long time to come.'[21]

World food summit 1996 There are presently some 840 million undernourished people. 'World agricultural growth is likely to be slower in the future compared with that of earlier decades . . . In many developing countries, per caput food supplies may remain stubbornly inadequate to allow for significant nutritional progress.' Net imports of cereals will grow to over 160 Mt by 2010. 'Without deliberate changes from the normal course of events, many of the food security problems of today will persist and some will become worse. This need not be so if action is taken now.' If policies and action continue as at present, the number of undernourished will fall by 2010 to 680 million; a realistic target, adopted at the summit, was to reduce the number to half the present level by 2015.[22]

By the World Bank:

Resources and global food prospects: supply and demand for cereals to 2030 A 25% production increase over the present could be met from new land, now under range or forest. Land degradation is less severe than commonly claimed. Wider application of present knowledge will not be sufficient; but new technologies, developed by research, could fill the demand–supply gap. 'If the demand scenario is met at acceptable economic and environmental costs, it will be because the supply of knowledge farmers are able and willing to use increases enough to compensate for the insufficient supplies of land and water resources.'[23]

By the International Food Policy Research Institute (IFPRI):

Population and food in the early twenty-first century On a baseline scenario of best assumptions, by 2010, developing countries will have a cereal demand of 1392 Mt, production of 1231 Mt, leading to a net import requirement of 161 Mt; this can be met by the surplus from developed countries. Under four alternative scenarios, net developing country cereal import requirements range from 152 to 200 Mt. The necessary 1.5–2.0% growth in production of the major cereals could just be achieved on an aggregate basis by a sustained commitment to all aspects of the agricultural sector; even then, there will be severe regional problems in Sub-saharan Africa and South Asia.[24]

This group may be called the middle or orthodox view, meaning that it covers studies by the three major international institutions concerned with food and agriculture: FAO, the World Bank, and the International Food Policy Research Institute (IFPRI). The two later FAO studies include detailed climatic and soil inventories, assessments of potential production and consideration of a wide range of questions, including land degradation, non-agricultural land use, and problems of distribution. The IFPRI study is based on a computerized model, which takes projections of demand and production from different regions as given variables, and analyses consequences for international food trade; there is a matching analysis by the World Bank in the same volume.[25]

The findings of this group are complex, but two features stand out. The first is that large differences in food availability between countries will inevitably continue, with a resulting need for immense transfers of cereals from the few exporting countries, largely in the temperate zone. The second is that serious problems of food shortage can only be avoided if the commitment to research, extension and development is substantially greater than at present.

The first FAO study, *Agriculture: toward 2000*, is of interest in that it will shortly be possible to compare its projections with the actual outcome. The estimate of the number of people undernourished, 600–650 million by 2000, was overoptimistic; already by 1995 this is believed to be 800 million. To the extent that any credence can be given to data on arable land (pp. 57, 240), this has increased less than was assumed; a 1990 forecast of 823–841 Mha (on alternative scenarios) for developing countries excluding China compares with the reported 1990 value of 775 Mha inclusive of 97 Mha for China.[26] For the slightly less unreliable data on individual crops, the subsequent study, *Towards 2010*, includes a retrospective comparison. For the three major cereals, wheat, rice, and maize, the increase in harvested area was lower than projected, but the growth in yields was higher, giving a close agreement between forecast and observed production.[27] (There is an element of circular argument in such data, see 'Statistics on crop production', p. 211 above.) After smoothing year to year fluctuations, the continuous increase in net imports of cereals was close to that forecast.

The innovative feature of *Land, food and people*, also called the study of potential population supporting capacities, was to break away from a world or broad regional basis, and consider the food needs and potential production country by country. Besides climatic and soil inventories, it was based on three levels of agricultural inputs, low, intermediate, and high. Low meant traditional farming methods, using no external inputs (fertilizers, chemical biocides), and high was 'full use of all inputs . . . equivalent to Western European levels of farming'.[28] It was implied that countries would progress from the low level to the high. This assumption is no longer entirely valid, universal use of high inputs being limited by reasons of supply, economics, and environment. What was then called the intermediate input level, with limited levels of fertilizers and pesticides, is more nearly equivalent to what is now called low-input sustainable agriculture.

The basis was to consider food needs of countries, at their 1975 populations and as projected for 2000, and assess the extent to which these could be met from their own resources. The methodology is complex, involving climatic and soil limitations, rainfed and irrigated agriculture, crop suitability, effects of soil degradation, and the need for fallows (called 'soil rest periods') to maintain soil fertility.[29] The fallow period requirement has a large effect on land availability at different levels of inputs. At the low input level, some soils, such as strongly leached soils in the rain forest zone, can only be cultivated one year in seven if fertility is to be maintained, whereas with higher input levels cultivation is possible one year in two.[30]

As often happens in numerically based analyses, the extremely detailed and precise computerized calculations made up to this point were then subjected to a large and uniform adjustment, called 'the one third deduction'. It was realized that no account had been taken of a number of other factors, including land requirements for non-food crops, fuelwood and forestry, and 'some degree of inequality' in food distribution (i.e. poor people will receive less food). It was not practicable in the time available to make adjustments for each of these, so a deduction of one third was made from the food production capacities of all countries. Not surprisingly, 'this adjustment makes a considerable difference to the results, often swinging the balance of the land resource from a sufficiency to a deficiency'.[31]

This analysis gave countries which would be in food deficit if all agriculture were at the low-, intermediate-, or high-input levels. In the year 2000, for example, Ethiopia and Uganda would be unable to support their population with low inputs; Haiti, Burundi, and Somalia would fall short at intermediate inputs; while Rwanda and Yemen would be unable to feed their populations even at high inputs. Countries such as Kuwait or Singapore which are in food deficit but have assured exports of oil, minerals or manufactures are classified as low risk. Countries in the most serious situation are those which are in food deficit and at the same time have exports that are heavily dependent on agriculture, and which therefore cannot spare land from export crops for food crops.

Again, some appraisal of the outcome of the forecasts is possible. Of the 64 countries forecast to be unable to feed their populations in 2000, every one of them by 1995 had substantial imports of cereals - but then this applies to almost every developing country except Thailand. More striking is the list of countries most seriously at risk: Rwanda, Yemen, Afghanistan, Haiti, Burundi, Somalia, the Comoros Islands, Ethiopia, and Uganda.[32] The coincidence of these with recent incidence of famine and civil strife is a sad confirmation of the forecast.

The more recent FAO project, *World agriculture: towards 2010*, in fact contains two studies employing different methods, conducted more or less independently in the first instance and then partially combined. Chapter 3 is based on informed projection of trends of the period 1990–2010; it covers the whole world, and includes trade. It is comparable with the World Bank and IFPRI studies. Chapter 4, on the other hand, is an updated and methodologically improved version of the population supporting capacity study, based on land inventory, crop suitability, and potential production. It is confined to developing countries, and gives attention to land requirements for purposes other than food. Forestry, fisheries, trade, poverty, research, and education are covered in later chapters. Taking the two approaches together, this is the most detailed and thorough assessment to date, and the source for many other estimates.

Earlier FAO studies are linked and extended in the technical background documents to the second *World food summit*, 1996. Taking reduction in the number of undernourished people as the objective, this is notable in casting the net across a wide range of policy aspects: the political and social environment, poverty, the role of women, the importance of the rural sector, participation and sustainability, trade, and disaster relief. The reviews of land availability and projections of crop yield increase and food imports are largely drawn from *World agriculture: towards 2010*. Land degradation receives proportionally very little attention, and research is not one of the 'commitments', or priority areas for action. The overall conclusion is that it is not plausible to eliminate hunger in the foreseeable future. Only if governments commit themselves to measures for improvements in the rural sector, and match their commitments by budgetary allocations, might it be possible to halve the present number of 840 million undernourished people by 2015. To meet population increase and achieve well-balanced diets by 2050, developing countries would have to increase their plant-derived food energy by 174%, a formidable target.

The FAO 2010 study is also taken as the primary basis for the following discussion. The demand side of the food equation is considered first, leading to an assessment of the requirements for food security. Discussion of the supply side begins by considering two alternative sources, food from the seas and food from rangelands, together with a solution of a different kind, importing food for developing countries from the temperate zone. This leads to the primary question, the land availability and potential productivity of lands of the developing world.

Food requirements, effective demand, and food security

A population basis

A target year of 2025 will be taken as the basis for the following discussion, within the range of long-term planning and development objectives of the present. Population change is considered in the following chapter, but for present purposes the three UN estimates, called high, medium, and low variants, will be taken. What matters is not the total world population but the populations of developing countries. The nearly constant number of 1200 million people in the developed world will continue to produce enough food to feed themselves and have a surplus for export. The UN estimates for developing countries, rounded to the nearest 50 million people, are:

Population, all developing countries, 2025

UN variant	Million	Percent increase on 1995
High	7650	67
Medium	7050	55
Low	6450	42

In the past, the UN medium variant forecast has proved to be accurate. Developing countries will probably need to feed at least 7000 million people in 2025. It could be 600 million more, or less.

Food needs and undernutrition

Although the SI unit of food energy supply is the joule, most international statistics employ the kilocalorie (kcal), equal to 1000 calories.[33] The minimum level to maintain body weight and support moderate activity is 2000–2300 kcal per day, varying with ethnic group. In detailed nutritional balance sheets, adjustments are made for sex, age, activity, and climate. In addition to energy, the primary requirement for a healthy diet is a protein supply of 60 g per day, nearly all of which can be vegetable protein. In constructing food balance sheets, however, it is common to convert this into energy terms by adding 10% as a 'protein constraint'. After making further allowance for inequalities in distribution, the energy requirement to avoid undernutrition[34] in tropical countries is about 2400 kcal per person per day. As there are inevitable losses of at least 10% in storage and processing, this converts to a minimum dietary need of 1 million kilocalories a person per year.

Many calculations of food production are carried out in terms of cereals or cereal equivalents, production from roots and pulses being converted on an energy basis. The calorific value of the major cereals is similar, about 3500 kcal per kilogramme or 3.5 M kcal per tonne. Hence one person needs on average 285 kg of cereal equivalent per year (this agrees with what people whose diet is mostly rice or maize actually eat, about 1 kg a day). Over a year, one tonne of cereals actually consumed could theoretically support 3.5 people, but making further allowance for losses, the conversion is nearer to 3 people per tonne of cereal equivalent. In broad terms, therefore, for a minimum adequate nourishment the 1995 population of developing countries requires 5700 ÷ 3 or 1900 Mt of cereal equivalent.

As with many statistics, it must not be supposed that the existence of data on the number of people undernourished means that this value has actually been measured. The procedure is to estimate the average dietary energy supply at national level, and make an allowance for inequality in distribution, usually taken as a standard deviation of 20% of the average. This procedure 'makes it possible *to draw inferences* about the *approximate* proportions of the population with access to food below a given nutritional threshold.'[35]

The average food supply for developing countries is only just above the critical limit of 2400 kcal per day – meaning that for half their population it falls below this. In eight countries of Africa the average is under 2000 kcal, and for Bangladesh only just above that level. The proportion of the population undernourished is estimated at 20% for all developing countries. The percentage is highest for Africa, 37%, but absolute numbers are greatest in East and South Asia, 230 and 220 million respectively. An appalling situation, with over 30% of the population undernourished, is believed to exist in Afghanistan, Mongolia, Bangladesh, Peru, Bolivia, Haiti, the Dominican Republic, Madagascar, and almost all countries of Sub-Saharan Africa other than South Africa. There are about 190 million children underweight for their age. With respect to nutritional balance, the most widespread critical problem of malnutrition is vitamin A deficiency in children, estimated to affect 40 million children; besides being found over much of Sub-Saharan Africa, clinical deficiency is found in some humid countries in which total food energy is adequate, such as India, the Philippines, and Vietnam. The 1996 World Food Summit resolved to halve the number of hungry people within 20 years, but it is unlikely that this target will be achieved. In round numbers, about 800 million people in developing countries are chronically undernourished. All of them are poor.

Hence, projections of food requirements for the future start from an already severe deficit position. To the 1995–2025 population increase of 55%, an additional increase in food supplies is needed if undernutrition is to be lowered. To make even a small contribution to reducing hunger would require an increase in food supplies of at least 60%.

Dietary change and waste

Western countries have an average daily consumption of 3400 kcal per person, well in excess of the nutritional requirement. This is not for the most part a case of actually eating more, nor of throwing away food or feeding pet animals, much as these are contributory causes. The main reason is diets high in meat. Owing to the nature of the food chain, the calorific value of meat is much less than that of the plant matter used to produce it. The world's cattle population is 1300 million; adding the equivalent body weight of other kinds of livestock brings this close to 2000 million livestock units.[36] If the world's domestic livestock were converted to humans on a basis of bodyweight they would amount to 11 000 million 'human-equivalents', twice the human population. Whilst part of their feed comes from grass, over one third of world cereal production is fed to animals.[37]

Food and the resources needed to produce it would indeed be saved if these affluent Western diets were reduced to nutritionally adequate levels. Any such change is regrettably unlikely to happen. Whilst some individuals choose to follow a wholly or partly vegetarian diet, a major change would only come about through a rise in the price of meat.

Set against any such change is the fact that, in those developing countries where incomes are rising, the first change in consumption patterns is to eat more meat.

There is a small modification in the diet of rural areas, together with a change among the more affluent of city dwellers towards the meat consumption of Western diets. Even a small dietary change, say to the level now found in South America, would increase the food requirement of developing countries by at least 5%.

There is opportunity for increasing efficiency of food use by reducing post-harvest losses. Direct waste, the throwing away in the West of food uneaten, may be regrettable but is not the most significant cause. Where the major losses occur in developing countries is in food storage, processing and distribution. Losses from household grain storage bins from pests may be 10–15%. By comparing production with consumption data, one estimate has put apparent post-harvest storage losses as 13% of the original crop, and distribution and household losses as an additional 17%.[38] Some of these losses, however, are inevitable consequences of processing and marketing, and the scope for their reduction is less than has sometimes been claimed.

It is right that Western countries should be made to recognize their excess food consumption; once there is awareness, action marginally to reduce it might follow, but this is not where the main problem lies. A reduction of even 5% in total post-harvest losses in developing countries is a direct gain in available food, and research and extension efforts are being made to achieve it.

Requirements and effective demand

Food requirements are based on physiological needs for adequate nutrition. Effective demand is an economic term, referring to willingness to buy food, and therefore ability to pay. Effective demand is very often shortened to 'demand', creating an unfortunate impression. A hungry person with no money exerts no economic demand.

Projection models of future food production, consumption and trade are nearly always conducted in terms of demand. They do not take account of hunger. Such economic models depend basically on the intersection of supply and demand curves, in terms of price. One of their main objectives is to forecast international trade and its cost. The starting-point for projections, effective demand at the present day, is taken as net imports, what a country has been willing and able to pay for; gifts, as food aid, may be added. Estimates for the future are then based on projections of the quantity and price curves for supply and demand.

Because of hunger and poverty, therefore, real food requirements are substantially greater than the values usually quoted as 'demand'. If 1 in 5 of the world's population have a 20% shortfall in food, this would add 4% to total food needed at the present day. The problem is more difficult than this figure would imply, because the food is not available where it is most needed.

FOOD SECURITY: COUNTRIES MOST AT RISK

Food security is lowest in countries with a low income, a food deficit, and a low capacity to pay for food imports. Insecurity is increased where the exports are themselves dependent on production from the land. On this basis, countries most at risk in 1994 are listed below.

Food imports more than 25% of total export earnings

Sub-saharan Africa	Middle East and North Africa	South and Central America	Asia and the Pacific
Agricultural exports more than 25% of total exports			
Burkina Faso		Dominican Republic	Afghanistan *FN*
Ethiopia *FN*		Nicaragua	Sri Lanka
Gambia *F*			
Guinea-Bissau *F*			
Mali			
Rwanda *FN*			
Somalia *FN*			
Agriculutual exports less than 25% of total exports			
Benin	Egypt	Haiti *FN*	Bangladesh *N*
Cape Verde	Yemen		Cambodia *F*
Comoros	Sudan *F*		Laos
Djibouti			Maldives
Lesotho			Nepal *F*
Mauritania			Samoa
Mozambique *FN*			
Senegal			
Sierra Leone *FN*			
Togo			

F = Countries which received emergency food assistance in 1995

N = Nutritional deficit, per capita food supply less than 2100 kg per day

Food security

Food security is more than an international jargon term for freedom from hunger. It means that in every country, over an extended period, all of the population should be adequately nourished. For this to be so, there are many aspects additional to having total food calorie supplies equal to the requirements of the population. Achievement of food security requires that:

total food supply is equal to dietary needs;

there is proper nutrition, an adequate balance to diets, including for the special needs of children and nursing mothers;

the present food supply is produced without compromising production in the future by land degradation;

food needs are available, through production or imports, in every country;

those who most need food, the poor, have adequate incomes to pay for it;

each country has effective mechanisms to meet shortages that occur in a year of drought, flooding, or other disaster;

the above requirements are not interrupted by war or civil strife.

Otherwise expressed, food security requires absence of undernutrition and of malnutrition, based on sustainable production; effective demand, both at national level and by individuals; measures in place for famine relief; and good governance to achieve these ends.

Food supply: alternative sources

Food from the seas

Marine fisheries provide less than 5% of human food supply but are important in diet as a source of protein. There are also signs that diets high in fish have benefits to health. Fish in all forms provide 16% of all animal protein but substantially more in areas of the developing world with low meat consumption, 21% and 28% in Africa and East Asia respectively, Marine fisheries (by capture) supply over three quarters of the total, inland fisheries 7%, and inland and marine aquaculture 15%.

With 70% of the earth's surface covered by the oceans, a rise in their food yield of only a few percent would greatly increase food supplies, but there is no ecological or technical likelihood that this can be done. On the contrary, there are clear indications that, as a result of technical advances in fishing methods, the offtake from marine fisheries has reached a sustainable limit at about 80 Mt. The all-time peak catch of 1989, 86 Mt, has not since been surpassed. Almost all ocean fisheries show signs of stress and in some, Atlantic and Pacific herring, Atlantic cod, and haddock, for example, the catch in recent years has fallen to 50–80% below the peaks of the 1960s and 70s. In addition to over-fishing, many coastal fisheries are affected by pollution. Many tropical lakes have also experienced declines. Fishing by capture provides the clearest illustration of a natural resource which has reached, and in many areas passed, the limit for sustainable offtake.

By contrast, in aquaculture, both freshwater and marine, there is opportunity to manage the resource on a sustainable basis. Production is increasing steadily and will continue to do so; fish ponds can form a valuable component of more intensive farming systems. Fisheries have an important and, through aquaculture, growing contribution to make to protein supplies and hence good nutrition, but there is no prospect of a substantial increase in their contribution to total food supplies.[39]

Food from rangelands

For purposes of assessing food production from land resources, it is convenient to separate livestock production systems into two groups: those based on managed pastures and feedstuffs (including stall feeding), and those based largely or entirely on grazing of open rangelands. The former systems are found predominantly in humid and subhumid climates, the latter in the semi-arid zones on either side of the subtropical deserts. The present and potential food supply position of these two groups of management systems are very different.

World meat production has risen steadily from 50 Mt in 1950 to approaching 200 Mt today. Annual production per person has risen from 20 to 33 kg and the recent trend suggests that it will level off below 40 kg. Much of this comes from two sources of primary plant production, grazing from temporary (planted) pastures and the use of cereals as animal feed. Analysis of livestock production is hindered by the fact that, whilst there are data on animal numbers and production, there are virtually no reliable data on either planted or natural pastures.[40] It is estimated, however, that 37% of world cereal production is fed to animals, or about 130 kg per person a year. The newly industrialized countries have greatly increased their meat consumption, and if China succeeds in economic growth it will massively add to this demand - in this instance, effective demand in economic terms. Besides the vital role of meat protein in diets, there are environmental benefits of short-term grasslands. Rotational grassland has for centuries been the primary means of maintaining soil fertility in the temperate zone. A striking feature of farms in the Indian subcontinent is the way that even very small holdings devote a proportion of land to fodder crops. There is scope to substitute planted grass–legume mixtures for natural fallows in other parts of the tropics. In terms of food energy, however, the inevitable dietary trend towards more meat consumption will divert cereals to animals. This lowers the amount available for direct consumption and therefore in food energy terms, reduces the food supply.

The position with respect to production from open rangelands differs in almost every respect. To the extent that they cannot be used for crop production, the use of natural pastures for grazing adds to food supplies. However much reports of 'desertification' may have been exaggerated, it is clear that degradation of pastures is extremely widespread. Contrary to what was formerly supposed, it appears that conversion from communal grazing systems to managed ranching, whilst it would improve the status of pastures, would not greatly increase livestock production (p. 192).

Because livestock scientists show such reluctance to make pasture surveys, the inventory approach, multiplying area by productivity, is impossible. Livestock production data do not separate animals raised from natural pastures, nor do they include most subsistence consumption. It is therefore necessary to fall back upon qualitative observation. All indications are that, as with the open seas, the limits of sustainable production from natural grazing in semi-arid environments are being approached. Continued research into livestock breeding, nutrition, and management will bring about gains in productivity, but only a change from full reliance on natural grazing to a mixture of grazing and planted feedstuffs will substantially increase production.

Imports from the temperate zone

A common basis in forecasts of food production potential is to estimate the productive potential of the world as a whole, and then to qualify the findings by pointing to 'problems of distribution', namely the massive transfers to the developing world from exporting countries in the temperate zone. A different approach is adopted here. The developed world produces, net, its own food requirements. It can therefore be treated as an alternative source of supply to the developing world, constrained not by the physical capacity of exporting countries to produce, but by the ability of importing countries to pay.

Reasons for this treatment are that developed countries have, and will continue to have, no problems in meeting their food requirements, exchanging a surplus production of cereals for tropical products. Farm output is below its potential, held down by subsidies for non-production. Output from this region responds to changes in world prices, increasing intensity of production if these rise. There is an unknown factor: how production from countries of the former USSR will settle down after its current transitional period.

The world food problem is a problem of the developing world. There will continue to be a large import of foodstuffs, in particular cereals, up to the limit of effective demand, i.e. what they can pay for (gifts, as food aid, have been 5–10% of purchased imports in recent years). One can therefore take forecasts of net trade and treat this as an alternative source of supply. In this treatment, imports are effectively subtracted from requirements for production.

A justification for this treatment is that forecasts of international trade are probably accurate, perhaps to within 10–20% over a 20-year projection. In the first instance, data on trade are relatively reliable. Secondly, since the ending of the cold war, forecasts have been based on modelling of supply and demand in a free market economy, for which tested technical methods are known. Finally, forecasts for 2010 by three major international agencies, FAO, the World Bank, and IFPRI, are in good agreement.[41] Projection to 2025 is more speculative. China is the largest source of uncertainty; if its economy succeeds in developing along the lines of the newly industrialized countries of Asia, it could create a demand larger than the total current imports of the developing world; current Chinese government policy favours of food self-sufficiency, yet its cereal imports reached 16 Mt in 1994.[42]

Taking cereals only, the position in the early 1990s on a highly generalized basis is as follows. World production is slightly under 2000 Mt, just over half of this from the developed world. There are net imports by developing countries of 80–85 Mt, nearly all from the temperate zone. Only five countries have exports of more than 10 Mt: USA, Canada, France, Argentina, and Australia. Kazakhstan may be emerging as an exporter from the former Soviet zone. The only large and consistent tropical exporter is Thailand with 5 Mt; it has recently been joined by Vietnam, but whether this surplus will continue remains to seen.

Forecasts for 2010 are consistent in suggesting a net import by developing countries of 160 Mt, double the present. There are formidable problems of distribution and payment, but this is what is realistically likely to happen. Adding 10% for non-cereals and 5% for food aid gives a total transfer to developing countries in 2010 of 185 Mt cereal equivalent. If the increase were projected to 2025 at the same rate it would give 220 Mt cereals or 250 Mt food supply as cereal equivalent, but this is speculative.

There is a problem with cereal imports: few of them reach the people who most need them. By definition, effective demand is, and will continue to be, met. Many accounts of world food trade blur the distinction between demand and requirements. Demand comes largely from the cities and the more affluent sections of the community. Other than through charitable and emergency distribution, food imports do not reach the rural poor. Food imports meet effective demand, but, because of poverty, bear no necessary relation to requirements.

Food requirements: from the 1990s to 2025

Future food requirements can be estimated by considering the present position and projecting into the future. However, a simple addition of the relative population increase underestimates the need in two respects: shortfalls at the present day, and per capita consumption changes in the future.

To recapitulate, the 'world food problem' is a problem of the developing world. Developed countries, largely in the temperate zone, will continue to feed themselves and to produce a net surplus of cereals. Food from the seas and from rangelands will continue to contribute to dietary balance, but their resources are now close to the limits for sustained production. Many developing countries now require large cereal imports for their food security, and this dependence is likely to increase, doubling over the next 30 years. There is scope for reduction of post-harvest losses, but the magnitude of this option is not as large as has been supposed.

In taking the 1990s as a basis for assessment, one needs to take account of three shortfalls already in existence:

the 800 million hungry people;

the massive cereal imports with their accompanying burden of debt;

the element of food production which at present is being obtained on a non-sustainable basis, degrading the land resources.

The projected population increase of 55% in 30 years makes by far the largest contribution to increased future requirements. Changes in diet, primarily eating more meat, could raise total food consumption considerably, but here there is a mitigating circumstance: to the extent that this change results from higher living standards, it will create effective, economic demand. When it comes to alleviating hunger, however, requirements fall short of effective demand. Considered on an overall basis, food needs for reduction of hunger might appear to be an additional 5%, but the real need is greater for two reasons: to get the food to where it is needed, and to make provision for bad years.

In summary, in the race into the future, we begin from behind the starting line. Food requirements for 2025 will exceed those of the mid-1990s not simply in proportion to the population. If population rises by 55%, then to reduce chronic undernutrition and recurrent famines will call for an increase in food supplies of at least 65%.

The food-production capacity of developing countries

The initial basis for assessment of food producing capacity is cultivable land area and average crop yield, giving production. Simplistic use of this formula, however, has led to estimates for the world as a whole such as '3000 Mha $\times$ 5 t ha^{-1} = 15 000 Mt, sufficient to support 45 000 M people'. Calculations of this kind are grossly misleading. By referring to the world as a whole, they do not take account of how food is to reach countries which need it. They make the highly unrealistic assumptions that all cultivable land is of good quality, and that all of it is used for food production. The effects of these and other factors are cumulative and large.

In addition to area of cultivable land and crop yield, assessments of food production potential must take account of:

non-agricultural uses of land;

non-food crops on agricultural land;

land requirements for conservation and maintenance of fertility;

lower yields obtained on marginal land;

input levels which are realistic, economically, environmentally and socially;

loss of land and lowering of yields by degradation;

post-harvest losses.

The FAO study, *World agriculture: towards 2010*, is the most thorough recent assessment. It is for developing countries (except China), employs the approaches of inventory and projection jointly, is based on immensely detailed primary data, and takes all the above factors into account. This will be taken as the main basis for discussion, adding estimates for China. The basic framework is threefold: land available for food production, crop yields on this land, and a comparison of the resulting production with food requirements.

Land available for food production

Three detailed inventories of potentially cultivable land have been conducted, two by FAO, the third independently.[43] They are based on the similar source material, the *Soil map of the world* and climatic inventories, and reach conclusions which are in good agreement (Table 8).[44] The areas include both rainfed cultivation and irrigated arid land. For the developing countries, there are said to be some 2700 Mha of 'potentially cultivable' land, of which 850 Mha or 31% is presently cultivated. This leaves a 'land balance' of 1850 Mha. There is a further 1000 Mha 'potentially cultivable' land in developed countries, of which 66% is under cultivation.

The terms 'potentially cultivable' and 'land balance' are highly misleading if quoted out of context, giving the impression of a large reserve of land which can be brought into use as future needs require. The figures for potentially cultivable areas are the interim results of a multi-stage assessment, obtained purely from data on climate, soil, and slope. The land balance is simply the result of subtracting from this the present cultivated area (annual and perennial crops). There is absolutely no possibility that anything like the whole of this 'balance' could be used for food production for two reasons: non-agricultural uses, and agriculture for non-food crops.

The non-agricultural uses are human settlements, protected areas, and land which it is considered desirable to retain under forest, productive or protective. There are statistical difficulties in taking these alternative uses of land into account. Data on the area covered by human settlements are available for only a few

Table 8. *Estimates of cultivable land*

FAO, World agriculture: towards 2010 (Alexandratos, 1995)

	Million hectares			
Region	With crop production potential	In use 1988/90	Land balance 1988/90	Estimated in use 2010
Sub-Saharan Africa	1 009	213	797	255
Near East and North Africa	92	77	16	81
East Asia excluding China	184	88	97	103
China*	[170]	[130]	[40]	[145]
South Asia	228	191	38	195
Latin America and Caribbean	1 059	190	869	217
Developing countries	2 743	887	1 856	995

*Data for China are not found in this source; they have been added, based on Fischer and Heilig (1998) and G. Fischer (personal communication).

Other estimates

	Potentially cultivable M ha	
Developing countries	2526	(FAO, 1982a, 1984b)
	2487	(Luyten, 1995)
	2759	(Fischer and Heilig, 1998)
World	3536	(FAO, 1982a, 1984b)
	3803	(Luyten, 1995)
'Expanded by one third'	1900	(Kendall and Pimentel, 1994)
'Expanded by 20%'	1700	(Smil, 1994; Bongaarts, 1996)
'Twice the present'	2900	(Waggoner, 1994)

countries, and that on forests subject to much uncertainty; data on protected areas are good,[45] but these overlap to an unknown extent with forest areas. Only rough estimates of the proportion of these three types of use which occupy land suited to cultivation can be made. FAO estimates are shown in Table 9. Human settlements, although frequently occupying good agricultural land, do not occupy a large area on a world scale although they are locally significant; in particular, they occupy 45% of

Table 9. *Land in non-agricultural uses: developing countries, excluding China, 1988/90*

	Area under land use Mha	Area of land balance Mha	Percent of land balance
Human settlements	94	51	2.8
Protected areas	385	201	12
Forest areas	1 690	774*	45*

*Minimum, 'probably much larger'.
Source: Alexandratos (1995), 155–8.

the land balance in South Asia. Protected areas are for the most part on non-cultivable land.

Forest land raises a question of principle: should it be confined entirely to land not suited to the growth of crops? The FAO estimate of 45% is based on the assumption that *all non-cultivable land is under forest*, so, in fact, well over half the land balance is occupied by forest. The functions of forests are vital to society, international opinion is strongly opposed to any further forest clearance, and national governments and local communities increasingly so. Forestry has always been the best use of steeplands, but its varied functions, productive and environmental, cannot possibly be fulfilled if it is excluded from potentially cultivable land. It would be unacceptable, ecologically and socially, to preserve the forests by taking into cultivation the whole area of the savannas. Expansion of cultivated land means clearance of forest.

A further deduction must be made for agricultural land under non-food crops. These occupy about one third of land in developing countries, and many such crops form the primary basis of their exports.

The supposed land balance is highly unevenly distributed. The area cultivated as a proportion of the potentially cultivable is 84% in South Asia, and in the Near East with North Africa, 48% in East Asia, but only 21% in Sub-Saharan Africa and 18% in Latin America. In relative terms, this is in accord with general observation. Out of the total land balance, 72% is found in only 15 countries, with Brazil having 27% and Zaire 9%. The uncultivated areas of Brazil and Zaire, together with those of 6 of the other 13 countries (Bolivia, Central African Republic, Colombia, Mexico, Peru, and Venezuela), are largely under highly valued rain forest. Nine countries are assessed as having negative land balances, present cultivation exceeding what is sustainably cultivable; these are again in agreement with qualitative observation.

Even when allowance is made for alternative uses, there appears to be a fundamental weakness in assessments of cultivable land by this method. Inventory

Table 10. *FAO estimates of cultivable land, selected countries*

	Thousand hectares			
	With crop production potential	In use 1988/90	Land balance	Balance as percent of potential
Tanzania	55 195	10 801	44 394	80
Ghana	15 902	4 876	11 026	69
Ethiopia	35 851	15 408	20 443	57
Malawi	5 992	2 674	3 318	55
Kenya	9 620	4 841	4 779	50
Jamaica	528	267	261	49
Vietnam	14 319	7 999	6 320	44
Sri Lanka	3 566	2 174	1 392	39
Philippines	16 261	11 608	4 653	29

Countries which contain 72% of the total land balance:
Brazil (27%)
Zaire (9%)
Angola, Argentina, Bolivia, Central African Republic, Colombia, Indonesia, Mexico, Mozambique, Peru, Sudan, Tanzania, Venezuela, Zambia.

Countries with negative land balances 1988/90:
Bangladesh, Haiti, Mauritius, Pakistan, Rwanda, Tunisia, Yemen.

Source: Alexandratos (1995).

data for nine countries are shown in Table 10. All are shown as having large balances of cultivable but presently uncultivated land; in four countries, the land balance exceeds the presently cultivated area and in three more it is approximately equal. These results are seriously in conflict with field observation. In Ethiopia, the cultivated land of one village is adjacent to the next, and land for grazing and fuelwood production grossly inadequate. Malawi still had some spare land in 1982; 10 years later, cultivation had extended up the hills and down onto steep rift valley dissected areas, land on which arable agriculture cannot be sustainable. Similarly, Jamaica is cultivating almost unbelievably steep hillsides. In Vietnam, encroachment into the already severely depleted forest remnants is continuing, and in the Philippines efforts are being made to protect such forested hill zones as remain. There is indeed some spare land remaining in Tanzania and Ghana, but it is inconceivable that it is 2–4 times the present cultivated area.

The analyses leading to these estimates have been conducted with great care and in immense detail, breaking the landscape down into small units and re-aggregating to country level. For the results to be so seriously in conflict with general

FIGURE 25 Land shortage: cultivation up a hillside, with incipient erosion, Malawi.

observation, there must be some basic problems in the method of assessment by inventory. Three reasons may account for the discrepancy: non-agricultural uses of land, underestimation of presently cultivated areas, and overestimation of cultivable land on small-scale maps.

Two of these reasons have already been considered. The essential use of substantial areas for settlements, conservation, and forest is noted above. The highly unreliable nature of data on land use, including the practice of governments ignoring illegal cultivation, was discussed in chapter 4. The *Towards 2010* assessment makes large 'adjustments' to cultivated areas arising from statistical discrepancies, for example raising the figure for Sub-Saharan Africa from 140 to 212 Mha, an addition of 51%.[46]

The third reason is inherent in maps at small scales, and specifically the 1:5 M

FIGURE 26 Land shortage: this coffee smallholding in Nyeri District, Kenya, is well managed, but on Kenya's humid zones one farm adjoins another, with no spare land.

scale of the *Soil map of the world*. Soil scientists give little attention to non-agricultural soils, and this has led to a serious underestimation of their extent. When mapping land from air photographs, lines are drawn around numerous areas of hills, scarps, rock outcrops, stony soils, shallow laterite, swamps, and sometimes areas of toxic soils; these often occupy some one third of the mapping unit. However, such areas can only be mapped individually at scales of 1:20 000 and larger. On smaller-scale maps they appear only as statements in the legend such as 'in association with lithosols'. When detailed surveys are reduced to national soil maps, and again to the international scale, these areas are often simply lost, and the whole of the generalized mapping unit assigned to its dominant or other usable soils. It is true that the world soil map includes 'phases' for such non-productive soils, but their areas are almost certainly underestimated.

As one who has long advocated soil surveys and their application through land evaluation, and taken part in the earlier FAO assessment, it is painful to come to the conclusion that findings based on the approach of inventory are fundamentally flawed. This is not because of problems intrinsic to the method itself, but stems from the seriously inadequate nature of the source material on which it is based, including maps both of land resources and current land use.

There is a means by which country assessments could be tested. Consultants visit the country and, through the national land use planning organization, ask to be shown in which regions there are areas of potentially cultivable land. They visit these regions, assess what proportion of the area is indeed cultivable, and, more importantly, find out what the land is being used for at present. There might be cases which required sensitive handling, such instances where influential landowners held land for speculative purposes. It is likely, however, that one of three options for the supposed cultivable land would be very widely found:

FIGURE 27 Terracing with hill irrigation of rice, Nepal; attempts to initiate terracing outside of Asia have rarely been successful.

that substantial areas were not in fact cultivable, consisting of hills and other such areas, not shown on national-scale maps;

that it was environmentally necessary that land should be retained under protection forestry or other forms of conservation;

that much of the land that remained, and had the physical potential for sustainable cultivation, was already being used for necessary purposes, such as production of fuelwood and other forest products, grazing, or support for minority ethnic groups.

Many other aspects are taken into account in the FAO assessment, for example the extent of land which is only marginally suited to cultivation, the contribution from irrigated land, and loss of land through degradation. It appears, however, that because of the problems with the inventory type of assessment, the estimates of future cultivated land are based more on the approach of projection.

Table 11. *Two estimates for developing countries (adding a proportional change for china)*

	Million hectares		Growth
	1988/90	2010	% per year[48]
Arable land			
Excluding China	757	850	0.6
China	95	107	–
Developing countries	852	957	–
Harvested area, cereals			
Excluding China	331	389	0.8
China	92	108	–
Developing countries	423	497	–

Source: Alexandratos (1995), 166 and 169.

Because of the lesser unreliability of the data, it is better to take the area for cereals as a basis. The growth rate for 1970–90 was only 0.6% per year, and this rate had shown a declining trend for the previous three decades; the prediction of 0.8% a year is therefore unlikely.[47] Abandonment of land following the more severe degrees of degradation is currently still taking place. A more reasonable extrapolation would be 0.5% a year to 2010, slowing to 0.3% for the period 2010–25. This would lead to cereal areas of 465 Mha in 2010 and 486 Mha in 2025.

Finding new land

In relative terms, the inventories are in agreement with observation. Land shortage is most severe in South Asia; the 'culturable wastelands' of early Indian taxation assessments have almost entirely disappeared. Two large and densely populated countries, Bangladesh and Pakistan, have negative land balances. Land is also in very short supply in the Middle East and North Africa, where water constraints limit the expansion of irrigation, and Tunisia and Yemen have negative land balances.[48] For China, the long-quoted figure for cultivated land of 96 Mha is now believed to be a 30% underestimate.[49] Part of the supposed land balance in East and South-East Asia is in fact already cultivated, much of it is under rain forest, and part of this is hilly; available land in this region is already very limited.

A problem with the land balance in Africa and Latin America is that much of it is under rain forest. Official statistics for present cultivation are again probably underestimates. For a number of ecological reasons, it is considered undesirable to cut down these forest areas, although substantial further clearance will certainly occur. One recent assessment adopts the view that rain forest should not be considered as land available for cultivation. If this is assumed, the implication is that the woodlands of the savanna zone should be almost entirely cleared, which is desirable neither on grounds of production nor conservation. Further expansion of cultivation in Africa will often be at the expense of the livelihood of pastoral peoples.

Two suggestions are sometimes made for the existence of large potential land reserves, based on extension of Asian practices to Africa and America: terracing of steeplands, and the conversion of the Amazon and Zaire (Congo) basins into swamp rice production. There is no physical reason why the spectacular flights of terraces found in such countries as Java or Nepal should not be constructed in mountain chains of the same age in other continents, but, quite apart from dubious claims about capacity for sustained labour, the time scale renders this impossible. Asian terrace systems were constructed progressively over hundreds or thousands of years. Attempts to introduce terracing for soil conservation purposes outside areas where it is traditional have repeatedly been unsuccessful There are local opportunities, particularly where there is water for hill irrigation; but, on grounds of labour and more particularly time, there is no chance that a large addition to cultivable land will come from terracing.

The vision of converting the Amazon and Zaire basins into Asian-type ricelands, sacrificing forests to the demands of food production, rests on a misunderstanding of physical potential. The Indo-Gangetic plains, the Mekong delta, and other alluvial plains of Asia are formed from recent alluvium derived from the Himalayan chain. Many are on young sediments of river terraces, with moderate fertility. The two great river basins of Central Africa and South America are downwarped areas of the ancient rocks of the basement complex, their sediments derived largely in Africa, and partly in South America, from similar material. The soils of these basins are highly weathered, the fertility of the topsoils lost within a few years of forest clearance, and with much less potential to restore this by mineral weathering. Rice schemes are locally successful in Africa, and political stability in Zaire and adjacent countries would bring some opportunities. The time-scale required for the complex measures of water control will severely restrict an already limited potential.

Water resources and the expansion of irrigation

In the 1960s and 70s, irrigation land grew at 2–4% per year. Irrigation made vast contributions to the expansion of food production, particularly in Asia. Cereal yields on irrigated land are typically twice as high as under rainfed. The assured water supply meant that the inputs of the green revolution could be applied without the risk of crop failure. Large dam and canal schemes with high investment were accompanied by tubewell expansion, both assisted by aid and through private capital. For a time, the potential for continued expansion appeared almost limitless.

In the 1980s, through a combination of factors, the rate of growth slowed to 1% per year. The best sites were becoming used up, and costs per hectare of irrigation development soared. Some serious failures on large-scale schemes, particularly in Africa, discouraged investment. Then, in the 1990s, a further constraint to expansion appeared in the form of water shortage, most serious in the semi-arid regions where irrigation is most needed. A recent assessment showed a balance of 110 Mha of potentially irrigable land in developing countries, more than 50% of the present irrigated area.[50] At recent rates of growth, this would become exhausted by 2050 (in Asia by 2025), but no one supposes that anything like this total area could in practice be developed.

There are now some 250 Mha of irrigated land, three quarters of this in developing countries. Much of the further expansion will be of supplementary irrigation to land already under rainfed cultivation. This will contribute to food production in the form of crop yields that are more assured, and thus higher on average. The potential for bringing further arid lands into cultivation is by now quite small, because in most of these regions the limits to water availability have been reached.[51] Irrigated land in developing countries is projected to rise from 123 Mha in 1990 to 146 Mha by 2010, which as a percentage of arable land is a rise only from 16 to 17%.[52]

Yields and production

Following land, the second factor in the food production equation is crop yield. Oversimplified, over optimistic views on potential for raising yields are sometimes encountered. It is true that, for any crop, the gap in average yields between countries is large: rice yields of more than 5 t ha^{-1} in Egypt, Japan, and North and South Korea, for example, compared with 2.7 t ha^{-1} in India and Bangladesh, or for maize

Table 12. *Forecasts of cereal production, 2010, developing countries*

			Growth 1988/90–2010		
		Production 2010 Mt	Total %	% per year (compound)	Net imports Mt
FAO: Estimate I		1 318	56	2.1	162
Estimate II			58	2.2	–
Excluding China	995				
China pro rata	149				?
		1 144			
IFPRI: Baseline projection		1 231	38	1.5	161
Four scenarios		1 187–1 264			152–200
World Bank		1 300	45	1.8	210

Sources: Alexandratos (1995); IFPRI (1995); Crosson and Anderson (1992). For China, see note 44.

an average for the whole of Africa of 1.2 t ha^{-1} compared with Asia's 3.4 t. More extreme estimates rest on the attainment of maximum yields over 10 t ha^{-1} on research stations, and suppose that, under pressure of need, these could be widely attained.

Such views rest on many misconceptions. Besides the biological potential resulting from research, the levels of crop yields which can be obtained in practice are limited by a succession of constraints. Land is rarely optimal for a crop, nearly always having limitations of climate, water, and soil: out of all land with crop production potential, 38% has low natural soil fertility and 23% sandy or stony soils, whilst two thirds of the total has one or more soil constraints.[53] For a variety of reasons, economic and environmental, inputs cannot be applied at levels which give optimal yields. There is certainly further scope for raising levels of fertilizer use in Africa and South America, about 20 and 60 kg ha^{-1} respectively, closer to the Asian level of 120 kg. But countries with high fertilizer usage are experiencing decreasing responses, and in some instances seem to be approaching a plateau in yields comparable to that already present in the temperate zone.

There is no way of predicting changes in yield other than by projection. It is widely agreed that it will become harder to obtain future increases than it was in the non-repeatable circumstances of the green revolution in the past. For cereals as a whole, FAO proposes that the 2.2% annual rise over 1970–90 will be reduced to 1.4% for 1990–2010, raising the average yield for developing countries, excluding China, from 1.9 to 2.6 t ha^{-1}. China, however, with 38% of world production, has an average yield of 4.4 t ha^{-1}. Adding this as a weighted average to present data, and

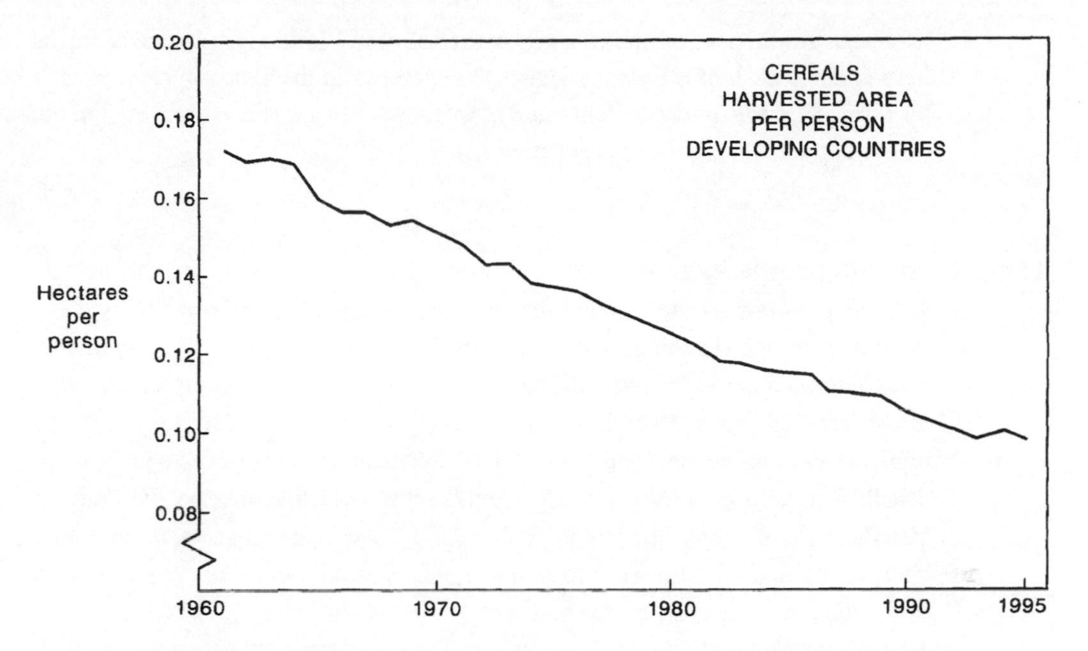

FIGURE 28 Harvested area under cereals, per person, developing countries. Source: FAO statistics.

projecting proportionally into the future, adjusts these values upwards, making the average yield for all developing countries rise from the 1990 level of 2.5 t ha^{-1} to a 2010 level of 3.3 t ha^{-1}. Beyond that period, a more speculative average increase of 1.0% per year would bring the average cereal yield for all developing countries in 2025 to 3.8 t ha^{-1}.

Combining increases in land area and crop yield, together with projections of trade, Table 12 shows estimates of cereal supplies to developing countries in 2010, from three international agencies. For FAO there are two estimates, obtained by different approaches, and for IFPRI a baseline or best estimate and four alternative scenarios, sets of assumptions. There is a reasonable degree of consensus on a total production of 1200–1300 Mt, 40–60% above the 1990 level, to be supplemented by net imports to developing countries of 160–200 Mt; a rise in living standards in China could greatly increase the figure for imports.

There are signs that 1990 may prove to have been a watershed year for world food, and particularly cereal production. After climbing almost non-stop from 1950 to 1990, world cereal production has remained static for the six years 1990–5. As compared with the projections in *World agriculture: towards 2010* there is a shortfall of 68 Mt in world cereal production 1994–6, but this is mostly due to a downward trend in countries of the former Soviet Union, and production in developing countries has exactly followed the forecast. In relation to population, the 1995 harvest of 293 kg per person is the lowest for 30 years. World grain carryover stocks have fallen in each of the years 1992–5. These straws in the wind appear against a background of the most constant trend in world statistics, the inexorable fall in the cereal cultivated area per person (Figure 28).[54]

Land, food, and people in 2025

The predicted population increase for the developing world 1995–2025 of 55% is a starting-point for assessing food needs. It makes by far the largest single contribution to the increased requirements, but is by no means the whole story. A reduction in chronic undernutrition will call on an overall calculation for a further 5%, but a total increase of nearer to 10% would be necessary because of inequalities of distribution. Whether means can be found to finance the forecast doubling of already massive cereal imports is problematic,but even such an increase would contribute less than a tenth of the food supplies of developing countries. Overall, a production increase of 65%, in cereal production matched by other food supplies, is a minimum target if increase in hunger and associated problems are to be avoided.

For land area, it appears unlikely that the FAO forecast of an average 0.8% annual increase to 2010 can be attained. The lower rates assumed above would give an increase in cultivated area, for cereals and *pro rata* for other food crops, of 10% to 2010 and a further 4–5% to 2025, net of land loss through degradation. This is 15% above the actual area under cultivation at present, not the lower areas shown by official statistics.

For crop yield increase, the FAO estimate of 1.4% a year gives a rise of 28% to 2010; this could be reduced by effects of land degradation, say to 24%. Three factors will then make it hard to maintain this rate: genetic advances will become more difficult, supplies of inputs may cause constraints, and limits to response to inputs will certainly be reached in some areas. The increase for the period 2010–25 might therefore fall to 0.8–1.0%. This would give a total yield increase for the period 1995–2025 of between 36 and 43%.

Combining a 15% land increase with this range of estimates for crop yields gives a total production increase above 1995 levels of 56–64%. This matches the increase in food requirements only for the most optimistic set of assumptions. It applies to the developing world as a whole, not to the already severe problems of many countries. The overall position is well summarized in a conclusion to the 1995 IFPRI study:

> **If governments and the international community maintain (or renew) their commitment to agricultural growth through policy reform and sustained, cost-effective investment in agricultural research, extension, irrigation and**

> **water development, human capital, and rural infrastructure, pressure on aggregate world food supplies from rising populations and incomes will not be overwhelming . . . However, problems will continue in getting food to those people who need it most.**[55]

This is a very large 'if', the implications of which are serious in the extreme. Forecasts made by the international agencies responsible for agricultural development, what has been called above 'the orthodox view', necessarily assume a set of favourable conditions concerning research, improved land management, and the investment necessary for these; they are bound by their mandates to do so. They take no account of interruptions to production brought about by war and civil strife. Although the skills for aid from developing countries are available, unless there is a quite unforeseen change in attitudes, there will be no increase in their level of funding support. This leaves the obligation in the hands of governments of the developing world.

The implications of continuing food deficits for land resources are equally serious. In the countries with lowest food security, there is already little suitable land remaining to take into cultivation. Growing populations will lead to increasing pressures on a finite land resource, so people will be forced to exploit land in ways which cannot be sustained. The nexus of population increase, poverty, and land degradation will be reinforced.

An increased level of research, extension, resource assessment, better land management, education, and rural infrastructure is needed if even the present unsatisfactory position over land, food, and population is to be maintained. There is no hope of achieving these aims without a greatly increased commitment to the agricultural sector by governments of developing countries. What is more disturbing is that even the most optimistic range of projections is only marginally able to meet requirements. The target of food requirements set by population growth, as it will be if action is not taken, may prove to be too high.

14 Population, poverty, and conflict

The present rate of world population increase, 240 000 people a day, poses immense problems, not least for land resource management. Future growth will nearly all take place in the developing world. Some countries, such as Bangladesh, Ethiopia, Malawi, Pakistan, and much of the African Sahelian zone, are already close to their limit for sustainable support to their people. Population increase will augment problems of landlessness, land degradation, and food security. Poverty and undernutrition add to these problems. There are also major consequences for human welfare or its opposite, suffering: direct links between population pressure and famine, and indirect but strong connections with civil conflict and war.

Scientists and governments have each made major international policy statements on population. Both accept the integral nature of population, economic growth, and environment, but there is one critical difference. The scientists state clearly that global social, economic, and environmental problems cannot be solved without an early approach to zero population growth, if possible within the lifetime of today's children. The governments could not commit themselves to such a statement. All are agreed on a set of ethically acceptable measures for reducing population growth, based on provision of family planning and reproductive health services, and measures to improve the welfare, education, and status of women. An immediate and major effort to apply such measures would contribute more than any other form of development to human welfare. Policy, planning, and action to check population growth is not merely an associated factor, but in the wider perspective is an integral part of land resource policy and management.

Much has been written on population, both at scientific and popular levels. International conferences have been held, forecasts of future world population made, and policy statements issued. There is a general awareness that the continued growth of world population at present rates would ultimately lead to some kind of disaster, although views differ on the urgency of the problem, and whether positive action needs to be taken or whether it will solve itself through economic development. In a book on land resources, no more than a brief indication of these questions can be given, so some justification is needed for writing about population at all.

The reason is that many aspects of land policy, including requirements for land use, food production, and land degradation, are strongly influenced by population. Population has impacts upon water, soils, forests, pastures, and nature conservation, as well as upon the pollution aspects of the global environment. It would be quite unrealistic to take population as an external variable, with its future changes as a given factor, to be allowed for. The links between population and land resource management are strong and close, to the extent of being virtually integral.

It is not just population numbers that are involved, but their adjunct, poverty. The intimate links between population, poverty, and land degradation, the so-called 'nexus', have been discussed in chapter 7. Finally, out of the joint consideration of land, food, population, and poverty, difficult but vital questions arise, of famine, civil strife, and conflict.

Population

World population, past and future

The past history of world population is well known: the long period of very slow growth to about 1800, a gradual acceleration through the next 150 years, and the modern period of rapid growth since 1950. One of the most striking ways to express this is by showing the years taken to add successive increments of 1000 million:[1]

Year	World population (million)	Interval
c. 1830	1 000	All human history
1930	2 000	100 years
1960	3 000	30 years
1975	4 000	15 years
1987	5 000	12 years
1998–9	6 000	11–12 years

Many people alive today were born when the population was half its present size, a situation which has never happened before this century. The 6000 million mark will be reached before the year 2000, so the twentieth century will go down in history as the only century in which the world population more than tripled. The UN will publicize the reaching of the 6000 million mark some day in 1998 or 1999, but, as there is certainly an error margin of 5% in the population estimates for most developing countries, this can be taken to all intents and purposes as the population 'now'.[2]

The most graphic figure is the rate of population growth, which from 1986 onwards has been close to 88 million a year. This is an increase of 240 000 a day, or 10 000 an hour, nearly 3 *extra* people every second. Every day, that additional quarter of a million people has to be fed, clothed, housed, and found work. Every year, the governments of the world must order their affairs so as to facilitate these tasks for the extra 88 million.

In forecasting future changes, the reliability differs between two phases. For the first 20–25 years, one generation, the uncertainties are relatively small, because the

FIGURE 29 Populations of the future will have even greater needs for land resources.

women who will bear the children are already born. The forecasts depend only on the direct effects of changes in the birth and death rates. This is the momentum built into population change. After 25 years, two greater sources of error arise. First, estimates of the birth rate at that time become very much less reliable, since major changes in customs, attitudes, and practices may have taken place. Secondly, any early errors are transmitted by a multiplier effect, as the new birth rates apply to the number of women, fewer or more numerous, in the first generation.

As a consequence, forecasts based on alternative assumptions about changes in birth rates start to diverge greatly between 30 and 50 years hence. The UN makes high, medium, and low variant forecasts, together with others such as 'medium-high' and also constant fertility, the assumption that present birth rates will continue. These effects can be seen as follows:[3]

	World population, million			
Year	Low variant	Medium variant	High variant	Constant fertility
2025	7 591	8 504	9 444	10 978
2050	7 813	10 019	12 506	21 161
2100	6 009	11 186	19 156	109 405

The low and high variants differ from the medium by about 1000 million for 2025, but diverge greatly for later years. The constant fertility assumption leads (as Malthus foresaw) to impossibly high values, 21 000 million by 2050; this becomes 694 000 million, over a hundred times the present population, in the year 2150.

With these reservations, it must be said that the actual course of world population over the past 50 years has followed the UN medium variant forecasts quite closely. The 1957 forecast of the 1990 and 2000 populations were 5140 and 6280 million respectively, compared with an actual 1990 value of 5285 million and the most recent forecast for 2000 of 6158 million.

Nearly all of the increase will take place in developing countries:

	Population, million, UN medium variant		
	1995	2025	Increase %
Developed regions	1 236	1 354	10
Developing regions	4 534	7 150	58
Percent in developing regions	79	84	

At present, almost 4 out of 5 people live in the developing regions. By 2025, nearly 6 out of 7 people will do so, and about 90% of all children. China and India together have 38% of the population of developing regions, and this is made up to 50% by five other countries: Indonesia, Brazil, Pakistan, Bangladesh, and Nigeria. Other population giants in the developing world, each with more than 1% of its total, are Mexico, Vietnam, Iran, the Philippines, and Egypt. Only 3 of the 10 countries with over 100 million people, the USA, Russia, and Japan, are in the developed world.

Consequences of population change

It is hard to appreciate the significance of totals for the world or major regions, but figures for individual countries are more meaningful (Table 13). Ethiopia is densely crowded today in its humid, highland regions, with famines recurring in several provinces every year that the rainfall is below average; yet, from the base of this already critical situation, it is expected to more than double its population. Kenya was certainly an 'empty' country, with spare land, until the 1950s, but today its people have fully occupied the humid regions and are forced into the marginal semi-arid zone. Kisii District in the west is under severe environmental stress, with fragmentation of farms already as small as one hectare, encroachment into the limited areas of remaining forests, and outward migration. With not only one of the highest birth rates in the world, but also a high proportion of children, there is little chance to check a further doubling of population; yet with few opportunities

Table 13. *Past, present, and future populations of selected countries*

	Population, million		
	Actual		Estimated
	1950	1995	2025
Ethiopia	18	55	127
Kenya	6	28	63
Malawi	3	11	22
Tanzania	8	30	63
Mexico	28	94	137
Brazil	53	162	230
Bangladesh	42	120	196
China	555	1221	1526
India	358	936	1392
Pakistan	39	140	285

Source: UN data and Fischer and Heilig (1997).

remaining to expand farmland, there is already a serious problem of unemployment for school leavers.

Bangladesh lives from its high rainfall, the waters of the Ganges-Brahmaputra system, and alluvial soils, permitting harvests of two and three crops a year. The last 10 years have seen a rise in cereal production of 25%, achieved entirely by raising yields as the cultivated area has not, and cannot, change - *and stands at an average of 0.1 hectares per person*. This higher production has been entirely taken up by population increase, with food supplies remaining at 2000 kcal per person per day, 20% below the average for Asia. Cereal imports are 1.4 Mt a year plus food aid of more than 1 Mt, making Bangladesh (jointly with Egypt) one of the two highest recipients of food aid in the world. On top of this chronic problem, the country is subject to natural disasters caused by flooding. A future population of nearly 200 million will massively increase the effort needed to meet such disasters.

Taken as a continental average, the situation is less critical for South and Central America: arable land per capita is higher and population growth has fallen below 2% per year. It is said that there is some 'spare' land to take into cultivation, but much of this is sloping, forested land, with a high degradation hazard and valued for its forest cover. Whilst this is the overall situation, crowded areas with dense populations and no spare land occur, particularly in the Caribbean. It is unbelievable that the small and poverty-stricken island of Haiti could support twice its present population.

Such examples could be multiplied many times. The specific problems differ

MALAWI: PAST THE LIMITS

Malawi is a small, land-locked country of 94 000 km^2 in Central Africa. Its varied relief gives to it much beauty, and indirectly leads to a good rainfall. It also means there are large areas of sloping land, both in the Rift Valley and the hills which rise above its plains. It has substantial areas of fertile soils, as on the plains around the old and new capitals, Zomba and Lilongwe.

These favourable natural resources attracted Bantu peoples from the south, and, by 1960, with 3.5 million people, it was already densely inhabited. On the more fertile plains, shifting cultivation had long-since disappeared and even short fallows had become rare; the agricultural system was basically continuous maize monoculture. But cultivation stopped short at the foot of the hills, and there was no serious erosion.

By 1975, with 5 million people, the arable land was fully occupied. Cultivation had extended up many hillsides and become extensive on the steep slopes of the Rift Valley scarp zone, with visible sheet and rill erosion. The *dambos*, valley-floor pastures, were becoming gullied. Civil strife in neighbouring Mozambique led to a vast influx of refugees, more than 1 in 10 of the population in 1992.

By 1995, the population had passed 11 million and was increasing at over 3% a year. The remaining forests had been widely invaded for cultivation, at a rate of 50 000 ha per year. Crop yields had *declined* since 1980, in part through recent drought years but also because of a low level of fertilizer use and soil fertility decline. There was clear evidence of lower yields under the same crop varieties and inputs. The cereal exports of the 1970s had given place to imports of over 100 000 t and food aid of nearly twice that amount, together supplying 20% of food needs. Per caput food supplies were barely above 2000 kcal a day, well below the accepted minimum for health.

The *Agriculture towards 2010* study estimates there are 6.0 million hectares with crop production potential, 64% of the total land area; with a presently uncultivated 'balance' of 3.3 million hectares. To anyone who knows Malawi, these figures are ludicrous: something is seriously wrong with the methodology which produced them. Nearly a million families are living off farms of less than one hectare. Large areas of land are in use that cannot be sustainably cultivated, and the resource potential is being lowered by erosion and soil fertility decline.

The long period of rule by President Banda, for all that its human rights record has been criticized, was at least one of stable and relatively efficient government. Roads were kept in repair, agricultural services functioned, and marketing systems operated. It is also a favoured country for foreign aid, $50 per head per year, accounting for a quarter of the gross national 'product'. Malawi has not suffered serious civil strife.

There is scope to raise crop yields substantially, but this will need more than fertilizer supplies and demonstration plots. A widespread change to farming based on land husbandry methods is needed, and it is hard to see how this can be done on such small farms. Malawi is permanently dependent on aid, and requires external assistance if famine is to be avoided whenever there is a dry year. It is a country that is well past its sustainable limit of population.

from one country to another, but common features recur. Among problems which are being created, or made more severe, by continued population increase are:

Reduction in farm size: subdivision, on inheritance, of family farms that may already be less than one hectare.

Movement onto marginal land: the cultivation of steep slopes, or migration to semi-arid zones, with consequent risk of land degradation.

Increased landlessness in rural areas.

Reduced food security: smaller carry-over stocks from year to year, both at farm and national levels; declines in per capita food production and consumption.

Resort to the uncontrolled exploitation of open-access resources, e.g. forest degradation, overgrazing of communal pastures.

Out-migration from rural areas, driven by the push factor of lack of rural opportunities, when there may already be urban unemployment.

An increased scale of effort needed to meet disaster, whether natural or linked with political causes.

Whilst they are cumulative at the national level, these problems are felt primarily in villages. Under the situation of land and population reached in most developing countries, all further increases in population will bring about greater pressures upon land resources and, except with massive and ever more difficult efforts, a decline in the quality of life.

Population policy

The causes of the population explosion since 1950 are well known: the spread of advances in medical science and health services to developing countries, bringing large reductions in death rates whilst birth rates initially remained high. In the Western world, this stage was followed by a fall in birth rates, leading to the present stable or falling populations of most European countries, and slow rates of increase in other parts of the developed world. The falls in death and in birth rates both took place during a period of continuing economic growth. It is a widely held article of faith that developing countries will follow the same sequence, leading to a levelling-off of population with low death and birth rates, the so-called demographic transition. A few former or present developing countries, for example South Korea, Singapore, and Cuba, have passed through this transition to achieve current growth rates of below 1% a year.

Considerable progress has been made in reducing the total fertility rate, the

PAKISTAN: IRRIGATION AND POPULATION

Pakistan's population could not have grown to anything like its present size without the vast irrigation systems of the Punjab and Sind, based on the Indus and its tributaries. Out of a total of 21 Mha of arable land, 17 Mha are irrigated. Without this water, almost all of which has its sources outside the country, it would be semi-desert, as can be seen in patches of land too high to be reached by irrigation water. The extent and complexity of the dam-and-canal system is illustrated by the names of successive subdivisions in the water supply system: main canal, branch canal, major distributary, minor distributary, watercourse, all this before the water passes through a *nakkar* (weir) to enter the farm system of distribution channels.

Vast sums have been spent over the past 40 years, not in improving the irrigation system but in reclaiming salinized land. Cereal yields and production in Pakistan have increased by 30% since 1980, yet production per head is 15% below its 1980 level, transforming the country from a cereal exporter to a net importer. Like Bangladesh, it has achieved this rise in yields through increasing levels of fertilizer, now averaging 100 kg ha^{-1}; but the yield response per unit of added fertilizer is falling steadily. In a recent FAO appraisal, the country is assessed as having a negative land balance, the presently cultivated land exceeding that which is considered to have cropping potential.

What would the present population have been if the large-scale irrigation systems had not been developed? Nothing like the present 140 million could have been supported. So would there have been deaths in great numbers from recurrent famines, large-scale peaceful migration to more humid regions of India, or war? Or - and such a possibility is rarely discussed - under conditions of severe land shortage, would birth rates have been lower? This may seem to be academic speculation, yet comparable situations are going to arise in many countries in the future.

Of more direct concern is present high rate of population increase. With the 1995 population of 136 million predicted to double by 2025, even on the UN low variant prediction, there will be an ever more frightening degree of dependence on stable government, avoidance of land degradation, and continuing crop yield increases if even the present low level of food supply is to be maintained.

average number of births per woman. In developing countries as a whole, this is now about 3.5, compared with 6.0 in the early 1960s.[4] The reduction has been greatest in eastern Asia, intermediate in Latin America and South Asia, and least in Africa. Almost all individual countries, even in Africa, have experienced a decline. There is a strong correlation between increased use of contraceptives and lower total fertility rates. Thus, in Thailand, an increase in contraceptive use from 15% to 70% of the population was accompanied by a halving of the total fertility rate to 2.8 children per woman. It is this fact that has led to one of the cornerstones of current population policy, the approach of 'children by choice'. As it is largely women who make this choice, the improvement of their education, status, and economic position is fundamental.

A few early attempts were made to reduce birth rates by means such as compulsory sterilization, for example in India. This approach was ethically unacceptable and also widely circumvented. Financial disincentives to having more than two children have been used, as in Singapore. The major example of a coercive approach is the continuing policy of China in applying a combination of legal, financial, and moral pressures for one-child families. This has brought China's total fertility close to the replacement rate of 2.1 children per woman, which, if continued, will stabilize its population at about 1500 million after 2025. The methods employed are not considered acceptable in the rest of the world, although justified by the Chinese government on grounds of the colossal magnitude of their population problem.

Views of the scientists

The current approach to population policy, formulated over more than 10 years, found expression in the Third UN Conference on Population and Development, the Cairo conference of 1994.[5] It was preceded by a Population Summit of the World's Scientific Academies, held in New Delhi in October 1993. This was only the second occasion that top-level scientific organizations of the world have undertaken a large-scale collaborative activity intended to influence governments.[6] The objective of this meeting was to influence governments and international decision-makers at the Cairo Conference, calling on them to 'adopt an integrated policy on population and sustainable development on a global scale'. The statement issued after the meeting was signed by 58 scientific academies, of which 26 were from developing countries, 10 ex-USSR, and the remainder from developed countries.[7] A high degree of consensus was achieved, with only one dissenting statement.[8]

A starting-point for this document was the three alternative UN population scenarios:[9]

Optimistic Average world fertility declines to 1.7 children per woman early in the 21st century; population peaks at 7800 million in the mid-21st century.

Medium Fertility declines within 60 years from its current rate of 3.3 children per woman to the replacement rate of 2.1. Population levels off at 11 000 million near the end of the 21st century.

Pessimistic Fertility declines only to 2.5 children per woman. Population reaches 19 000 million by the end of the 21st century and continues to grow.

Many of the views expressed were essentially the same as those subsequently included in the outcome of the Cairo meeting: the integral nature of population, development, and environment; the threat to environmental resources caused by population growth; the dependence of quality of life on conservation of natural

POPULATION, ECONOMIC GROWTH AND THE ENVIRONMENT: VIEWS OF THE SCIENTISTS

'The relationships between human population, economic development, and natural environment are complex and not fully understood. Nonetheless, there is no doubt that the threat to the [global] ecosystem is linked to population size and resource use.'

'Our common goal is the improvement of the quality of life for all, both now and for succeeding generations. By this we mean social, economic, and personal well-being . . . and the ability to live harmoniously in a protected environment.'

'Family planning could bring more benefits to more people at less cost than any other single technology now available to the human race.'

'Success in dealing with global social, economic and environmental problems cannot be achieved without a stable world population . . . We must achieve zero population growth within the lifetime of our children.'

Extracts from the Population Statement issued by the World's Scientific Academies, New Delhi, 1993.

resources; and the opportunities which exist to reduce rates of population growth in an ethically acceptable manner, through provision of family planning services, and improving the education and rights of women. They called for the 'incorporation by governments of environmental goals in legislation, economic planning, and priority setting', and for 'pricing, taxing, and regulatory policies that take into account environmental costs'. As would be expected, they stressed the potential of science for improving human welfare, and hence called for strengthening the indigenous scientific capacities of developing countries, not just in research, but more generally in its efficiency to manage the socio-economic and natural environment. They mentioned the fact that you cannot achieve any of these desirable aims without good governance and management.

Views of governments

The Third UN Conference on Population and Development, the Cairo conference of 1994, was in many ways a strange meeting, in that there was comparatively little discussion of population as such. Instead, the focus was upon health, education, women's rights, and poverty, together with family planning services. The major advance over earlier approaches was a recognition that population was integrally linked with conservation of the environment and sustained economic growth. This is most concisely set out in the UN General Assembly resolution adopting the Programme of Action of the conference which, 'acknowledges that the factors of population, health, education, poverty, patterns of production and consumption,

POPULATION, ECONOMIC GROWTH, AND THE ENVIRONMENT: THE VIEW OF GOVERNMENTS

'Efforts to slow down population growth, to reduce poverty, to achieve economic progress, to improve environmental protection, and to reduce unsustainable consumption and production patterns are mutually reinforcing. Slower population growth has in many countries bought more time to adjust to future population increases. This has increased those countries' ability to attack poverty, protect and repair the environment, and build the base for future sustainable development. Even the difference of a single decade in the transition to stabilization levels of fertility can have a considerable positive impact on quality of life.'

Extract from Programme of Action of the 1994 International Conference on Population and Development, paragraph 3.14.

empowerment of women, and the environment are closely interconnected and should be considered through an integrated approach'. The extract from the Programme of Action quoted above says, in effect, that if population growth can be slowed down, then a whole range of major economic and environmental problems will become less difficult to combat.

The new consensus was on acceptable measures, intended to promote human welfare and at the same time check increase of population. Whilst ranging widely over economic development and reduction of poverty, the elements which bear most directly upon population are:

- to make family planning services, including provision of contraceptives, available to everyone;
- to improve reproductive health, both of mothers and children;
- to improve the education, rights and status of women.

The Cairo Conference statement requested governments of donor countries to devote more aid resources to projects with these aims, and many have done so. Thus, the UK Overseas Development Administration (ODA) launched a 'Children by choice not chance' initiative in 1991, leading to a 33% increase in aid to population and reproductive health programmes between 1990 and 1993.[10] The US aid programme similarly supports 'reducing population growth rates to levels consistent with sustainable development', through promotion of family planning services, reproductive and child health, and education of girls and women.[11] As countries have not increased total development aid budgets at the same time, this has effectively meant diverting aid from projects in other fields.

A comparison of views

There is a crucial difference between the New Delhi statement issued by the scientists and the Cairo statement issued by governments. The scientists made no reservations about saying:

> Success in dealing with global social, economic and environmental problems cannot be achieved without a stable world population . . . We must achieve zero population growth within the lifetime of our children.[12]

The nearest that the governments got to this is a sentence in the Preamble, that 'Intensified efforts are needed . . . in a range of population and development activities, bearing in mind the crucial contribution that early stabilization of the world population would make towards the achievement of sustainable development.'[13] No such statement is to be found among the 15 principles, nor in the subsequent programme of action.

The outcome of the World Food Summit (1996) illustrates how population is still treated as a taboo subject at government policy level. The technical background documents for this meeting clearly state that, 'Population growth is probably the single most important global trend influencing food security.' They demonstrate that whether the high, medium, or low variant of UN population projections is followed makes huge differences to food requirements over the period 2020–50. In particular, 'the demographic transition in Africa would facilitate the process of achieving food security', requiring an annual growth rate in agricultural production of 2.6% for the low population variant, compared with 3.3% for the high variant. Yet population appears in the high-level document, the Rome declaration, only once, in a phrase 'within the framework of . . . early stabilization of world population', and does not appear at all, neither as a commitment nor an objective, in the Plan of Action.[14]

The same attitude is found in the 1996 'Mission Statement' of the UN Fund for Population Activities (UNFPA), the body charged with implementing UN population programmes.[15] The original, earlier, UNFPA mandate refers frequently to 'population problems'. In the recent statement, this phrase is no longer found. The main sections cover poverty eradication, environment, food security, empowerment of women, education, and health. There are explicit goals for infant mortality, life expectancy, education and reproductive health, but there is no goal for birth rates. The nearest the statement comes to expressing an opinion on desirable population change is to refer to 'the universally accepted aim of stabilizing world population'.

National population policies

There is a still more politically sensitive aspect. If the world population is to stabilize, then the populations of most individual countries must do so.[16] With the notable exception of China, no politician has ever expressed a desirable target population for their country. It is almost universally considered that women should be free to choose the number of children they have, the role of governments being to give them the opportunity for choice. Where stability or a decrease has been achieved, as will soon be the case in Denmark, Germany, Italy, Japan, Portugal, and Spain, it has happened mostly through the choice of life styles by women, without ever having been regarded as a national objective. The hope of optimists is that this will simply 'happen' as a corollary of economic growth by developing countries. Without enormous efforts, international and above all national, it will not do so.

An exception to the principle of freedom of individual choice is China, where the government has taken forceful measures to control births, through financial and other penalties for more than one child (two in rural areas).[17] Recent government statements confirm that 'the Chinese government has made it a basic state policy to carry out family planning and population control.'[18] They justify this attitude on the grounds of the extreme seriousness of problems caused by further population growth, notably the food problem, linked to the fact that, owing to the mountainous nature of the country, only one tenth of its area is cultivable. The policy has achieved its aim of reducing the total fertility rate to below the replacement rate of 2.1 births per woman, although, because of the youth of the existing population, a levelling off of numbers cannot be reached below 1500 million. China's coercive methods are ethically unacceptable, although it remains to be seen whether their undoubted harshness will prove to reduce suffering in the longer term.

There are encouraging signs to the population scene. The fall in fertility in most parts of the developing world is continuing, and at slightly faster rates than previously estimated. The total fertility rate for 1990–5 is 3.5 for less developed regions as a whole, falling from 6.1 in 1960–5 although still well above the 2.1 replacement rate. It is 5.8 for Africa, 3.1 for Central and South America, and 3.0 for Asia. The fall in fertility rates is found in all continents and almost all countries of the developing world, including those which formerly had high rates of increase or which, until recently, opposed or took no action on birth control. For example, between 1975–80 and 1990–5, the total fertility rate fell in Kenya from 8.1 to 6.3, in Pakistan from 7.0 to 6.2, and in Iran from 6.8 to 5.0.[19]

The actual 1995 population of the developing world is 1% lower than had been forecast three years earlier, and in the UN 1996 revision, the forecast for 2050 is 5% lower than in the earlier estimate prepared for the Cairo conference in 1994. It is impossible to separate the effects of international assistance, government policy, urbanization, and economic trends in bringing about this reduction; in the final analysis, women are deciding to have fewer children.

Poverty

The meaning and reality of poverty might be thought too obvious to require definition. The poor are those who are chronically or recurrently hungry, cannot afford medical services or education for their children, have housing of the most minimal standard, go cold in winter, are frequently landless or otherwise insecure, and have no provision for old age other than the charity of their families. If they are women, they may be working 14 or more hours a day to obtain basic water, food, and fuel for the family. The poor are marginal economically and sometimes, even in a spatial sense, crowded around the edges of villages, forming long queues for water at a single standpipe, or crowded into shanty towns. The poor have always formed part of human society, but in recent years there are more of them, and the disparity in income between rich and poor is growing greater.[20]

However, there is an official UN definition of poverty, and rightly so, as it provides a basis on which to monitor change. The poverty line is defined as the income level below which a minimum nutritionally adequate diet plus essential non-food requirements are not affordable.[21] There are more refined indices which take into account such factors as food security, underweight children, births attended by midwives, life expectancy (as a measure of adequacy of health services), and education of women.[22] Using the simpler measure of income-based poverty, at the money values of the 1990s and the food prices of the poorer developing countries, an income of about $1 a day, or $400 a year, approximates to the poverty line. Much more is needed at the food prices of richer countries.

Depending on definition, estimates of the number of people living in poverty in the mid-1990s range from 900–1300 million. The proportion is highest in Sub-Saharan Africa, but absolute numbers are greatest in Asia. In Latin America, poverty exists within economies which are, on average, richer. The poor form just under 50% of the total population of South Asia and Sub-Saharan Africa, 33% of the Middle East and North Africa, 22% of Central and South America, and 10% of East Asia. By the year 2000, these proportions are expected to fall in Asia but remain similar in other regions. Coupled with overall population growth, absolute numbers of the poor will increase. In summary terms, it can be said that, in today's world, there are about 1000 million very poor people. It seems regrettably likely that by 2025 there will be nearly 2000 million.

Most of the poor, about 90%, live in rural areas.[23] They are found predominantly in areas of lower potential, with less favourable climate and soils. Many are landless, in some countries making up over half the rural population. Some have inherited farms of one hectare or less, subdivided from those of their parents. Others are surviving by cultivating marginal land, dry or sloping. Rural poverty exerts a strong 'push' effect, forcing migration to cities where there may be no jobs. Economically and politically, the poor are weak. In the classic illustration, a man with no shirt on his back and no money exerts no 'demand' for a shirt. Politically, in autocratic forms of government, their influence is small, exerted only through the wasteful method of a *coup d'état* by a new, supposedly populist, dictator. Their position is better in a well-functioning democracy where, because of their numbers, albeit limited by lack of education, the poor can have a substantial voice.

International aid efforts, which in the 1950s and 60s were directed primarily at economic return, are now strongly focused on reduction of poverty. 'The World Bank's overarching objective is to reduce poverty in the developing world.'[24] The main strategy is still that broadly based economic growth generates income opportunities for the poor, but expenditures are analysed from the perspective of poverty reduction, sector assistance is said to be focused on help for the poor (although a fall in agricultural sector aid would appear to conflict with this aim), and efforts are made to monitor poverty by countries. The International Fund for Agricultural Development (IFAD), a UN financial institution, has a specific mandate to combat rural poverty and its concomitant, hunger: 'About 800 million people in the world are persistently hungry and most of them are hungry because they are poor.'[25] Most non-governmental organizations have long directed their assistance towards the poor, and in recent years increasingly towards two disadvantaged groups, women and children. 'No combination of war or natural disaster inflicts suffering or destroys human potential on the scale of the "silent emergency" of poverty . . . Stated in cold figures, the scale of global deprivation retains the power to shock. But . . . they cannot convey the tragedy, for example, of the one-in-six African children who will not live to see their fifth birthday'.[26]

This international action directed towards the poor is not yet matched by a similar level of concern and action by most developing countries. For their governments, the poor do not appear to contribute greatly to economic growth. Ethnic groups, pastoral peoples, and other such minorities in particular are frequently marginalized in national policies. There are worthy exceptions, such as the efforts made by the Indian government to overcome social discrimination against the harijan (untouchable) caste.

Investment in poverty reduction can be an uphill task, not producing the early and direct economic returns of many other forms of aid. The previous view on development was that growth in the overall economy would 'trickle down' to the poor. The new approach, or development paradigm, is that it is not enough simply to get people across an income threshold; there must be 'an integration of the poor into the process of growth'.[27] Population increase, however, is making this objective increasingly hard to fulfil. It is a fundamental tenet of reduction in rural poverty that the poor should have better access to benefits from land; but with one third to half of rural populations already landless, and many existing landholdings already unable to provide support for full-time farming, this basic need becomes impossible to meet. In many parts of the world, the position is currently being made worse through the vicious circle between population increase, poverty, and land degradation (p. 129).

Famine

From biblical times onwards, famine has always been present in history with the same two causes, natural disaster and conflict, primarily drought and war. Because of its variable monsoonal rainfall, India has always been affected and the situation has been relatively well recorded, although estimates of the number of deaths are necessarily very uncertain. Between the Bengal famine of 1769–70, when between a tenth and a third of the population died, and the famine of 1943–4 in the same region, with one and a half million deaths, the Indian subcontinent experienced 10 major famines in 200 years, with total deaths probably in excess of 20 million. Allowing for the non-existence of subsequent multiplication from these lost lives, it could be estimated that recurrent famines have reduced the population by over 60 million, or 5% of the present population of the subcontinent of 1200 million. This is the manner of operation, through untold and inconceivable suffering, of the Malthusian check on population. China has been similarly afflicted, with famines less frequent but deaths on the worst occasions on an even larger scale.[28]

A famine hazard is inevitable in the semi-arid zone, occurring whenever there is more than one successive year of drought, as in the Sahel belt of Africa in 1968–74. Growth of population has also led to recurrent famine becoming endemic in some more humid areas. In the Ethiopian highlands there is high average rainfall and relatively fertile soils; an early traveller reported that the land was so bounteous that the people need only cultivate one year in two to have enough food. From 1970 onward, some provinces, such as Wollo, have become so heavily populated that famine, or a need to bring in food relief supplies, occurs every year that the rainfall is substantially below average. The Red Cross *World disasters report* estimates that, in Ethiopia over the 25 years to 1995, 1.2 million people have died from disasters, mostly famine.[29] This is equal to 2.2% of the population in 1995 - which on present rates of growth is projected to become theoretically 127 million by the year 2025.

The difference in modern times is the effort made, on humanitarian grounds, to prevent starvation by famine relief. Countries themselves frequently ship reserve food supplies to affected areas, and where this is done in a timely and efficient manner it attracts no outside attention. Most of the larger famines of recent years have been brought about by a combination of natural causes with misgovernment or civil war. Some receive wide publicity through the media, such as the disasters of the Sahel, Somalia, Ethiopia and Rwanda. The worst famine of recent years, the 'three terrible years' of 1959–61 in China, was kept hidden from the outside world;

this was of such magnitude as to cause a noticeable check to the growth of world population. In 1997, mass starvation struck North Korea, refugees reporting deaths of the order of one million; in January 1998, to alleviate this disaster, the World Food Programme launched 'the largest food aid programme in its history . . . to distribute to seven million people in danger of starvation.'[30] It has been said, with some degree of truth, that starvation has never occurred in a functioning democracy.

Since the Second World War, major famines have been met by international emergency relief efforts, the UN working jointly with voluntary agencies, and no one would suggest that such actions should not be continued. Population increase is affecting famines in two respects, their frequency and magnitude; in 1995, 25 countries, 15 of them in Africa, received emergency food assistance. In the first place, there are areas where population is so close to the capacity of the land to supply food that a danger of famine arises whenever, through drought, flood, or other natural hazard, food production is substantially below average. Secondly, where disaster strikes, the sheer magnitude of the effort needed to meet it becomes ever greater. Thus, disasters struck the area that is now Bangladesh in 1944 and 1974, with populations of 40 and 80 million respectively. If, as is bound to happen, a flood or other natural calamity occurs there again, the relief operation will have to be of a magnitude to feed 120 million people.

Recently there has been an unintended reversal in aid priorities. Since the early days of foreign assistance programmes, it was a tenet that famine relief was a last resort and aid was more effectively spent on promoting the agricultural development that would avoid the need for it. Pressure of population upon land resources, however, has led to a rise in the number of areas chronically prone to famine, in frequency of local food shortages, and in the magnitude of disasters when they occur. The overriding humanitarian concern means that, where disaster does occur, relief must take priority. As a consequence, an increasing proportion of aid funds are now being forced to go towards famine relief, and so are necessarily diverted from objectives of development.

Conflict

The link between environmental pressures and conflict – war, civil war, and civil unrest – is complex yet potentially strong. The warlike reputation of pastoral peoples is, in part, a response to their need to move to neighbouring lands in time of drought. Post-war conflict over the water of the Indus nearly brought about war between India and Pakistan. A serious potential for war exists over competing demands for the water of the Nile, Mekong, and other major international rivers. The proximate cause of conflict, as in the recent examples of Somalia, Sudan,

Rwanda, Afghanistan, and Haiti, is political, but population pressures are a contributory factor; and where food supplies are already marginal, any disruption of the rural economy leads to hunger. The recent anarchy and genocide in Rwanda has as its apparent causes long-standing ethnic enmities and a recent breakdown of governance, but there is a case to be made that it fundamentally stems from overpopulation.[31]

Perhaps the most serious consequence of the combination of famine and conflict has been the rapid growth in the number of refugees. From 1960 to 1976, the total recorded by the UN remained in the range of 1–3 million. Since 1977, these numbers have risen to over 20 million in 1993.[32] Numbers are approximately equal in Africa, Asia, and Europe (largely from the former Yugoslavia conflict), but proportional to population are highest in Africa, where 7 million, 1 in 100 of the population, are refugees. A refugee group will cut down every tree, and farm all available land, obviously without any possibility of conservation practices. The refugee population must then be largely supported, often for many years, by food aid.

It has been argued that environmental pressures will become the principal cause of conflict in the twenty-first century.[33] We stand on the verge of an entirely new set of security issues, taking the predominant place formerly occupied by the cold war. The link is indirect, the proximate cause being usually political strife or a breakdown in law and order. Countries with good governance can achieve far higher levels of sustainable production and thus of food security – a factor not taken into account in estimates of population-supporting capacity. In political history, one cannot argue in a scientific manner for a direct cause and effect between environmental pressures and conflict; many factors of personalities and chance intervene. But population pressures upon resources, and associated land degradation, will certainly continue to form a dangerous and possibly tragic underlying source of conflict.

The integral nature of population change and land resource policy

An increase in world population to between 7500 and 8500 million by 2025 is inevitable, and further growth to at least 8000 million, and possibly 10 000 million, by the middle of the twenty-first century is highly likely. Almost all of this increase will take place in developing countries. Only strong and sustained efforts, not primarily at international level but by developing countries themselves, could lead to the lower of these estimates. Because of the generational carry-over effect, action taken now and over the next 20 years will strongly influence possibilities for the longer term. Most human suffering affects the poor. They lack purchasing power, go hungry, and have a weak political voice. Their children suffer in physical and mental development from malnutrition.[34]

The arguments linking population, poverty, and land resources may seem extended, but the links are close to the extent of being virtually integral. There is a clear connection, or self-reinforcing nexus, between population pressure, poverty, and land degradation. Most developing countries now have endemic food shortages, and close to one fifth of the world's people are undernourished. Some areas have reached the point where any drought year brings about potential famine. The prospect over the next 20 years is for increasingly frequent and severe famines, placing a huge strain on efforts at relief. The connections between land shortage and civil strife or war are indirect and linked strongly to efficiency of governance, but, on an underlying basis, environmental pressures lead to a strong underlying potential for conflict.

All of these problems are rendered more severe by population increase. There is now an ethically acceptable set of measures to reduce birth rates. It is in the interests of all, the developed and developing world, their governments and peoples, to devote greater funds and efforts to this aim.

All that can be done in the field of land resources – better land use planning, higher productivity, checking degradation, and other measures – could be nullified in its effect by continuing population increase. It is not sufficient to say the development efforts in agriculture, forestry, and environmental conservation should merely be combined with efforts in the population sector. In the wider perspective, policy, planning, and action to check population increase is an integral part of land resource policy and management.

15 Awareness, attitudes, and action

Improvements in land resource management can only come about if they are preceded by awareness of the problems, and recognition of the need for action. At international level, there is a strong measure of agreement on priorities, including poverty reduction, avoidance of land degradation, research, and people's participation in decision-making. At national level, land resource policies cannot be applied in isolation. There first needs to be avoidance of civil conflict, good government, and attention to development of the rural sector, leading to a recognition of the role of land resources. This provides a framework for a set of national land resource policies, including improved survey and evaluation, efforts to combat land degradation, the effective linking of research with extension, a national land use plan, and monitoring of the national heritage of land resources. This will require a strengthening of institutions, with improvements in education and training. None of these measures will be fully effective unless accompanied by greater efforts to reduce rates of population growth; demographic policy is an integral part of rural land development.

Land resources play a critical role in human welfare. Land is no longer abundant, and its productive potential is being reduced by degradation. Sustainability, the combination of production with conservation of resources to meet the needs of future generations, is the key to land management. Whilst a valuable contribution can come from international co-operation, the ultimate responsibility lies with the people and governments of developing countries. Awareness, and with it the will to bring about change, can only come from within.

In discussing different aspects of land resources and their management, some recurrent themes appear in the conclusions. The first is the key role of sustainability, in the sense of production combined with conservation. Sustainable land use is the critical element which links the use of land resources to meet present needs with their conservation for use by future generations. Sustainability should by no means be regarded as a jargon term. It is the most basic concept, which incorporates an essential principle of management, that land resources are to meet the needs of people both now and in the future.

A second theme is the need for better information. This applies to all aspects of land resources, including basic surveys, but the deficiency is greatest in two areas, land degradation and land use. In order to justify action to reverse loss of natural resource potential, there needs to be better knowledge of the extent and severity of land degradation in all its forms: soil erosion and degradation, water resource degradation, deforestation, and degradation of pasture resources. As a basis for national land resource policy, a greatly improved knowledge of existing land use is essential.

Another common element is the close link between research and improved methods of land management. Research does not just mean scientists running controlled trials, no more than improved land management means farmers doing what they are told. There is a continuous spectrum from research scientists via extension services to farmers. Every time farmers try out something different, they are conducting research. Based on feedback from farmers, extension staff should build up experience of what works best in their area. Basic research is a foundation for improved methods, but indigenous knowledge helps in adapting them to the great diversity of local environmental conditions.

People and institutions are fundamental to good management. Governments can supply a political and economic environment. It is also obligatory for them to support the institutions which in turn will help land users. But, ultimately, it is only people who can effectively manage natural resources. Planning and policy should be directed towards improving the incentives and opportunities for communities and individuals to practise sustainable land management, which farmers have always sought to do.

A more controversial but recurrent theme is the pervasive effects of population increase. Land shortage, with resulting competition for its use, is now endemic over large parts of the developing world. If the gains in productivity that have been achieved over the past 50 years had been accompanied by a stable population, living standards in rural areas would have risen many times. In fact, income gains and reductions in undernutrition have been reduced to a fraction of this rate of growth by population increase. Where endemic land shortage is combined with poverty, the vicious circle of overexploitation, land degradation, and falling production is often seen.

A final theme is that the primary responsibility for resources lies with national governments. Valuable as international action can be, through financial aid and technical assistance, there are severe limits to what it can accomplish. Externally funded development projects are restricted in space and time; they cover only a fraction of the country and come to an end, so far as foreign assistance is concerned, after 5 or at most 10 years. If growth is to be sustained, and affect the whole of the rural sector, developing countries need to move forward into an era in which they more effectively manage their own resources. A special role for governments lies in exercising stewardship of resources for future generations. Sustainable management of resources at national level calls for changes in policies, institutions, education, and practice.

Effective action can only be brought about if there are improvements in awareness and attitudes. The first step is for people, institutions, and governments to become aware that problems exist: that forest cutting is exceeding regrowth, for example, or soil fertility is declining. Awareness is of no use if accompanied by a *laissez-faire* attitude, that the problems are unimportant or will solve themselves. There must be a recognition of the seriousness of their effects, and a resolve to do something about them. Once these two critical stages have been achieved, action can sometimes be remarkably rapid and effective, as shown in the rehabilitation of areas devastated by natural disasters or war. Awareness, attitudes, and action will be examined first at international level, and then at the more important level of national governments.

The international context

At international level, there is no lack of awareness of the problems of the rural sector and natural resources. Since its foundation in 1945, FAO has continually sought to give guidance in the form of technical services to agriculture, water resource management, forestry, and fisheries, and to promote improvements in nutrition. It has also been the focus for international inventories of climatic and water resources, soils, and forests. The International Fund for Agricultural Development (IFAD) has the specific aim of improving the food security and well-being of the world's rural poor. The major investment institutions, UNDP and the World Bank, were earlier criticized for neglecting the environmental consequences of development. Since the 1980s, this situation has been much improved, and projects in natural resource management now form a major field of investment. There remains, however, a bias towards evaluating and comparing projects largely in economic terms.

A way of assessing international attitudes is to compare the policy priorities of major recent international statements, four of which have been selected for Table 14. *Agenda 21* is the outcome of the most valuable of all recent international meetings, the UN Conference on Environment and Development (UNCED). *A 2020 vision for food, agriculture, and the environment* represents views of the research community, as the outcome of a conference organized by the International Food Policy Research Institute (IFPRI), but with wide participation. *An agenda for development* has the widest scope of the documents selected; it is a progress report to the General Assembly of the United Nations, with priorities for the future. The *Rome declaration on world food security* was the policy outcome of the second World Food Summit, prepared by technical experts and approved by governments.[1]

Given their differing objectives, there is substantial measure of agreement over international priorities for the rural sector. Reduction of poverty and avoidance of land degradation are emphasized in all four documents, whilst good government,

Table 14. *International policy priorities in agriculture, food and land resources*

	UN: *Agenda 21* (1992)	IFPRI *A 2020 vision* (1995)	UN *An agenda for development* (1995)	World Food Summit *Rome declaration* (1996)
Primary Objectives:	Environment and Develop-ment	Food and Agri-Culture	Economic Develop-ment	Food and Nutrition
The Political environment				
Peace	X		X	
Good government		X	X	X
The Economic environment				
International trade	X			X
National policies		X	X	X
Social justice				
Reduction of poverty	X	X	X	X
Importance of the rural sector			X	X
Sustainable land management				
Avoidance of degradation	X	X	X	X
Research and technology transfer	X	X		
Efficient inputs and markets		X		
People's participation	X			X
The role of women	X			X
Demographic policies	X		X	
Strengthening of institutions	X	X	X	
National responsibility	X			X

Absence of a symbol does not mean that an aspect is ignored, but that it has not been chosen for emphasis at the level of principles or summary.

national macro-economic policies, and strengthening of institutions appear in three. Nearly all the aspects which are not itemized at the highest level appear in the supporting detailed accounts of all four documents. An international economic environment of relatively free trade, based on the Uruguay round of the General Agreement on Tariffs and Trade (GATT), forms the international complement to market-based national economic policies. Where formerly regarded as undiplomatic to mention openly, international advocacy of good government, meaning efficient, responsible, and incorrupt government and institutions, has recently become widespread.

International aid

Whilst there is a high level of awareness and a desire to assist among institutions, the international contribution to rural development is severely hindered by the low and recently falling level of funding. This applies in some degree to bilateral aid, assistance of one country by another, but particularly to multilateral aid, the funding of international organizations such as the UNDP, World Bank, and FAO. Total international aid, or net official development assistance, is presently close to $70 billion. In the earlier part of the development era, the nominal target was 1% of gross national product, and many countries exceeded 0.75%. The current UN target of 0.7% of GNP 'has become increasingly illusory'.[2] In recent years it has been exceeded only by Denmark, Norway, Sweden, and the Netherlands; next in order of generosity are France, Finland, and Canada. The average for the major donors (OECD countries) is 0.3%, with 11 countries falling in the range 0.35–0.25%. Until recently, the United States was the largest donor in absolute terms (although the smallest as a proportion of GNP), but in 1993 it was overtaken by Japan.[3]

Development budgets for international organizations are extremely small as compared with the budgets of governments and commercial organizations in Western countries. The Director-General of FAO recently observed that his total budget for activities in world agriculture and food was one fifteenth of the amount spent on slimming products in 'a single country'. International research to improve rice production receives one tenth of the agricultural research budget of France or the UK.

Furthermore, external assistance to the agricultural sector has fallen considerably in recent years. Expressed in constant 1990 dollars, after reaching a peak of $18–19 billion between 1982 and 1986, it has fallen to $10 billion. This is channelled in approximately equal amounts through the World Bank, bilateral assistance, and other intermediaries including the regional development banks. Non-governmental organizations (NGOs) receive in excess of $2.5 billion, within which the rural sector receives a substantial share.[4] One reason given for the fall is the generally poor record of the agricultural sector in post-project evaluations. Disaster relief is also now taking an increasing proportion of scarce aid funds. Out of a United Nations budget of $5.2 billion in 1992–3, $2.5 billion was spent on humanitarian assistance and disaster management.[5]

There needs to be recognition of the wider role of agriculture in national economies, in reducing the problems stemming from urban migration for example (in economic terms, an externality). More hopefully, the fall in assistance to agriculture may be linked to a rise in assistance directed at reducing rates of population growth, directly through family planning and reproductive health services, or indirectly through programmes such as poverty alleviation and better education for women.

All institutions know that real progress calls for a rising budget. However, barring a radical change in attitudes, no substantial increase in aid funds from developed countries can be expected. People respond readily to charitable appeals for disaster relief, and some make regular donations to non-governmental organizations. But the greater proportion of aid funds comes from governments, and the same people regard paying taxes in quite a different light. Western governments are presently unwilling to do more than maintain the low proportions of their budgets allocated to foreign aid. In default of official assistance, attention is turning to the role of the private sector, but opportunities for profitable investment are least likely to be found in rural areas.

Much can be gained by more efficient use of funds, reducing the waste, corruption, and development failures that were common in the past. Sound government and efficient administration are prerequisites for this, but enough has been said to make it clear that a sound knowledge of resources, matching land use to land and avoiding environmental hazards, can contribute greatly.

The greater part of official aid comes not as gifts, but nominally in the form of loans. There has always been an air of unreality to this practice, the supposition that countries will one day repay these loans. For the great majority of poor countries there is no possibility that this will be done. By 1993, 26 developing countries had external debts in excess of $10 000 million; in many of them, this debt exceeded the annual gross national product. In 20 countries, debt service, or paying the interest on these loans, exceeds current borrowing, meaning that these countries are locked into a spiral of increasing debt. This situation makes it increasingly difficult to channel development assistance to areas where expenditure by government is most needed, such as rural infrastructure and conservation.

National priorities for land resource policy

The national context

Valuable as international effort is in providing technical and economic assistance, it is far less important to resource management than policies and action at national level. The sustainable management of land resources is complex; it requires consistent policies, translated into action which is maintained over a long term. This can only be achieved within a supportive national environment. The prerequisites for an effective land resource policy begin with the political and economic context, leading towards a recognition of the role of the rural sector.

There are two elements to the political context, avoidance of civil conflict, and good government. Just as, internationally, peace is the foundation for welfare, so, at the national level, avoidance of civil conflict is the priority. In recent years, a high proportion of world conflicts have been civil rather than international. Besides

NATIONAL LAND RESOURCE POLICY: A HIERARCHY

The international context
- Peace
  - A market-orientated system of trade
  - International co-operation and assistance.

The national policy framework
- The political context
  - Avoidance of civil conflict
  - Good government
- The economic context
  - Appropriate macro-economic policies
  - Attention to development of the rural sector
- Recognition of the role of land resources.

National land resource policies
- Improved knowledge: survey and monitoring
- Efforts to combat land degradation
- Research, linked with extension, into land management
- A national land use plan
- Effective environmental legislation
- Monitoring of the national heritage of land resources, including environmental and economic accounting
- Linking of scientific information with social and economic analysis

founded upon
- Strengthening of institutions
- Improved education and training

and integral with
- Demographic policy.

actual destruction of the rural infrastructure, the existence of conflict precludes investment and enforces short-term, often destructive, land management. In 1997, almost all the armed conflicts in the world were within, not between, states. Military spending by governments of developing countries is twice the total amount of development aid.

The second element of the political context is good government (also referred to as 'governance'). Besides the ministerial government and politicians, this covers the civil service, parastatal organizations, and private companies which have sometimes taken over their responsibilities, such as commodity marketing boards. Good government implies that national bodies are responsible and answerable to the people. It involves a relatively low level of corruption, which in turn requires that civil servants should be adequately paid, and receive promotion from doing their job well. 'Improving and enhancing governance is an essential condition for the success of any agenda or strategy for development . . . It means ensuring the capacity, reliability and integrity of the core institutions of the modern State . . . accountability for actions and transparency in decision-making.'[6] In practice, good government is usually based on a functioning democracy. Political changes which bring about better government are invariably followed by striking improvements in economic development, agriculture, and food security; among recent examples are Bangladesh, Chile, Ethiopia, Indonesia, the Philippines, and Uganda.

The economic context is founded on macro-economic policies. Governments are not the primary agents in economies, but provide the framework of price structures and institutions within which economic progress can take place. Details of such policies lie beyond the scope of the present discussion, but elements include avoidance of excessive inflation, a market-price currency and hence exchange rate stability, and non-distortion of market prices, although a degree of protection and subsidy is recognized to be necessary for countries at an early stage of development. Also to be avoided are policies which subsidize urban consumers at the expense of penalizing farmers. More generally, economic policies must be such as to provide incentives to producers. The structural adjustment programmes of the 1980s onward, implemented through the World Bank, had as their objective economic reforms directed towards a more open and competitive economic environment.[7]

A further element in the economic context is that governments should recognize the special role of the rural sector in developing-country economies. Burkina Faso, Costa Rica, India, South Korea, and Thailand are examples of countries which have achieved advances in agricultural productivity and food security through recognizing the role of the rural sector in overall economic development.[8] The argument for this key position is complex, but includes, first, a link with the poverty reduction, and, secondly, the avoidance of economically enforced migration ('push' migration) to cities. There also needs to be recognition that measures for development cannot be confined to high-potential rural areas. Indeed, since the high-potential areas are most easily made productive by private efforts, governments should give special attention to low-potential areas, those with low or unreliable rainfall, soils with fertility limitations, and environmental hazards. Some of these areas are occupied by pastoral peoples or ethnic minority groups.

These political and economic conditions lead towards the focus of the present argument, that governments should give greater recognition to the role of land resources. The sustainable use of natural resources is central not only to the welfare of people in rural areas, but to national economies as a whole. Such recognition is the final element in the hierarchy of political and economic prerequisites for policies on land resources.

National land resource policies

Provided that the contribution of land resources to national welfare is recognized and there is a wish to ensure their sustainable use - once there are the right awareness and attitudes - then governments are in a position to take action. This must be promoted through a set of national land resource policies, with steps for their implementation. Details have been outlined in previous chapters.

There should be a better knowledge of the national land resource base, its changes over time, and its productive potential. This can be obtained through improved survey and inventory, monitoring of change, and land evaluation. Knowledge of resource potential forms the foundation for strategic planning of development, including project identification, and provides a framework for more detailed local studies where they are necessary. Survey and monitoring will form the physical basis for an ongoing national resource inventory, showing both losses to degradation and land improvements. A further necessary element is the ongoing survey of changes in land use, to provide much needed data on the extent of arable land, permanent crops, forest cover and its utilization (for production, protection, etc.), grasslands, and, of much importance, the extent of urban and industrial land.

There should be a national land use policy, comprising a statement of objectives, policy guidelines, and a plan of action. It should set out the respective roles and obligations of government, institutions, and individuals, and the scope of actions needed to fulfil these. Among the policy guidelines are likely to be:

The government, on behalf of its people, will seek to increase production, based on land suitability for different kinds of use and on the needs of the people.

It will seek to give all sections of the community, including the poor and minority ethnic groups, appropriate access to land resources, whilst protecting these from exploitation.

It will encourage participation by communities, farmers, and other land users in decision-making on matters of land use planning and land management.

Recognizing its obligation to protect the land resource heritage for future generations, the government will take measures to promote sustainable land use.

Among the policy objectives included in the national land use plan should be a commitment to combat land degradation in all its forms: soil erosion and fertility decline, water resource degradation, deforestation, and pasture resource degradation (in dry zones, desertification). Efforts must be made to improve knowledge of the extent and severity of degradation, and its effect upon productivity, but action should not be held back by uncertainties over this.

There needs to be a stronger national system of applied and adaptive research, linked with extension, into methods of sustainable land management suited to the environmental conditions and productive potential of the country. Depending on the economy, this will cover crop production, livestock production, forestry, fisheries, and management of the nation's water resources.

As an element in the national land use plan, the system of environmental legislation should be revised and made effective. Its scope should be restricted to areas where it is most needed, such as protection of sensitive watershed areas, or planning controls for urban encroachment upon agricultural land. Elsewhere, for example in checking soil erosion, approaches other than legal compulsion should be employed. This element recognizes the special role of government in conservation of resources.

The government should institute a programme to monitor changes in the national heritage of land resources: water, landforms, soils, forests, pastures, and biological resources. Monitoring will be conducted initially in physical terms, such as soil condition, forest area and status, groundwater and river flow, and the condition of grazing lands. These data will also be converted to economic form, in a system of environmental and economic accounting.

Consistent with the statements of principle on the needs of the people, all the activities related to land resources should take account of social and economic considerations (just as economic policy should involve resource aspects). Whilst there are difficulties in practice, every effort should be made to integrate physical and economic aspects. The objectives of land resource planning and management are to encourage sustainable land use, which requires:

promoting land uses which are suited to the resource potential of different areas, economically viable, and socially acceptable;

conserving the resource potential for the needs of the future.

The statements of principles will be translated into a proposed plan of action, specific to the land resource issues of the country. Guidelines to governments on necessary action are set out in *Agenda 21*, particularly chapter 10, 'Integrated approach to the planning and management of land resources', and the succeeding chapters on combating deforestation, managing fragile ecosystems, and promoting sustainable agriculture.[9]

Strengthening of institutions

This vision of national land resource policies and actions may seem highly unrealistic to those who are familiar with the institutions in developing countries. With respect to the present situation, they are right. Forestry departments, soil surveys, agricultural research services, and other land resource institutions are generally understaffed and receive weak support. Often they have capable senior staff but a totally inadequate operating budget, so that even basic necessities such as vehicles and equipment are lacking. One international review, *A 2020 vision for food, agriculture, and the environment*, places as its number one priority, 'Strengthening the capacity of developing-country governments to perform their appropriate functions.'[10]

It is impossible to generalize about institutional structures, since governments have a number of equally effective ways of arranging these. Some aspects of land resource policy will be primarily the responsibility of sectoral agencies, for example forest monitoring by the Forestry Department, agricultural research as a branch of the Ministry of Agriculture.

There needs to be an institution to act as a focus for land resource inventory, monitoring and planning. This could be developed from the former soil survey, transformed into a natural resources institute with wider responsibilities. Alternatively, a new natural resources or land use planning authority could be established, of which the soil survey would become a branch. This central body would itself conduct studies of climatic, soil, and vegetation resources, and also, most importantly, of land use. It would also act as a co-ordinating centre for data collected by sectoral departments such as forestry and water resources. Survey of pasture resources, including, in dry lands, monitoring of desertification, could either be made integral with soil survey activities, or made the responsibility of a body such as a range management division or drylands development authority. The central land resource body would incorporate, or develop close links with, existing institutes for soil conservation and reafforestation. It would also act in liaison with national parks authorities and conservation authorities.

There need to be better linkages between survey, monitoring, and research, on the one hand, and field extension work in agriculture and forestry, on the other. This will require changes in attitudes, institutional structures and procedures. Soil survey, agricultural research, and similar activities should not be regarded as scientific luxuries, but should become an integral part of extension and rural development. Staff from the land resource institutes would be seconded to development projects, not only during planning but during implementation. Extension services should include regional or district technical advisers, for example in soil conservation, agroforestry, and pasture management, to serve as the link between research and land management.

The staffing of these bodies should not be limited to natural scientists. In order to achieve more effective integration with economic analysis and planning procedures, their staffing should include social scientists. In plain terms, one or more economists should be included in the staff of land resources survey organizations, together with other social scientists such as farm systems specialists where appropriate.

A difficult problem arises with respect to a national land use planning authority. Land use planning requires good advice on resource potential, yet its major decisions are matters of high-level politics. Some form of multi-level structure will be needed, for example a technical advisory board, which reports to a high-level committee with ministerial representation, such as a natural resource planning committee. A land use planning authority is needed to handle day-to-day decisions, such as applications for urban extension.[11]

Institutional capacity rests upon three foundations: capable staff, adequate funding, and an appropriate place in decision-making. Strengthening this capacity therefore requires improvements in education and training at all levels, professional and technical.[12] Developing countries differ greatly in this respect, but two widespread needs may be noted. First, changes in approaches and methods for land resource management call for re-education of existing staff; examples include guiding soil conservation staff in the approach of land husbandry, or forestry staff in multiple-use management. Secondly, research and planning is only effective if it can be implemented by an efficient cadre of technical staff, operating at the district level, a level of training that is often neglected.

The integral nature of national demographic policy

> To achieve sustainable development and a higher quality of life for all people, States should eliminate unsustainable patterns of production and consumption and promote appropriate demographic policies.[13]

> Unsustainable and unsupportable population growth can have adverse effects on development efforts globally. These effects in turn have profound implications for the use of natural resources such as water, wood, fuel and air. They affect the ability of Governments to supply the basic services that people require.[14]

These views, the first forming Principle 8 of the Rio Declaration of the UN Conference on Environment and Development (UNCED), the second from the UN *Agenda for development*, are among many statements on the link between population and environment which have increasingly been made in recent years. For reasons of

FIGURE 30 Action is needed now for the welfare of future generations.

diplomacy they are expressed as 'demographic policy'. What this refers to, above all, is an effort to reduce the rate of population increase. This must be done by means which are ethically, socially, and politically acceptable, central to which are provision of family planning services and improved education of women.

If this view has only recently come to be expressed openly at international level, it is still more sensitive as a national political question. It would be unrealistic and politically unthinkable for any country to propose a desirable maximum population. The most that can be realistically hoped for is to agree on a target date for stabilization of population. A very few developing or former developing countries, Singapore, Thailand, South Korea, Uruguay, and China, have reached the point in the demographic transition when stabilization is in sight, after the increase deriving from past high birth rates has been passed. For most countries, the earliest at which a stable population could conceivably be attained is 2050, and for most there will still be some increase to come.

A reduction in the rate of population growth is absolutely essential to sustainable land resource management. If it is not achieved, improvements in knowledge, planning and methods of management will constantly be counteracted by the decline in the ratio between population and resources. This has been seen throughout Africa in the 1980s, with rises in total national wealth, but falls in per capita income. For once, statistics are consistent with what can be seen on the ground, in the form of ever-decreasing farm areas. The nexus between population increase, land shortage, poverty, and land degradation is the most serious expression of this problem.

Population policy is an integral part of rural development. It would be impractical actually to include measures such as family planning clinics in agricultural development projects. What can be done more realistically is to take account of population change, and measures to influence it, in planning at national level. In many countries, giving priority to population-related services over agricultural investment could be justified, both economically and in terms of longer-term welfare. Having recently gained international recognition of the integral nature of demographic policies and development, the more difficult task of achieving this at national level remains.

The need for action

Action by the international community

Motivation for action by the developed world covers development aid as a whole, through international institutions and directly to the developing countries. A wider recognition of the role of the environment as a productive resource, as compared with pollution aspects, is still needed.

Reasons why development aid should be increased above its present low level have often been rehearsed, notably in the Brandt and Brundtland Reports. The arguments are currently meeting with little success. They are of two kinds, ethical reasons and self-interest. The ethical argument, whether based on religious belief or otherwise, is that it is wrong for Western peoples to continue living their generally comfortable and often affluent existence without making efforts to reduce poverty and malnutrition. It is based on the huge and increasing gap in income levels and rates of resource use between rich and poor peoples. The ethical argument stands or falls on whether people feel a moral obligation to help those less fortunate than themselves.

One possible reason derived from self-interest is that food would become in short supply world-wide, and rise in price. This aspect was first raised following the shortages of the Second World War, and has been revived from time to time on grounds of population increase. On the basis of the self-interest of Western nations, this argument is not valid. Other than through war, food will not run short for them; money will give them first choice from the temperate-zone exporting nations, and the ability to buy from the tropics.

Hence, the argument of self-interest rests on the concept of 'only one world'. It is that poverty and unrest in one part of the world will indirectly affect welfare as a whole. Civil war is now frequently met by international efforts to bring about peace, and to alleviate the suffering that it brings, particularly in the form of refugees, by disaster relief. Wars are the result of states seeking to widen their area of control; conflicts over environmental resources have been a cause in the past and may become increasingly so in the future. Atomic weapons will inevitably become more widely available, and the worst scenario is that sooner or later they will be used in warfare.

The direct threat of armed invasion of rich nations by poor presently appears remote, but the position could appear different if the wealth gap is bridged, for we shall soon live in a world in which the present developing countries outnumber those now developed by six to one. The second part of self-interest lies in the strong pressures for immigration from poor nations to rich. In places these have reached the point of continued attempts at illegal immigration, such as along the USA–Mexico frontier, by sea from North Africa to southern Europe, and, formerly, by the refugee boat-people from Vietnam. It would reflect better on human nature if ethical reasons proved more forceful than arguments based on self-interest.

Action by national governments

Why should governments take action to promote the rational use and sustainable management of their land resources? In positive terms, the answer is that this is an essential element in improving the welfare of their people, not only in rural areas, but to benefit the country as a whole. In negative terms, reasons are that, if such action is not taken, then countries will suffer to differing degrees from:

increasing problems in meeting food needs, both as chronic shortages and increasingly frequent and severe famines;

shortages of fuelwood, water, and other products and services from forests;

continued land degradation, lowering the productive capacity of resources, including responses to inputs;

continued and increasing rural poverty;

a push-induced migration of the poor to cities, leading to unemployment and problems of health, and law and order;

ultimately, if problems continue and become more serious, civil unrest, possibly leading to overthrow of government.

A summing up

'We must be careful not to succumb to despair, for there is still the odd glimmer of hope.'[15] For a statement such as this to be made, in 1993, by the retiring Director-General of FAO, there must be a serious need for improvement.

Land resources play a critical role in human welfare, supplying not only food, but fuel, water, and other necessities. All countries need to promote better management of this resource heritage, but the need is greatest in the developing world. There must be better information, more research, and measures to bring about improved management. Sustainable land use, meeting the needs of the present whilst conserving resources for the future, is the basis of good land resource management.

Land is not still abundant. Past estimates of land available for cultivation have neglected the many other necessary and competing uses for land. A high proportion of the land not yet under productive use possesses limitations which reduce its potential and means that there are hazards of degradation.

Land degradation is widespread. It takes many forms, including soil erosion and fertility decline, loss of water resources, forest clearance and degradation, and reduction of the productive capacity of pastures. For two aspects, erosion and desertification of dry lands, good information is lacking on the exact degree of severity, but this should not be a reason for failing to take action. The appropriation of resources for production and settlement has also led to great losses in biodiversity. The causal link between population growth, food needs, rural poverty, and land degradation is in danger of reaching crisis point in one country after another.

Continued increases in population can nullify improvements in land management. Rates of increase are falling, but a continuation of present trends will not be sufficient to avert serious pressures in many regions. There needs to be greatly increased awareness of the dangers of this position, leading to action to bring rates of population increase below those of the current best estimates. Action to check population increase is an integral part of land resource policy.

Much has been done internationally, in awareness, aid, and technical assistance. A range of methods to assist developing countries with land resource management is available, but international institutions for land resource development are handicapped by their low levels of funding. The international scene is marked by good technical co-operation, but low levels of development aid. A change in this situation could only be brought about through reasons of ethics or self-interest. As expressed by a former Secretary-General of the United Nations, 'Preserving the availability and rationalizing the use of the earth's natural resources are among the most compelling issues that individuals, societies and states must face.'[16]

But sustained action to promote better resource management can only come about through the efforts of developing countries themselves. There needs first to be awareness of the role of land resources, followed by recognition of the need to improve their conservation and management. At present, most developing countries fail to recognize the importance of preserving their national heritage of land resources. Only when there is a change in awareness and attitudes will the conditions be met for action to promote sustainable land management. International co-operation can assist, but the ultimate responsibility lies with the people and governments of developing countries. Awareness, and with it the will to bring about change, can only come from within.

Notes

1 Concern for land

1 Brown and Kane (1994).
2 Brandt Commission (1980); Brundtland Commission (1987).
3 The official document is UNCED (1992); a condensed version has been published by Sitarz (1993).
4 In *Agenda* 21 (UNCED, 1992), land resources occupy chapters 10–14 and 18, pollution chapters 19–22, biological diversity chapter 15, and the atmosphere and oceans chapters 9 and 17 respectively.
5 World's Scientific Academies (1993), 1.
6 FAO (1996a). In the three-page *Declaration*, the contribution of women is mentioned five times; the adjective 'man-made' appears twice: man-made disasters and man-made emergencies!
7 In the *Plan of action*, Commitment 3 (out of 7) is for governments to pursue participatory and sustainable policies; and Objective 3.2 is to combat environmental threats to food security, including land degradation (FAO, 1996a, 17, 20). In the 15 technical background documents (FAO, 1996b), the role of water forms the whole of Document 7; but in Document 11, on environmental impact, soil degradation occupies less than one page.
8 The Publications cited are, in the order listed: FAO-UNESCO (1970–80); FAO (1976); Doorenbos and Pruitt (1977); FAO (1978–81); Lanly (1982) and FAO (1995a); Chambers (1981); FAO (1982a, 1984b); Shaxson et al. (1989); Magrath and Arens (1989); Pearce et al. (1990); Solorzano et al. (1991); Cruz and Repetto (1992); Lutz (1993); Lutz et al. (1994); Oldeman et al. (1990); Turner et al. (1991); Jazairy et al. (1992); FAO (1993a); Alexandratos (1995); Pieri et al. (1995).
9 The distinction between developed and developing countries has recently become less clear-cut, with some overlap in income levels. In particular, some of the central Asian republics of the former USSR should be considered as developing countries. In the classification employed for FAO statistics (up to 1992, when the distinction was discontinued), the developed countries are taken as Europe, the United States, Canada, Australia, New Zealand, Japan, and Israel, together with the 15 countries of the former USSR (in earlier years called 'centrally-planned economies'). Where statistical comparisons are made, this is the base that has been taken. In the text, 'developing countries' refers to low-income countries in a more generalized sense. To reduce repetition, 'the developing world', 'less-developed countries', and 'the Third World' are employed as synonyms.

2 Land resource needs

1 In terms of the most widely used climatic classification, Köppen, the humid zone covers classes Af, Am, Cf and Cm, the subhumid zone classes Aw and Cw, with the bimodal rainfall area identified as Cw″, and the dry lands are BS (semi-arid) and BW (arid). On the basis of length of growing season for annual crops, the humid zone has over 270 days, the subhumid zone 75–270 days, and the dry lands 1–75 days in the semi-arid zone and no growing period in the arid zone.
2 That is, nutrient losses from the soil-plant ecosystem are less than 10% of the nutrients recycled between soils and plants; examples are given in Young (1989, chapter 10). It was an early classic work, *Les pays tropicaux* (Gourou, 1946), which first set out the fragility of this environment once the forest cover is removed.
3 For estimates of soil rest period requirements, for sustainable use, under major climates and soil types, see Young and Wright (1980) and FAO (1982a, 20).
4 These are not identified as a major soil type in either the FAO or US classification systems. An attempt to distinguish them as 'weathered ferrallitic soils' was not widely adopted. They have also been called 'leached, pallid soils' and 'plateau sandveld soils' (Young, 1975, and 1976, 142).

5 As Chapter 12 of *Agenda* 21 (UNCED, 1992).
6 In an astonishing passage, Plato wrote that in Attica (Greece), there were once abundant trees and rich soil on the mountains, supplying timber and feed for animals. Only remnants of these trees survive today [4th Century BC]. Rainfall was stored in the soil and fed springs and rivers, not running off barren ground as it now does. By comparison with the former landscape, only the skeleton of a body is left. I have paraphrased, but these remarkably perceptive observations are all present (Plato, *Critias*, Section IIIb-d, quoted by Harrison (1992), 114).
7 As chapter 13 of *Agenda* 21 (UNCED, 1992).
8 Hugh Brammer, personal communication.
9 Australia is of special interest in showing tropical environments, from humid to semi-arid, managed wholly within the context of a developed, high labour cost economy. Among developing countries, China carries this socio-economic situation furthest into temperate regions. The future course of the dual economy of South Africa, with rich and poor economies superimposed in the same country, will be of much interest.
10 Agricultural Advisory Council (1970).

3 **Resource survey and land evaluation**

1 An exception is Karen Blixen's coffee estate, which failed not because it was too cold, as the film *Out of Africa* implies, but too dry.
2 The earliest tropical soil maps were of the Dutch East Indies (Indonesia), the earliest in 1901. Elsewhere, most date from 1928 onwards, see Young (1976), 328.
3 Wood (1950).
4 Dent and Young (1981), 97.
5 For a discussion of soil survey procedures, see Dent and Young (1981).
6 FAO (not dated, *c.* 1960).
7 The natural classification school was led by European scientists, the artificial by Australians. It was written at the time, 'Scientists who are otherwise reasonable and unemotional are liable to behave quite differently when discussing this topic' (Mulcahy and Humphries, 1967). For a discussion of soil classification systems see Young (1976), 235.
8 Soil Survey Staff (1975, 1992).
9 FAO-UNESCO (1974); FAO (1988a); Spaargaren (1994).
10 The extent of the 30 major soil groups in 8 world climatic zones is given in FAO (1991c), 30.
11 Buol and Couto (1981).
12 The basic rule, that a soil map was of no use unless the internal variance of soil unit A plus that of unit B is less than the variance of the area as a whole, was first stated by Beckett and Webster (1971). For a discussion of soil variability in relation to mapping, see Dent and Young (1981), 92.
13 Young and Goldsmith (1977).
14 The pioneering work was Christian and Stewart (1953). It was followed by 40 surveys of Australia and Papua New Guinea. For accounts of the land systems method see Christian and Stewart (1968); Young (1968); and Dent and Young (1981, chapter 7).
15 In England and Wales in the 1960s, conventional soil survey was found to be proceeding at a rate which would have taken 150 years to complete. When a landscape approach was adopted in the 1980s, coverage of the country was accomplished in 10 years.
16 Young and Brown (1962, 1964).
17 Hendricksen (1984).
18 Zinck (1994), 4.
19 Moss (1968).
20 An unusual opportunity arose when a senior soil scientist realized that samples taken for his PhD were still held in storage, with notes on their exact sampling locations. Resampling 40 years later showed a trend towards acidification of soils in Scotland, although effects of acid rainfall and afforestation with conifers could not be separated from possible long-term post-glacial soil development (Billett et al., 1988).
21 Young (1991).
22 Climatic classifications are reviewed in Young (1987), 43.
23 Doorenbos and Pruitt (1977); Doorenbos and Kassam (1979); FAO (1993c).
24 The growing period refers only to water availability for annual crops, in the top 2 m of soil. Trees stay

green and continue to grow outside this period, an important reason for the potential of agroforestry.

25 FAO (1978–81); databases of soil water balances and growing periods are now available.
26 Rodda (1995), 360 and 367.
27 FAO (1993b), 6.
28 A comprehensive account of vegetation mapping is given by Küchler and Zonneveld (1988); for mapping at international scale, see Young (1987), 53.
29 Verboom and Brunt (1970).
30 FAO (1991a).
31 Hendricksen (1984).
32 ISRIC (1993).
33 FAO (1994b).
34 FAO (1976).
35 FAO (1983a, 1984a, 1984d, 1991a); Young (1993b).
36 For technical purposes an artificial term, land utilization type, is employed, defined as any system of land use, existing or potential, which is taken as the subject of land evaluation. A land utilization type is an idealized description, in contrast with what is actually practised on the ground, called the land use. In this book, land use type is employed in both senses.
37 Young (1985).
38 Rossiter (1995).
39 Smyth and Dumanski (1993).
40 FAO (1976), 4.
41 Kassam et al. (1993).
42 Taken as a set, the Land Resource Studies are one of the greatest contributions to geographical knowledge ever made by the UK. For a review, see Young (1978).
43 The use made of information from land resource surveys is reviewed by Dalal-Clayton and Dent (1993).
44 Examples of the effective integration of land resource information with economics and development planning are found in four of the Land Resource Studies produced by the UK Natural Resources Institute (formerly the Land Resources Development Centre), feasibility studies of the development in Nepal (Land Resource Study 17), Ethiopia (21), Gambia (22), and Belize (24).
45 Dalal-Clayton and Dent (1993), 4.

4 Competition for land

1 Sombroek and Sims (1995), adapted.
2 FAO (1986a); FAO (1993b).
3 This section is based mainly on Young (1994c) and UNEP/FAO (1994a), in which many further references to sources will be found.
4 Korotkov and Peck (1993).
5 Anderson et al. (1976).
6 UNDP-FAO (1994).
7 EEC (1993).
8 A healthy attitude to statistics was taken by the British politician Dennis Healy. During the Second World War, he was stationed at Swindon and ordered to count the service personnel leaving by train, arriving by train, and arriving and leaving again. With six platforms this was impossible, so he counted the number leaving, made up the number arriving and leaving, and asked the ticket collector to count the number arriving. 'After a few weeks I discovered he was making up his numbers as well. This gave me a life-long scepticism about the reliability of statistics, which served me well when I became Chancellor of the Exchequer. It also taught me never to take for granted any information I was given unless I was able to check it from another source' (Healey, 1989, 50).
9 Quotations in this sections are from 'Notes on the tables' in the *FAO production yearbook* (FAO, annual, a) and *World resources* (World Resources Institute et al., biennial).
10 Alexandratos (1995), 161, 166.
11 As FAOSTAT, formerly AGROSTAT (FAO annual, c).
12 Heilig (1994), 846–7.
13 FAO (1996b), Documents 1, page 29, and 11, page 17.
14 Rodenburg (1993). The joke is abstruse, requiring both literary and environmental knowledge. Easier to appreciate is the title of a data review by the International Institute of Environment and Economic Development, *O.K., the data's lousy, but it's all we've got* (Gill, 1993).
15 The official figure for world urban population in 1995 is 45%, but this, 'is best considered not as a precise percentage . . . but as being between 40 and 55 per cent, depending on the criteria used to

define what is an "urban centre".' (Habitat, 1996,) 14.

16 Reviewed in Habitat (1996), UN (1995b) and *Part I: the urban environment* of *World Resources 1996–97* (World Resources Institute et al., biennial).

17 Potts (1995).

18 Ghosh (1984); Lowry (1991); UN (1995b); and Habitat (1996).

19 Bhadra and Salazar (1993) and sources cited therein.

20 Grübler (1992); an earlier estimate gave 200 M ha (Buringh and Dudal, 1987).

21 Alexandratos (1995), 152–6.

22 The Lilongwe Series, and its catenary associates (Young and Brown, 1964).

23 Sombroek and Sims (1995), 11.

24 There are many fine collections of views of the world from space, for example Sheffield (1981).

25 Turner et al. (1991).

26 One cannot update these data to the 1990s without statistical juggling, since the FAO values for 1980 are 6% lower for world cropland and a mere 17% lower for forests.

27 FAO (1982b). It was paralleled by a UNEP version, called the *World soils policy* (UNEP, 1982).

28 UNEP (1995). Putting its own precept on the need for communication into practice, it commissioned a simplified version, *Down to earth* (Lean, 1995).

29 FAO (1985a). The original name, 'Tropical forests . . .' was later changed to 'Tropical forestry . . .'.

30 UNCED (1992), 389, This statement is given as a separate annex, rather than as a chapter in the main text, presumably for some reason of internal politics.

31 UNEP/FAO (1994c).

32 UNEP/FAO (1994b).

33 World Bank (1993a, 1995a).

5 Working with farmers

1 In 1970, the University of East Anglia joined forces with a consulting company in a land development project in Johor, Malaysia. The plea by the University to include one sociologist was accepted only with difficulty, and he was initially set to work counting population.

2 FAO (1980, 1988c).

3 Sundaram (1985).

4 Scoones and Thompson (1994), 104.

5 Collinson (1981, 1987).

6 A farm system has also been called a farm-household system, partly in recognition of the large proportion of farms in which off-farm activities of the household contribute to its livelihood.

7 There is no better collection than in Ruthenburg (1980).

8 Also called rapid rural appraisal (RRA); Chambers (1981), Longhurst (1981), Chambers et al. (1989), Hudson and Cheatle (1993), Scoones and Thompson (1994).

9 The original statement of diagnosis and design for research in agroforestry is Raintree (1987). The procedures were adapted to the general case of land systems improvement by Young (1985).

10 Young (1985).

11 Chambers (1993); Hudson and Cheatle (1993); ABLH (1995).

12 The Gusau Agricultural Development Project (ADP) was one of a series of World Bank ADPs in Nigeria. I obtained the information given from unpublished documents when working on the Sokoto State ADP which succeeded, and enlarged, the Gusau project.

13 Richards (1985); Davis (1993); Scoones and Thompson (1994).

14 Critchley et al. (1992); Kerr and Sangh (1992).

15 Agarwal (1994). For a reasoned discussion of women's position in the land-population-environment debate, see Arizpe et al. (1994).

16 Hudson (1992), 57.

17 World Bank (1994).

18 The first two examples are from personal information from Jules Pretty, the Haitian case from my own observation. A small UNDP/World Bank project costs of the order of $10 million, an NGO-based village participatory project $100 000 - $500 000. The total budget of NGOs is less than 5% of all development assistance.

19 In former times, well-structured advisory services, locally staffed at the field level, were operated (on very low budgets) in British colonial territories.

20 Benor et al. (1984); Swanson (1984).

6 Land use planning

1 But by no means all agricultural development projects are in land use planning; improved marketing facilities or veterinary services, for example, are basically planned by sector-specific procedures, although they certainly have impacts upon land use.
2 A difference of terminology exists. Land evaluation means assessing what is likely to happen in advance. In the project cycle, evaluation refers to assessing the success of a project during and after its implementation.
3 UNCED (1992), 84.
4 Sombroek and Sims (1995).
5 Projects to large countries, such as India or Indonesia, or to a province of China, have a budget of more than $150 million, plus an equivalent contribution from the national government in services and local currency.
6 FAO (1986b); Easter et al. (1986).
7 Sundaram (1985).
8 Sources: FAO (1991b); Kwakernaak (1995); World Bank (1993a); World Bank project listings; and personal experience.
9 FAO (1991b), 97.
10 J. Pretty, personal communication.
11 FAO (1988b).
12 Quoted from memory from a World Bank evaluation report, c. 1983.
13 Kwakernaak (1995), 26.
14 Beatty et al. (1979); Cloke (1989).
15 Cocks et al., (1986).
16 Patricros (1986).
17 Van Staveren and van Dusseldorp (1980).
18 FAO (1993a).
19 Information from A. Savary, Publications Promotion, FAO.
20 Robert Burns (1759–1796), *To the Reverend John M'Math*.
21 Chidley et al. (1993).
22 FAO (1994b).
23 Cocks et al. (1986).
24 Kwakernaak (1995); Sombroek and Sims (1995).
25 Sombroek and Sims (1995), 34.
26 I had early experience of this on the Jengka Triangle land settlement scheme in Malaysia. I had drawn up a system of land suitability evaluation for oil palm. A colleague, an older Dutchman, did not agree with the resulting map. 'I have been manager of an oil palm plantation. I should know which land is suitable, isn't it?' One could only defer.
27 During the Second World War, a regulation was in force in the UK requiring farmers to use land productively and efficiently; this was a reserve measure, to back up advice and exhortation, and cases of prosecution were rare.
28 FAO (1985c).

7 Land degradation

1 Lewis (1987).
2 Hudson (1993); Evans (1995).
3 Wischmeier and Smith (1978), 2; Smith and Stamey (1965).
4 Young and Saunders (1986). Rates can reach 10 t ha^{-1} per year in mountainous areas, but this is inclusive of river-bed erosion and there is no continuous soil cover.
5 Erosion models are reviewed in Young (1994b).
6 US Department of Agriculture (1989).
7 Stocking (1988).
8 As these quotations are deliberately selected to illustrate their exaggerated nature, it would be unfair to give citations.
9 Lal and Greenland (1979).
10 Young (1994a), 37 and references there cited; Pagiola (1995).
11 Stoorvogel and Smaling (1990); Van der Pol (1990). Some nutrient balance studies fail to take account of input of phosphorus by rock weathering.
12 Douglas (1994).
13 Umali (1993); Ghassemi et al. (1995).
14 Oldeman et al (1990); Oldeman (1994).
15 UNEP (1992). Despite being called *World atlas of desertification*, this shows GLASOD results for all climatic regions, highlighting the dry zones.
16 Sehgal and Abrol (1992).
17 Young (1994a).
18 *The state of food and agriculture 1993* (FAO, annual, b), 237.
19 Young (1994a), 45.
20 FAO (1996b), Document 7; Falkenmark (1998).
21 Skole and Tucker (1993).

22 Myers (1980), Grainger (1993), and FAO (1993b, 1995a); the quotation is from FAO (1993b), page x.
23 FAO (1993b, 1995a), summarized in Singh (1993).
24 FAO (1985a).
25 Research by the World Wide Fund for Nature, reported in *The Times*, 9 October 1997.
26 Grainger (1995).
27 De Montalembert and Clément (1983).
28 FAO/UNESCO/WMO (1977); subsequently revised and more correctly entitled *Map of desertification hazards* (UNEP (1984).
29 Young (1984).
30 UNEP (1995).
31 Details of the desertification story are from Thomas (1993) and Thomas and Middleton (1994).
32 Warren and Agnew (1988).
33 FAO (1984c), Bie (1990).
34 Biot et al. (1992); Sillitoe (1993).
35 Tiffen et al. (1994); English et al. (1993).
36 Boserup (1965); Phillips-Howard and Lyon (1994); Harris (1996).
37 Blaikie (1985); Blaikie and Brookfield (1987).
38 FAO (1980, 1988c).
39 *The state of food and agriculture* 1991 (FAO, annual, b), 106; World Bank (1992), 7.
40 Ahmad and Kutcher (1992).
41 Young (1994a).

8 Global issues: climatic change and biodiversity

1 Bazzaz and Sombroek (1996), 5.
2 Bazzaz and Sombroek (1996), page xi; the debate on wording reported in the *International Herald Tribune*, 2 December 1995.
3 Rosenzweig and Parry (1994); *Food Policy* (1994).
4 FAO (1994c).
5 Walker and Steffen (1996). The positive effects of carbon dioxide on growth mainly affect plants with the C3 photosynthetic pathway; these include rice, wheat, barley, potatoes, most leguminous crops, and many horticultural crops.
6 Vitousek (1994).
7 Bazzaz and Sombroek (1996), 12.
8 Colborn et al. (1996), Cadbury (1997), *New Scientist* 11 January 1997.
9 Bazzaz and Sombroek (1996), 329.
10 IUCN (1994).
11 Lovelock (1979).
12 Constanza et al. (1997).
13 IUCN/UNEP/WWF (1991); WCMC (1992); Royal Geographical Society (1996).
14 Noss (1990), Reid et al. (1993).
15 Pellew (1995).
16 Norton-Griffiths and Southey (1995).

9 Monitoring change: land resource indicators

1 This chapter is summarized from Pieri et al. (1995), which draws upon previous work, notably Hammond et al. (1995). In Pieri et al. (1995) the term 'land quality indicators' is used, but I have replaced this with land resource indicators, first, because 'land quality' is ambiguous, and could refer either to the inherent value of land or to its present condition, and, secondly, because land quality has an established meaning in the procedures of land evaluation.
2 World Bank (1995b).
3 O 'Keefe et al. (1984).
4 Noss (1990); Pieri et al. (1995).
5 Lewis (1987).
6 For a review of modelling soil changes, with references, see Young (1994b).
7 See the estimate of the cost of nutrient replacement for Zimbabwe, made on the basis of extrapolation from experimental data by Stocking (1988).
8 Young (1991).
9 WCMC (1992).
10 Noss (1990).
11 Reid et al. (1993), 10.
12 Kremen et al. (1994).
13 Heywood and Watson (1995); Royal Geographical Society (1996).
14 GTOS is linked with the International Geosphere–Biosphere Programme (IGBP) of the International Council of Scientific Unions (ICSU), under the auspices of UNESCO (Heal et al., 1993).
15 Information on environmental monitoring activities is contained in a UNEP newsletter, *Earthviews*.
16 The title of a review of data sources by Rodenburg (1993). Gaia is another name for Erda, the Earth Goddess.

17 Pieri et al. (1995), 49.
18 Estes et al. (1995).
19 UNCED (1992) 86.
20 UNDP (annual).

10 Costing the earth: the economic value of land resources

1 For the application of this procedure to land evaluation, see FAO (1984d), 86–90 and 106–7.
2 The fact that the 'rent' value of agricultural land derived from characteristics of its natural resources was recognized by Ricardo in classical economics.
3 Stocking (1988).
4 Java: Repetto et al. (1989) and Magrath and Arens (1989); Costa Rica: Solorzano et al. (1991); the Philippines: Cruz and Repetto (1992).
5 Among many sources used for the following discussion, some of the key concepts are drawn from Hufschmidt and Hyman (1982); Peskin (1989); Schramm and Warford (1989); Pearce et al. (1990); Pearce (1993); Pearce and Warford (1993); Turner (1993); Price (1993); and Bartelmus (1994).
6 Rates are similar for Norfolk, UK, and upstate New York. Since suppliers are offering horticultural quantities, and include profit, I have halved the quoted rates.
7 This was found to be the case for depletion by commercial forestry in Indonesia, where in some years the standing biomass fell but its 'value' rose (Repetto et al., 1989).
8 Space does not permit discussion of the evaluation of intangible benefits, a topic widely covered in economic texts.
9 Also called option value or option use value.
10 Leslie (1987), 56.
11 Still more strangely, this does not substantially alter the economic viability of the original scheme. A reclamation expenditure of $100 000 in each of Years 21–25, discounted at 10%, has a 'present value' totalling $56 000 for the whole period of reclamation.
12 These comments are based on cogent arguments by Price (1993).
13 Price (1993), 288.
14 This proposal was made in outline over 25 years ago, but in such an inappropriate publication that it received no attention (Young, 1973).
15 Bojö (1992); Lutz et al. (1994).
16 Bojö (1992).
17 Bishop and Allen (1989), page ii.
18 FAO (1991a), 111.
19 Repetto et al. (1989).
20 UN (1993).
21 For more extensive discussion see Peskin (1989); Abaza (1992, 1996); UN (1993); Lutz (1993); and Bartelmus (1994).
22 Repetto et al. (1989); Cruz and Repetto (1992).

11 Land management: caring for resources

1 Klingebiel and Montgomery (1961); Dent and Young (1981), chapter 9.
2 This arose from a convergence of ideas. The essentially socioeconomic underlying causes of erosion were forcefully set out by Blaikie (1985). The contribution of traditional methods of conservation was recognized (Critchley et al., 1992; Reij et al., 1996). Scientists in field projects worked out practical approaches and spread the concepts through training courses; these included Malcolm Douglas, Hans Hurni, David Sanders, Francis Shaxson, Michael Stocking, and, remarkably, a former leading figure in the conventional approach, Norman Hudson. The present author developed the approach with respect to agroforestry (Young, 1989). The land husbandry approach was brought together in Shaxson et al. (1989); Hudson (1992); and Norman and Douglas (1994). It has also been called conservation farming and, in Canada, stewardship of the land.
3 Pimentel and Pimentel (1979); Giampetro et al. (1993).
4 Reijntjes et al. (1992); Sanchez (1994); Greenland et al. (1994); FAO (1994a).
5 Minimum (zero) tillage is more suited to well-structured soils. It requires chemical weed killers, leading to both economic and environmental problems; yet it is undoubtedly highly effective as a means of soil conservation (Lal, 1974). Despite reading many accounts and questioning those presumed to know about it, I still find myself unable to form an appraisal of the magnitude of its potential.

6 As a consequence of the new emphasis, the former Fertilizer Service of FAO became the Plant Nutrition Service.
7 Irrigation is defined as human intervention in natural hydrological systems with the objective of bringing water onto agricultural land. Hence retention of rainwater by bunds in swamp rice systems, without bringing in additional water by channels or pumping, is not irrigation.
8 Most data originate from FAO. By comparison with other statistics, those for irrigation are probably relatively reliable, being based on a well-defined and identifiable practice, and showing internal consistency. It should be noted, however, that data for nearly all developing countries carry the tell-tale letter 'F', meaning this is an estimate by FAO, not a figure reported by the country.
9 Jones (1995).
10 Sanmuganathan and Bolton (1988).
11 Turner (1994).
12 Dambos have long fascinated natural resource scientists, although some of the best discussions are published in obscure sources; see, for example, Ackermann (1936); Milne (1947); Lamerton (1962); Ipinmidun (1970); Ferreira and Heery (1975–8); and Turner (1986).
13 Hardin (1968). The only publication on land development which matches this in frequency of citation is Boserup (1965).
14 This argument is taken mainly from Cossins (1985) and Behnke et al. (1993), which summarizes earlier work.
15 Tom Dunne, at an FAO meeting on land evaluation for extensive grazing.
16 Behnke (1995).
17 This account draws upon FAO (1985b); De Montalembert (1991); Young (1993b); and Lanly (1995).
18 Young (1993b).
19 This section is greatly expanded in Young (1989, 1997).
20 I believe this widely quoted joke originated with Oscar Fugalli at a meeting of the International Union of Forestry Research Organizations (IUFRO) in Peradeniya, Sri Lanka, 1985.
21 Kiepe (1995).
22 Young (1989, 1993a, 1997); Kiepe and Rao (1994); Kiepe (1995).
23 Originated by Sanchez (1987) and modified by Young (1989, 1997).
24 It is unfortunate that alley cropping has been up to now the more widely used name, as it is an agriculturalist's term, referring only to the crops; hedgerow intercropping, a name owing to Peter Huxley, is more appropriate, naming both the distinctive feature, the hedgerows, and the cropping which takes place between these. I continue to fight what is possibly a losing battle over this (Young, 1989, 1997).

12 Research and technology

1 FAO (1995b), 4. This proposal is not basically different from demonstration farms to be seen in the British colonies of Africa in the late 1950s.
2 A number of recent annotated bibliographies have sought to remedy this fault.
3 Farmer (1979).
4 Sanchez (1994), 69.
5 I came across a remarkable example of this. In 1958 the 'Birch effect', an accelerated mineralization of nitrogen when soil is first moistened after a dry season, was hailed as a major scientific discovery (Birch, 1958). But in the (handwritten) Annual Report of the Department of Agriculture, Nyasaland (Malawi) for about 1906, it is reported that early planting of maize, to germinate at the very beginning of the rains, produces a large increase in yield, for reasons unknown.
6 In practice, having done very precise calculations, engineers than add a highly approximate but large safety margin.
7 Scientists should occasionally test 'way-out' ideas; most are unsuccessful, but the pay-off if one works is large. It is said that Charles Darwin tried playing the trombone to tulips; the result of this experiment was negative.
8 I am indebted to Peter Huxley for this concept.
9 Greenland et al. (1987).
10 By the International Centre for Research in Agroforestry (ICRAF). For details of these networks, see ICRAF annual reports.
11 Powlson and Johnston (1994).

12 Steiner and Herdt (1993).
13 Data from *FAO production yearbook* (FAO, annual, a) and Alexandratos (1995), 172.
14 Often referred to as biotechnology, although this properly means any technological process that uses biological methods, such as conventional plant breeding or wine-making.
15 World average yields can be misleading, since they are influenced by the response to prices of farmers in the major, temperate-zone, producing countries. A fall in grain prices will be met by lower inputs.
16 Through chance elements of research politics, IITA is not located in the humid tropics nor ICRISAT in the semi-arid zone.
17 ODI (1994).
18 FAO (1996b), Document 9, page vi.
19 Ruttan (1994).
20 Eyzaguirre (1996); UNEP/FAO (1994c).
21 CIAT, ICRISAT, IITA, ICARDA, ICRAF, and IRRI.
22 Report of a CGIAR meeting, Washington DC, 1990, chapter 9.
23 Woomer and Swift (1994).
24 FAO (1983b).
25 Bunting (1987).
26 Bonte-Friedheim et al. (1994).
27 Tribe (1994), 261.
28 Tribe (1994), 118; FAO (1996b), Document 9, page 10.
29 Bonte-Friedheim et al. (1994).

13 Land, food, and people

1 Malthus (1798, 1803).
2 Ehrlich (1968) Meadows et al. (1972). The latter was also known as the 'Club of Rome' study. These books had immense sales.
3 President's Science Advisory Committee (1967), 44. A contemporary study was made in Britain (Hutchinson, 1969).
4 Ehrlich and Ehrlich (1990).
5 In the order cited: P. Bauer, *The Times*, 3 January 1994; W. Oddie, *The Sunday Telegraph*, 4 September 1994; Simon (1994), 388.
6 Cohen (1995), 213. Cohen plots these estimates against the dates they were made, showing that there is no upward or downward trend over time; indeed, the earliest, made by Leeuwenhoek in 1679, is 13 400 million, close to many of today's estimates.
7 For political reasons, most FAO studies prior to 1995 excluded China.
8 Buringh et al. (1975).
9 Smil (1994), 281.
10 Waggoner (1994).
11 Bongaarts (1996).
12 Luyten (1995), 3.
13 A leading proponent of this view is Simon (1994).
14 This study considers only a nitrogen constraint, assuming that phosphorus and potassium are not limiting.
15 Luyten (1995), 144.
16 Ehrlich et al. (1993), 27.
17 Kendall and Pimentel (1994), 203–4.
18 Brown and Kane (1994), 197–8, drawing also upon Brown (1994).
19 FAO (1981), 124 (my italics).
20 FAO (1984b), summarizing FAO (1982a).
21 Alexandratos (1995), 5, 6, 8, 18, 170.
22 FAO (1996a); FAO (1996b), Document 1, pages viii-ix, and Document 14, pages v-vi.
23 Crosson and Anderson (1992), 111.
24 Agcaoili and Rosegrant (1995).
25 Mitchell and Ingco (1995).
26 There are problems of comparability between countries included, but making all possible allowances the outcome is very much less than the forecast.
27 Alexandratos (1995), 180.
28 FAO (1984b), page x.
29 FAO (1982a), Annex Figure 1.
30 FAO (1982a), 20; Young and Wright (1980). As one of the consultants responsible for this section of the study I can attest that these requirements for rest periods (fallows, etc.) are far from exaggerated; indeed, mindful of the effect they would have in reducing land areas, they were made as short as was consistent with evidence of needs to maintain soil fertility.
31 FAO (1984b), 39.
32 FAO (1984b), 25.
33 Kilocalories are sometimes written, confusingly, as Calories or even simply calories. The SI unit, the joule, is still more inconvenient dimensionally: 1

calorie = 4.19 kilojoules.

34 In technical use, undernutrition means a deficiency in basic food energy requirements, malnutrition an inadequately balanced diet.

35 FAO (1996b), Document 1, page vii.

36 A livestock unit is a mature cow of 350 kg, the body weight equivalent of 6 human beings.

37 In addition to consumption of cereals harvested green for feed or silage, which the *FAO production yearbook* excludes from crop statistics.

38 Smil (1994).

39 *World resources* 1996/97, 295 (WRI/UNEP/UNDP/World Bank, biennial); *The state of food and agriculture* 1992 (FAO, annual, b); *The Times*, 27 July 1994.

40 Temporary grasslands are included in statistics for arable land; they presumably account for the fact that, for the developed world, the reported arable area is double the sum of areas under all crops; for developing countries it is 85%, possibly because double cropping nearly compensates for short-term fallows. Data on natural grasslands are highly unreliable (p. 60).

41 Alexandratos (1995), 145; Agcaoili and Rosegrant (1995); Mitchell and Ingco (1995); FAO (1996b), Document 1, page 19.

42 Brown (1995).

43 FAO (1982a, 1984b); Alexandratos (1995); Luyten (1995). The FAO estimates (Alexandratos, 1995) have been re-analysed, with additional data, by Fischer and Heilig (1998), who reached similar conclusions.

44 Statisical comparisons in this chapter are rendered difficult by many factors: differences in base years; data taken from FAO or US sources; changes in definitions of developing countries; the omission, at the time of the earlier forecasts, of China from some FAO statistics; and the fact that reports frequently do not make clear whether changes over a period, e.g. 1990–2010, are given as simple or compound percentages.

45 Data are good for protected areas of a nature conservation type. For obvious reasons, information is not available on areas protected for military reasons.

46 Alexandratos (1995), 166.

47 Growth rates in this section are percentages on the base period, not compound rates.

48 In Yemen, this may be because the flights of spectacular hill terraces are classed by the inventory as uncultivable.

49 G. Fischer, personal communication, based on recent information from Chinese sources.

50 Yudelman (1994), 12, quoting a World Bank estimate.

51 Postel (1989).

52 FAO (1996b), Document 1, page 29.

53 Alexandratos (1995), 155.

54 Area under cereals is a better indicator than total arable area, as data are slightly less unreliable.

55 Agcaoili and Rosegrant (1995), 77.

14 Population, poverty, and conflict

1 When writing of population, I deliberately do not use 'billion', as it is a shorthand which removes any sense of the vast magnitude involved. It is hard even to envisage the real meaning of one million. A large sports stadium may contain 100 000 people. One can just imagine flying over ten such arenas. But the real meaning of even six million people, 0.1% of the world's population, defies the imagination.

2 It is a mistake to suppose that, unlike other data, those for population are highly accurate. Follow-up studies have shown underenumeration of 3% in US censuses, and even in the UK, underenumeration in the 1991 census is believed to be 1–1½%. In developing countries, errors are unlikely to be less than 5–10%, and some countries have recently postponed censuses or suppressed the results.

There is no better illustration of the danger of uncritical acceptance of statistics than the population table in *World Resources* 1996–97 in which (at least in the first printing) the populations given for all countries of the world are one thousand times too large.

3 UN (1995a).

4 Sadik (1991).

5 UN (1994).

6 The statement says this was the first time that scientists had acted collectively to influence international policy, but the 'Pugwash' conferences on

nuclear armaments, beginning in 1957, also had this aim.

7 World's Scientific Academies (1993). This statement is exemplary in its clear and concise language, and absence of jargon, leading to astonishing brevity: a document of only 13 pages with a summary in 3 pages.

8 The dissenting statement was by the African Academy of Sciences. It takes the view that, for Africa, 'population remains an important resource for development' and 'the contribution of the North [developed world] to Africa's population predicament must be acknowledged.'

9 UN (1992).

10 ODA (1994).

11 USAID (1994).

12 World's Scientific Academies (1993), Statement page 1 and Summary page 2.

13 UN (1994), paragraph 1.11.

14 FAO (1996b), Document 3, page 9, and Document 4, pages viii and 30; FAO (1996a).

15 UNFPA (1996).

16 I cannot refrain from relating an interchange which took place at the 1987 Dahlem (Berlin) workshop on resources and world development. Speaker: 'It has been said that the desirable population of the world is about one billion.' Dennis Greenland (from the floor): 'Including us.'

17 Kane (1987).

18 Peng Yu (1994); Chinese government (1995).

19 UN (1995a).

20 A graphic description of the Indian poor is given in *Blossoms in the dust* (Nair, 1961), and the attitudes you are forced to take if you are poor are set out in *The peasant view of the bad life* (Bailey, 1966). The best literary insight into the mind of a farmer at the lowest level of security is given in Leonard Woolf's novel (1913), *The village in the jungle.*

21 UNDP, annual, 1996 edition, 222.

22 Jazairy et al. (1992), 376; UNDP (annual, 1996 edition), 27.

23 Jazairy et al. (1992).

24 World Bank (1993b), page v.

25 IFAD (1995).

26 Watkins (1995), 2. The scale, nature, and consequences of poverty, together with the stark contrasts with growth and consumption in the developed world, are powerfully marshalled in reviews by the IFAD (Jazairy et al., 1992) and in *Human development report* (UNDP, annual). The 1996 edition of the last includes the astonishing statistic that the world's 360 richest people (the ones that they know about) own more assets than the annual income of 45% of the world's people.

27 Jazairy et al. (1992).

28 Data from the 15th edition of *Encyclopedia Britannica* (1988), 675.

29 Red Cross (annual), reports for 1996 and 1997.

30 BBC report, September 1997, based on information from the German Red Cross; and *The Times*, 2 and 6 October 1997 and 7 January 1998.

31 Prunier (1995).

32 Data from UN statistics quoted in Brown et al. (1995), 103.

33 Rather emotively by Kaplan (1994), and in a more reasoned manner by Homer-Dixon et al. (1993); Myers (1993); Westing (1989); and by eleven writers from Sahelian countries in Bennett (1991).

34 Brown and Pollitt (1996).

15 Awareness, attitudes, and action

1 UNCED (1992); IFPRI (1995); Boutros Ghali (1995); FAO (1996a). Fourteen countries entered reservations to their approval of the Rome declaration on world food security (FAO, 1996a).

2 OECD (annual).

3 OECD (annual). These data cover 21 OECD donor countries. There is additional aid from Arab countries.

4 FAO (1996b), Document 10, page 35; ODI (1995). Inevitably, doubts have been expressed about the accuracy of these statistics.

5 Boutros-Ghali (1995), 78.

6 Boutros-Ghali (1995), 45.

7 Space does not permit discussion of controversial elements in structural adjustment programmes, which include possible adverse effects upon the poor. It has also been held that they made land degradation worse, by encouraging countries to exploit natural resources in order to reduce foreign debt, but the link, if any, is complex. For a dis-

cussion of this aspect, with case studies, see Reed (1992) and Cruz and Repetto (1992).

8 Twelve *Success stories in food security* as given as Document 2 of the 1996 World Food Summit (FAO, 1996b).

9 UNCED (1992), chapters 10–14. Further guidelines are given in *Planning for sustainable use of land resources* (Sombroek and Sims, 1995).

10 IFPRI (1995), 24.

11 For further discussion of institutional aspects, see Sombroek and Sims (1995), 45.

12 Falvey (1996).

13 UNCED (1992), 10.

14 Boutros-Ghali (1995), 38.

15 Saouma (1993), page D2.

16 Boutros-Ghali (1995), 32.

Land use statistics

The situation with respect to land use statistics has recently deteriorated. From 1996 onwards, the FAO *Production yearbook* 'presents a country data for a reduced number of land categories'. After arable use and permanent crops, the remaining area is given as '*all other land*', a class which combines pastures, forests, built-up areas, and barren land. It is noted that 'difficulties in standardization . . . may have led to some confusion among users'.

References

Abaza, H. (ed.) 1992 *The present state of environmental and resource accounting and its potential application in developing countries*. UNEP Environmental Economics Series 1. Nairobi.

Abaza, H. (ed.) 1996 *Environmental accounting*. UNEP Environmental Economics Series 17. Nairobi.

ABLH (Association for Better Land Husbandry) 1995 *Principles of good land husbandry*. Mimeo. London.

Ackermann, E. 1936 Dambos in Nordrhodesien. *Wissenschaftliche Veröffentlichungen des Deutschen Museums für Länderkunde Leipzig, Neue Folge* 4: 147–57.

Agarwal, B. 1994 The gender and environment debate: lessons from India. In Arizpe et al. (1994), q.v.: 87–124.

Agcaoili, M. and Rosegrant, M. W. 1995 Global and regional food supply, demand, and trade prospects to 2010. In Islam (1995), q.v.: 61–83.

Agricultural Advisory Council (UK) 1970 *Modern farming and the soil*. London: HMSO.

Ahmad, M. and Kutcher, G. P. 1992 *Irrigation planning with environmental considerations: a case study of Pakistan's Indus basin*. World Bank Technical Paper 166. Washington DC.

Alexandratos, N. 1995 *World agriculture: towards 2010. An FAO study*. Chichester, UK: Wiley, for FAO.

Anderson, J. R., Hardy, E. E., Roach, J. T., and Witmer, R. E. 1976 *A land use and land cover classification system for use with remote sensor data*. US Geological Survey Professional Paper 964. Washington DC.

Arizpe, L., Stone, M. P., and Major, D. C. (eds.) 1994 *Population and environment: rethinking the debate*. Boulder, USA: Westview.

Bailey, F. G. 1966 The peasant view of the bad life. *Advancement of Science*, December 1966: 399–409.

Bartelmus, P. 1994 *Environment, growth and development*. London: Routledge.

Bazzaz, F. and Sombroek, W. 1996 *Global climatic change and agricultural production*. Chichester, UK: FAO and Wiley.

Beatty, M. T., Petersen, G. W., and Swindale, L. D. (eds.) 1979 *Planning the uses and management of land*. Madison, USA: American Society of Agronomy.

Beckett, P. H. T. and Webster, R. 1971 Soil variability - a review. *Soils and Fertilizers* 34: 1–15.

Behnke, R. 1995 The limits on production and population growth in pastoral economies. *Tropical Agriculture Association Newsletter* 15(4): 2–4.

Behnke, R., Scoones, I. and Kerven, C. 1993 *Range ecology at disequilibrium*. London: Overseas Development Institute.

Bennett, O. 1991 *Greenwar: environment and conflict*. London: Panos.

Benor, D., Harrison, J. Q., and Baxter, M. 1984 *Agricultural extension: the training and visit system*. Washington DC: World Bank.

Bhadra, D. and Salazar, P. B. A. 1993 *Urbanization, agricultural development, and land allocation*. World Bank Discussion Paper 201. Washington DC.

Bie, S. W. 1990 *Dryland degradation measurement techniques*. World Bank Environment Working Paper 26. Washington DC.

Billett, M. F., Fitzpatrick, E. A., and Cresser, M. S. 1988 Long-term changes in the acidity of forest soils in north-east Scotland. *Soil Use and Management* 4: 102–7.

Biot, Y., Lambert, R., and Perkin, S. 1992 *What's the problem? An essay on land degradation, science and development in Sub-saharan Africa*. Development Studies Discussion Paper 222. Norwich: University of East Anglia.

Birch, H. F. 1958 The effect of soil drying on humus decomposition and nitrogen availability. *Plant and Soil* 10: 9–31.

Bishop, J. and Allen, J. 1989 *The on-site costs of soil erosion in Mali*. World Bank Environment Working Paper 21. Washington DC.

Blaikie, P. 1985 *The political economy of soil erosion in developing countries*. London: Longman.

Blaikie, P. and Brookfield, H. 1987 *Land degradation and society*. London: Methuen.

Bojö, J. 1992 Cost–benefit analysis of soil and water conservation projects: a review of 20 empirical studies. In *Soil conservation for survival* (ed. Kebede Tato and H. Hurni, Soil and Water Conservation Society, Ankeny, USA): 195–205.

Bongaarts, J. 1996 Population pressures and the food supply system in the developing world. *Population and Development Review* 22: 483–503.

Bonte-Friedheim, C., Tabor, S. and Roseboom, J. 1994 *Financing national agricultural research: the challenge ahead.* ISNAR Briefing Paper 11. The Hague: International Support for National Agricultural Research (ISNAR).

Boserup, E. 1965 *The conditions of agricultural growth.* Chicago: Aldine.

Boutros-Ghali, B. 1995 *An agenda for development.* New York: UN.

Brandt Commission 1980 *North-South: a programme for survival.* London: Pan.

Brown, J. L. and Pollitt, E. 1996 Malnutrition, poverty and intellectual development. *Scientific American*, February 1996: 38–43.

Brown, L. R. 1994 Facing food insecurity. Chapter 10 in *State of the world 1994* (ed. L. R. Brown, Worldwatch Institute, Washington DC).

Brown, L. R. 1995 *Who will feed China?* New York: Norton.

Brown, L. R. and Kane, H. 1994 *Full house: reassessing the earth's population carrying capacity.* New York: Norton.

Brown, L. R., Lenssen, N., Kane, H. and Starke, L. 1995 *Vital signs 1995.* New York: Norton.

Brundtland Commission 1987 *Our common future.* Oxford University Press.

Bunting, A. H. 1987 *Agricultural environments: characterization, classification and mapping.* Wallingford, UK: CAB International.

Buol, S. W. and Couto, W. 1981 Soil fertility-capability assessment for use in the humid tropics. In *Characterization of soils* (ed. D. J. Greenland, Clarendon, Oxford): 254–61.

Buringh, P. and Dudal, R. 1987 Agricultural land use in space and time. In *Land transformation in agriculture* (ed. M. G. Wolman and F. G. A. Fournier, Wiley, Chichester, UK): 9–43.

Buringh, P., van Heemst, H. D. J. and Staring, G. J. 1975 *Computation of the absolute maximum food production of the world.* Wageningen, The Netherlands: Agricultural University.

Cadbury, D. 1997 *The feminization of nature.* London: Hamish Hamilton.

Chambers, R. 1981 Rapid rural appraisal: rationale and repertoire. *Public Administration and Development* 1: 95–106.

Chambers, R. 1993 Participatory rural appraisal. In Hudson and Cheatle (1993), q.v.: 87–95.

Chambers, R, Pacey, A., and Thrupp, L. A. 1989 *Farmer first; farmer innovation and agricultural research.* London: Intermediate Technology Publications.

Chidley, D. R. E., Elgy, J., and Antoine, J. 1993 *Computerized systems of land resources appraisal for agricultural development.* FAO World Soil Resources Report 72. Rome.

Chinese government 1995 *White paper on family planning.* Translated in *Population and Development Review* 22, 1996: 385–90.

Christian, C. S. and Stewart, G. A. 1953 *General report on survey of Katherine-Darwin region, 1946.* Land Research Series 1. Melbourne: CSIRO.

Christian, C. S. and Stewart, G. A. 1968 Methodology of integrated surveys. In *Aerial surveys and integrated studies*, UNESCO Natural Resources 6 (UNESCO, Paris): 233–80.

Cloke, P. J. 1989 *Rural land-use planning in developed nations.* London: Unwin Hyman.

Cocks, K. D., Cole, R. P. Garrard, I. M., Ive, J. R., and Trethewey, S. V. 1986 *Using the LUPLAN package to assist in the assessment of crown lands near Lake Eucumbene.* Divisional Report 86/2. Canberra: Division of Water and Land Resources, CSIRO.

Cohen, J. E. 1995 *How many people can the earth support?* New York: Norton.

Colborn, T., Dumanowski, D. and Myers, J. P. 1996 *Our stolen future.* Boston, USA: Little Brown.

Collinson, M. P. 1981 A low cost approach to understanding small farmers. *Agricultural Administration* 8: 433–50.

Collinson, M. P. 1987 Farming systems research: procedures for technology development. *Experimental Agriculture* 23: 365–86.

Constanza, R., and 12 others 1997 The value of the world's ecosystem services. *Nature* 387: 253–60.

Cossins, N. J. 1985 *The productivity and potential of pastoral systems.* ILCA Bulletin 21. Addis Abeba: International Livestock Centre for Africa (ILCA): 10–15.

REFERENCES

Critchley, W. R. S., Reij, C. P. and Turner, S. D. 1992 *Soil and water conservation in Sub-saharan Africa*. Amsterdam: Free University, for International Fund for Agricultural Development (IFAD).

Crosson, P. and Anderson, J. R. 1992 *Resources and global food prospects: supply and demand for cereals to 2030*. World Bank Technical Paper 184. Washington DC.

Cruz, W. and Repetto, R. 1992 *The environmental effects of stabilization and structural adjustment programs: the Philippines case*. Washington DC: World Resources Institute.

Dalal-Clayton, B. and Dent, D. 1993 *Surveys, plans and people. A review of land resource information and its use in developing countries*. London: International Institute for Environment and Development (IIED).

Davis, S. H. 1993 *Indigenous views on land and the environment*. World Bank Discussion Paper 188. Washington DC.

De Montalembert, M. R. 1991 Key forestry policy issues in the early 1990s. *Unasylva* 42(166): 9–18.

De Montalembert, M. R. and Clément, J. 1983 *Fuelwood supplies in developing countries*. FAO Forestry Paper 42. Rome.

Dent, D. and Young, A. 1981 *Soil survey and land evaluation*. London: Allen and Unwin.

Doorenbos, J. and Kassam, A. H. 1979 *Yield response to water*. FAO Irrigation and Drainage Paper 33. Rome.

Doorenbos, J. and Pruitt, W. O. 1977 *Guidelines for predicting crop water requirements*. FAO Irrigation and Drainage Paper 24. Rome.

Douglas, M. 1994 *Sustainable use of agricultural soils: a review of the prerequisites for success or failure*. Development and Environment Report 11. Switzerland: University of Bern.

Easter, K. W., Dixon, J. A. and Hufschmidt, M. M. 1986 *Watershed management projects*. Boulder, USA: Westview.

EEC (European Economic Commission) 1993 CORINE *land cover: technical guide*. Brussels.

Ehrlich, P. 1968 *The population bomb*. New York: Ballantine.

Ehrlich, P. and Ehrlich, A. 1990 *The population explosion*. London: Hutchinson.

Ehrlich, P. R., Ehrlich, A. H., and Daily, G. C. 1993 Food security, population and environment. *Population and Development Review* 19: 1–32.

English, J., Tiffen, M. and Mortimore, M. 1993 *Land resource management in Machakos District, Kenya, 1930–1990*. World Bank Environment Paper 5. Washington DC.

Estes, J., Lawless, J. and Mooneyhan, D. W. (eds.) 1995 *International symposium on core data needs for environmental assessment and sustainable development strategies*. Nairobi: UNDP and UNEP.

Evans, R. 1995 Some methods of directly assessing water erosion of cultivated land - a comparison of measurements on plots and in fields. *Progress in Physical Geography* 19: 115–29.

Eyzaguirre, P. 1996 *Agriculture and environmental research in small countries*. Chichester, UK: Wiley, for International Support for National Agricultural Research (ISNAR).

Falkenmark, M. (1998) Meeting water requirements of an expanding world population. In Greenland et al. (1998), q.v., 929–36.

Falvey, L. 1996 *Food environment education: agricultural education in natural resource management*. Melbourne, Australia: Crawford Fund.

FAO not dated, *c*. 1960 *Guidelines for soil description*. Rome.

FAO 1976 *A framework for land evaluation*. FAO Soils Bulletin 32. Rome.

FAO 1978–81 *Report on the agro-ecological zones project*. FAO World Soil Resources Report 48/1–4. Rome.

FAO 1980 *The world conference on agrarian reform and rural development*. Rome.

FAO 1981 *Agriculture: toward 2000*. Rome.

FAO 1982a *Potential population supporting capacities of lands in the developing world*. Rome.

FAO 1982b *World Soil Charter*. Rome.

FAO 1983a *Guidelines: land evaluation for rainfed agriculture*. FAO Soils Bulletin 52. Rome.

FAO 1983b *A physical resource base. Map, scale 1:25 million*. Rome.

FAO 1984a *Guidelines: land evaluation for irrigated agriculture*. FAO Soils Bulletin 55. Rome.

FAO 1984b *Land, food and people*. Rome.

FAO 1984c *Provisional methodology for assessment and mapping of desertification*. Rome.

FAO 1984d *Land evaluation for forestry*. FAO Forestry Paper 48. Rome.

FAO 1985a *Tropical forests action plan*. Rome.

FAO 1985b *Intensive multiple-use forest management in the tropics*. FAO Forestry Paper 55. Rome.

FAO 1985c *The role of legislation in land use planning for developing countries*. FAO Legislative Study 31. Rome.

FAO 1986a *Programme for the 1990 world census of agriculture*. FAO Statistical Development Series 2. Rome.

FAO 1986b *Strategies, approaches and systems in integrated watershed management*. FAO Conservation Guide 14. Rome.

FAO 1988a *FAO–UNESCO soil map of the world: revised legend*. FAO World Soil Resources Report 60. Rome.

FAO 1988b *National parks planning: a manual with illustrated examples*. FAO Conservation Guide 17. Rome.

FAO 1988c *WCARRD 1979–1989. Ten years of follow-up on impact of development strategies on the rural poor*. Rome.

FAO 1991a *Guidelines: land evaluation for extensive grazing*. FAO Soils Bulletin 58. Rome.

FAO 1991b *Land use planning applications*. FAO World Soil Resources Report 68. Rome.

FAO 1991c *World soil resources*. FAO World Soil Resources Report 66. Rome.

FAO 1993a *Guidelines for land-use planning*. FAO Development Series 1. Rome.

FAO 1993b *Forest resources assessment 1990: tropical countries*. FAO Forestry Paper 112. Rome.

FAO 1993c *CLIMWAT for CROPWAT*. FAO Irrigation and Drainage Paper 49. Rome.

FAO 1994a *Cherish the earth: soil management for sustainable agriculture and environmental protection in the tropics*. Rome.

FAO 1994b *ECOCROP 1. The adaptability level of the FAO crop environmental requirements database. Version 1.0*. Rome.

FAO 1994c *Global climatic change and agricultural production*. Rome.

FAO 1995a *Forest resources assessment 1990: global synthesis*. FAO Forestry Paper 124. Rome.

FAO 1995b *Special programme on food production in support of food security in LIFDCs [low-income food-deficit countries] (SPFP)*. Rome.

FAO 1996a *Rome declaration on world food security and world food summit plan of action*. Rome.

FAO 1996b *World Food Summit: technical background documents 1–15*. Rome.

FAO annual, a. *FAO production yearbook*. Rome.

FAO annual, b. *The state of food and agriculture*. Rome.

FAO annual, c. *FAOSTAT*. [FAO statistical data on diskette.] Rome.

FAO-UNESCO 1970–80 *Soil map of the world 1:5 000 000*. Vols. 1–10. Paris: UNESCO.

FAO-UNESCO 1974 *Soil map of the world 1:5 000 000. Vol. 1: Legend*. Paris: UNESCO.

FAO/UNESCO/WMO 1977 *World map of desertification*. Nairobi: UNEP.

Farmer, B. H. 1979 *The 'green revolution' in south Asian rice-fields: environment and production*. Journal of Development Studies 15: 304–19.

Ferreira, R. E. C. and Heery, S. 1975–8 *Dambos: their agricultural potential*. Farming in Zambia 9(4): 37–8; 10(3): 39–42; 11(1): 14–17; 11(3): 19–21; 12(1): 36–40.

Fischer, G. and Heilig, G. K. 1998 Population momentum and the demand on land and water resources. In Greenland et al. (1998), q.v.: 869–90.

Food Policy 1994 Climate change and world food security. *Food Policy*, Special Issue 19(2): 97–208.

Ghassemi, F, Jakeman, A. J., and Nix, H. A. 1995 *Salinization of land and water resources*. Wallingford, UK: CAB International.

Ghosh, P. K. (ed.) 1984 *Urban development in the third world*. London: Greenwood.

Giampetro, M., Bukkens, S. G. F., and Pimentel, D. 1993 Limits to population size: three scenarios of energy interaction between human society and ecosystem. *Population and Environment* 14: 109–31.

Gill, G. J. 1993 *O.K., the data's lousy, but it's all we've got (being a critique of conventional methods)*. London: International Institute for Environment and Development (IIED).

Gourou, P. 1946 *Les pays tropicaux*. Paris: Presses Universitaires.

Grainger, A. 1993 Rates of deforestation in the humid tropics: estimates and measurements. *Geographical Journal* 159: 33–44.

Grainger, A. 1995 The forest transition: an alternative approach. *Area* 27: 242–51.

Greenland, D. J., Bowen, G., Eswaran, H. Rhoades, R., and Valentin, C. 1994 *Soil, water, and nutrient management research - a new agenda*. IBSRAM Position Paper. Bangkok: International Board for Soils Research and Management (IBSRAM).

Greenland, D. J., Gregory, P. J., and Nye, P. H. (eds.)

1998 *Land resources: on the edge of the Malthusian precipice?* London: CAB International and The Royal Society.

Greenland, D. J. and Szabolcs, I. (eds.) 1994 *Soil resilience and sustainable land use.* Wallingford, UK: CAB International.

Grübler, A. 1992 *Technology and global change: land-use, past and present.* IIASA Working Paper WP-92–2. Laxenburg, Austria: International Institute for Applied Systems Analysis (IIASA).

Habitat (UN Centre for Human Settlements) 1996 *An urbanizing world: global report on human settlements.* Oxford: University Press, for Habitat.

Hammond, A, Adriaanse, A., Rodenburg, E., Bryant, D., and Woodward, R. 1995 *Environmental indicators: a systematic approach to measuring and reporting on environmental policy performance in the context of sustainable development.* Washington DC: World Resources Institute.

Hardin, G. 1968 The tragedy of the commons. *Science* 162: 1243–8.

Harris, F. 1996 *Intensification of agriculture in semi-arid areas: lessons of the Kano close-settled zone, Nigeria.* Gatekeeper Series 59. London: International Institute for Environment and Development (IIED).

Harrison, P. 1992 *The third revolution: population, environment and a sustainable world.* London: Penguin.

Heal, O. W., Menaut, J.-C., and Steffen, W. L. 1993 *Towards a global terrestrial observing system (GTOS).* IGBP Global Change Report 26. Paris: UNESCO.

Healey, D. 1989 *The time of my life.* London: Michael Joseph.

Heilig, G. K. 1994 Neglected dimensions of global land-use change: reflections and data. *Population and Development Review* 20: 831–59.

Hendricksen, B. L. (ed.) 1984 *Ethiopia: a land resources inventory for land-use planning.* Rome: FAO.

Heywood, V. H. and Watson, R. T. (eds.) 1995 *Global biodiversity assessment.* Cambridge University Press, for UNEP.

Homer-Dixon, T. F., Boutwell, J. H., and Rathjens, G. W. 1993 Environmental changes and violent conflict. *Scientific American*, February 1993: 38–45.

Hudson, N. W. 1992 *Land husbandry.* London: Batsford.

Hudson, N. W. 1993 *Field measurement of soil erosion and runoff.* FAO Soils Bulletin 68. Rome.

Hudson, N. W. and Cheatle, R. J. 1993 *Working with farmers for better land husbandry.* London: Intermediate Technology Publications.

Hufschmidt, M. M. and Hyman, E. L. (eds.) 1982 *Economic approaches to natural resource and environmental quality analysis.* Dublin: Tycooly.

Hutchinson, J. (ed.) 1969 *Population and food supply.* Cambridge University Press.

IFAD (International Fund for Agricultural Development) 1995 *Conference on hunger and poverty, Brussels, November 1995: vision statement.* Rome.

IFPRI (International Food Policy Research Institute) 1995 *A 2020 vision for food, agriculture, and the environment.* Washington DC.

Ipinmidun, W. B. 1970 The agricultural development of fadama with particular reference to Bomo fadama. *Nigerian Agricultural Journal* 7: 152–63.

Islam, N. (ed.) 1995 *Population and food in the early twenty-first century: meeting future food demand of an increasing population.* Washington DC: International Food Policy Research Institute (IFPRI).

ISRIC (International Soil Reference and Information Centre) 1993 *Global and national soils and terrain digital databases (SOTER): procedures manual.* Wageningen, The Netherlands.

IUCN 1994 *Guidelines for protected area management categories.* Gland, Switzerland.

IUCN/UNEP/WWF 1991 *Caring for the earth: a strategy for sustainable living.* Gland, Switzerland.

Jazairy, I., Alamgir, M., and Panuccio, T. 1992 *The state of world rural poverty.* New York: University Press, for International Fund for Agricultural Development (IFAD).

Jones, W. I. 1995 *The World Bank and irrigation.* Washington DC: World Bank.

Kane, P. 1987 *The second billion: population and family policy in China.* London: Penguin.

Kaplan, R. D. 1994 The coming anarchy. *Atlantic Monthly* 273(2): 44–76.

Kassam , A. H., Van Velthuizen, H. T., Fischer, G. W., and Shah, M. M. 1993 *Agro-ecological assessments for national planing: the example of Kenya.* FAO Soils Bulletin 67. Rome.

Kendall, H. W. and Pimentel, D. 1994 Constraints on the expansion of the global food supply. *Ambio* 23: 198–205.

Kerr, J. and Sangh, N. K. 1992 *Indigenous soil and water conservation in India's semi-arid tropics.* Gatekeeper Series SA 34. London: International Institute for Environment and Development (IIED).

Kiepe, P. 1995 *No runoff, no soil loss: soil and water conservation in hedgerow barrier systems.* Tropical Resource Management Papers 10. Wageningen, The Netherlands: Agricultural University.

Kiepe, P. and Rao, M. R. 1994 Management of agroforestry for the conservation and utilization of land and water resources. *Outlook on Agriculture* 23(1): 17–25.

Klingebiel, A.A. and Montgomery, P. H. 1961 *Land capability classification.* Agriculture Handbook 210. Washington DC: US Department of Agriculture.

Korotkov, A. V. and Peck, T. J. 1993 Forest resources of the industrialized countries: an ECE/FAO assessment. *Unasylva* 44(3): 20–30.

Kremen, C., Merenlender, A. M., and Murphy, D. D. 1994 Ecological monitoring: a vital need for integrated conservation and development programs in the tropics. *Conservation Biology* 8: 388–97.

Küchler, A. W. and Zonneveld, I. S. 1988 *Vegetation mapping.* Dordrecht, The Netherlands: Kluwer.

Kwakernaak, C. (ed.) 1995 *Integrated approach to planning and management of land: operationalization of Chapter 10 of UNCED's Agenda 21.* DLO Winand Staring Centre Report 107. Wageningen, The Netherlands.

Lal, R. 1974 No-tillage effects on soil properties and maize (*Zea mays* L.) production in western Nigeria. *Plant and Soil* 40: 321–31.

Lal, R. and Greenland, D. J. (eds.) 1979 *Soil physical properties and crop production in the tropics.* Chichester, UK: Wiley.

Lamerton, J. F. 1962 Manda valleys in Tanganyika. *Journal of Ecology* 50: 771–4.

Lanly, J.-P. 1982 *Tropical forest resources.* FAO Forestry Paper 30. Rome.

Lanly, J.-P. 1995 Sustainable forest management: lessons of history and recent developments. *Unasylva* 46(182): 38–45.

Lean, G. 1995 *Down to earth: a simplified guide to the convention to combat desertification.* Geneva: Centre for Our Common Future.

Leslie, A. J. 1987 A second look at the economics of natural management systems in tropical mixed forests. *Unasylva* 39(9): 46–58.

Lewis, L. A. 1987 Predicting soil loss in Rwanda. In *Quantified land evaluation procedures.* ITC Publication 6 (ed. K. J. Beek, P. A. Burrough and D. E. McCormack, ITC, Eschhede, The Netherlands): 137–9.

Longhurst, R. (ed.) 1981 *Rapid rural appraisal.* Institute of Development Studies Bulletin 12(4). Brighton, UK: University of Sussex.

Lovelock, J. E. 1979 *Gaia, a new look at life on earth.* Oxford: University Press.

Lowry, I. S. 1991 World urbanization in perspective. In *Resources, environment, and population: present knowledge, future options.* New York: Population Council.

Lutz, E. (ed.) 1993 *Toward improved accounting for the environment.* Washington DC: World Bank.

Lutz, E, Pagiola, S., and Reiche, C. (eds.) 1994 *Economic and institutional analyses of soil conservation projects in Central America and the Caribbean.* World Bank Environment Paper 8. Washington DC.

Luyten, J. C. 1995 *Sustainable world food production and environment.* Wageningen, The Netherlands: AB-DLO.

Magrath, W. and Arens, P. 1989 *The costs of soil erosion on Java: a natural resource accounting approach.* World Bank Environment Working Paper 18. Washington DC.

Malthus, T. R. 1798, revised edition 1803 *An essay on the principle of population.* Edited, with an introduction, by D. Winch. Cambridge University Press (1992).

Meadows, D. H., Meadows, D. L., Randers, J., and Behrens, W. W. 1972 *The limits to growth.* London: Earth Island.

Milne, G. 1947 A soil reconnaissance journey through parts of Tanganyika Territory, December 1935 to February 1936. *Journal of Ecology* 35: 192–265.

Mitchell, D. O. and Ingco, M. D. 1995 Global and regional food demand and supply prospects. In Islam (1995), q.v.: 49–60.

Moss, R. P. 1968 Land use, vegetation and soil factors in south-west Nigeria: a new approach. *Pacific Viewpoint* 9:107–27.

Mulcahy, M. J. and Humphries, A. W. 1967 Soil classification, soil surveys and land use. *Soils and Fertilizers* 30: 1–8.

Myers, N. 1980 *Conversion of tropical moist forests.* Washington DC: National Academy of Sciences.

Myers, N. 1993 *Ultimate security: the environmental basis of political stability*. New York: Norton.

Nair, K. 1961 *Blossoms in the dust: the human element in Indian development*. London: Duckworth.

Norman, D. and Douglas, M. 1994 *Farming systems development and soil conservation*. FAO Farm systems Management Series 7. Rome.

Norton-Griffiths, M. and Southey, C. 1995 The opportunity costs of biodiversity conservation in Kenya. *Ecological Economics* 12: 125–39.

Noss, R. F. 1990 Indicators for monitoring biodiversity: a hierarchical approach. *Conservation Biology* 4: 355–64.

ODA (Overseas Development Administration) 1994 *Children by choice not chance*. London.

ODI (Overseas Development Institute) 1994 *The CGIAR: what future for international agricultural research?* London.

ODI 1995 *NGOs and official donors*. London.

OECD (Organization for Economic Co-operation and Development), annual. *Development cooperation*. Paris.

O'Keefe, P, Raskin, P., and Bernow, S. 1984 *Energy, environment and development in Africa. 1. Energy and development in Kenya: opportunities and constraints*. Stockholm: Beijer Institute.

Oldeman, L. R. 1994 The global extent of soil degradation. In Greenland and Szabolcs (1994), q.v.: 99–118.

Oldeman, L. R., Hakkeling, R. T. A., and Sombroek, W. G. 1990 *World map of the status of human-induced soil degradation*. Wageningen: International Soil Reference and Information Centre (ISRIC) and UNEP.

Pagiola, S. 1995 *Environmental and natural resource degradation in intensive agriculture in Bangladesh*. World Bank Environment Working Paper 15. Washington DC.

Patricros, N. N. (ed.) 1986 *International handbook on land use planning*. London: Greenwood.

Pearce, D. 1993 *Economic values and the natural world*. London: Earthscan.

Pearce, D, Barbier, E., and Markandya, A. 1990 *Sustainable development: economics and environment in the third world*. London: Earthscan.

Pearce, D. and Warford, J. J. 1993 *World without end: economics, environment, and sustainable development*. Oxford University Press, for World Bank.

Pellew, R. 1995 Biodiversity conservation - why all the fuss? *Royal Society of Arts Journal* 143(1): 53–66.

Peng Yu 1994 China's experience in population matters: an official statement. *Population and Development Review* 20: 488–91.

Peskin, H. M. 1989 *Accounting for natural resource depletion and degradation in developing countries*. World Bank Environment Working Paper 13. Washington DC.

Phillips-Howard, K. D. and Lyon F. 1994 Agricultural intensification and the threat to soil fertility in Africa: evidence from the Jos Plateau, Nigeria. *Geographical Journal* 160: 252–65.

Pieri, C., Dumanski, J., Hamblin, A., and Young, A. 1995 *Land quality indicators*. World Bank Discussion Paper 315. Washington DC.

Pimentel, D. and Pimentel, M. 1979 *Food, energy and society*. London: Arnold.

Postel, S. 1989 *Water for agriculture: facing the limits*. Worldwatch Paper 93. Washington DC: Worldwatch Institute.

Potts, D. 1995 Shall we go home? Increasing urban poverty and African cities and migration processes. *Geographical Journal* 161: 245–64.

Powlson, D. S. and Johnston, A. E. 1994 Long-term field experiments: their importance in understanding sustainable land use. In Greenland and Szabolcs (1994), q.v.: 367–94.

President's Science Advisory Committee 1967 *The world food problem*. Washington DC: The White House.

Price, C. 1993 *Time, discounting and value*. Oxford: Blackwell.

Prunier, G. 1995 *The Rwanda crisis, 1959–1994: history of a genocide*. London: Hurst.

Raintree, J. B. 1987 *D and D user's manual: an introduction to agroforestry diagnosis and design*. Nairobi: International Centre for Research in Agroforestry (ICRAF).

Red Cross (International Federation of Red Cross and Red Crescent Societies), annual. *World disasters report*. Oxford University Press.

Reed, D. 1992 *Structural adjustment and the environment*. London: Earthscan.

Reid, W. V., McNeely, J. R., Tunstall, D. B., Bryant, D. A., and Winograd, M. 1993 *Biodiversity indicators for policy-makers*. Washington DC: World Resources Institute.

Reij, C., Scoones, I., and Toulmin, C. (eds.) 1996 *Sustaining the soil: indigenous soil and water conservation in Africa*. London: Earthscan.

Reijntjes, C., Haverkort, B. and Water-Bayer, A. 1992 *Farming for the future: an introduction to low-external-input and sustainable agriculture.* London: Macmillan.

Repetto, R., Magrath W., Wells, M. Beer, C., and Rossini, F. 1989 *Wasting assets: natural resources in the national income accounts.* Washington DC: World resources Institute.

Richards, P. 1985 *Indigenous agricultural revolution: ecology and food production in West Africa.* London: Hutchinson.

Rodda, J. C. 1995 Guessing or assessing the world's water resources. *Journal of the Chartered Institution of Water and Environmental Management* 9: 360–8.

Rodenburg, E. 1993 *Eyeless in Gaia.* Washington DC: World Resources Institute.

Rosenzweig, C., and Parry, M. L. 1994 Potential impact of climate change on world food supply. *Nature* 367: 133–8.

Rossiter, D. G. 1995 Economic land evaluation: why and how. *Soil Use and Management* 11: 132–40.

Royal Geographical Society 1996 *Biodiversity assessment: a guide to good practice.* Vols. 1–3. London: HMSO.

Ruthenburg, H. 1980 *Farming systems in the tropics.* Oxford: Clarendon.

Ruttan, V. W. 1994 *Agriculture, environment, and health: sustainable development in the 21st century.* Minneapolis, USA: University of Minnesota.

Sadik, N. 1991 *The state of world population 1991.* New York: UN Fund for Population Activities (UNFPA).

Sanchez, P. A. 1987 Soil productivity and sustainability in agroforestry systems. In *Agroforestry: a decade of development* (ed. H. A. Steppler and P. K. R. Nair, International Centre for Research in Agroforestry (ICRAF), Nairobi): 205–26.

Sanchez, P. A. 1994 Tropical soil fertility research: towards the second paradigm. *Transactions of the 15th World Congress of Soil Science* 1: 65–88.

Sanmuganathan, K. and Bolton, P. 1988 Water management in third world irrigation schemes. *ODU Bulletin* 11: 4–10. Wallingford: Overseas Development Unit.

Saouma, E. 1993 *Statement by the Director-General. Appendix D to Report of the Conference of FAO, 27th session, November 1993.* Rome: FAO.

Schramm, G. and Warford, J. J. 1989 *Environmental management and economic development.* Baltimore, USA: Johns Hopkins, for World Bank.

Scoones, I. and Thompson, J. (eds.) 1994 *Beyond farmer first: rural people's knowledge, agricultural research and extension practice.* London: Intermediate Technology Publications.

Sehgal, J. and Abrol, I. P. 1992 Land degradation status: India. *Desertification Bulletin* 21: 24–31.

Shaxson, T. F., Hudson, N. W., Sanders, D. W., Roose, E., and Moldenhauer, W. C. 1989 *Land husbandry: a framework for soil and water conservation.* Ankeny, USA: Soil and Water Conservation Society.

Sheffield, C. 1981 *Earth watch: a survey of the world from space.* London: Sidgwick and Jackson.

Sillitoe, P. 1993 Losing ground - soil loss and erosion in the highlands of Papua New Guinea. *Land Degradation and Rehabilitation* 4: 143–66.

Simon, J. (ed.) 1994 *The state of humanity.* Oxford: Blackwell.

Singh, K. D. 1993 World forest resources assessment 1990: an overview. *Unasylva* 44(3): 2–9.

Sitarz, D. (ed.) 1993 *Agenda 21: the earth summit strategy to save our planet.* Boulder, USA: Earthpress.

Skole, D. and Tucker, C. J. 1993 Tropical deforestation and habitat fragmentation in the Amazon: satellite data from 1978 to 1988. *Science* 260: 1905–10.

Smil, V. 1994 How many people can the earth feed? *Population and Development Review* 20: 255–92.

Smith, R. M. and Stamey, W. L. 1965 Determining the range of tolerable erosion. *Soil Science* 100: 414–24.

Smyth, A. J. and Dumanski, J. 1993 *FESLM: an international framework for evaluation sustainable land management.* FAO World Soil Resources Report 73. Rome.

Soil Survey Staff 1975 *Soil taxonomy: a basic system of soil classification for making and interpreting soil surveys.* Agricultural Handbook 426. Washington DC: US Department of Agriculture.

Soil Survey Staff 1992 *Keys to soil taxonomy.* SMSS Technical Monograph 19. Blacksburg, USA: Soil Management Support Services.

Solorzano, R. and 9 others 1991 *Accounts overdue: natural resource depreciation in Costa Rica.* Washington DC: World Resources institute.

Sombroek, W. G. and Sims, D. 1995 *Planning for sustainable use of land resources: towards a new approach.* Rome: FAO.

Spaargaren, O. C. 1994 *World reference base for soil resources.* Wageningen: International Society of Soil

Science (ISSS)/International Soil Reference and Information Centre(ISRIC)/FAO.

Steiner, R. A. and Herdt, R. W. (eds.) 1993 *A global directory of long-term agronomic experiments. Vol. 1: Non-european experiments*. New York: Rockefeller Foundation.

Stocking, M. 1988 Quantifying the on-site impact of soil erosion. In *Land conservation for future generations: proceedings of the fifth international soil conservation conference* (ed. Sanarn Rimwanich, Department of Land Development, Bangkok): 137–62.

Stoorvogel, J. J. and Smaling, E. M. A. 1990 *Assessment of soil nutrient depletion in Sub-saharan Africa: 1983–2000*. Staring Centre Report 28. Wageningen, The Netherlands: Staring Centre.

Sundaram, K. V. 1985 *Toward improved multilevel planning for agricultural and rural development in Asia and the Pacific*. FAO Economic and Social Development Paper 52. Rome.

Swanson, B. E. 1984 *Agricultural extension: a reference manual*. Rome: FAO.

Thomas, D. S. 1993 Sandstorm in a teacup? Understanding desertification. *Geographical Journal* 159: 318–32.

Thomas, D. S. G. and Middleton, N. J. 1994 *Desertification: exploding the myth*. Chichester, UK: Wiley.

Tiffen, M., Mortimore, M. and Gichuki, F. 1994 *Population growth and environmental recovery: policy lessons from Kenya*. Gatekeeper Series 45. London: International Institute for Environment and Development (IIED).

Tribe, D. 1994 *Feeding and greening the world*. Wallingford, UK: CAB International.

Turner, B. 1986 The importance of dambos in African agriculture. *Land Use Policy* 3: 343–7.

Turner, B. 1994 Small-scale irrigation in developing countries. *Land Use Policy* 11: 251–61.

Turner, B. L., Clark, W. C., Kates, R. W., Richards, J. F., and Mathews, J. T. 1991 *The earth as transformed by human action: global and regional changes in the biosphere over the past 300 years*. Cambridge: University Press.

Turner, R. K. 1993 *Sustainable environmental economics and management*. Chichester, UK: Wiley.

Umali, D. L. 1993 *Irrigation-induced salinity: a growing problem for development and the environment*. World Bank Technical Paper 215. Washington DC.

UN 1992 *Long-range world population projections*. New York.

UN 1993 *Integrated environmental and economic accounting: interim version*. New York.

UN 1994 *Program of action of the 1994 international conference on population and development*. New York. Reprinted in *Population and Development Review* 21: 187–213.

UN 1995a *World population prospects: the 1994 revision*. Population Studies 145. New York.

UN 1995b *World urbanization prospects: the 1994 revision*. New York.

UNCED 1992 *Agenda 21: programme of action for sustainable development*. Rio de Janeiro: United Nations Conference on Environment and Development (UNCED).

UNDP, annual. *Human development report*. Washington DC.

UNDP-FAO 1994 *Cambodia: land cover atlas 1985/87–1992/93*. Rome: FAO.

UNEP 1982 *World soils policy*. Nairobi.

UNEP 1984 *Map of desertification hazards*. Nairobi.

UNEP 1992 *World atlas of desertification*. London: Arnold.

UNEP 1995 *United Nations convention to combat desertification*. Geneva.

UNEP/FAO 1994a *Report of the UNEP/FAO expert meeting on harmonizing land cover and land use classifications*. GEMS Report series 25. Nairobi.

UNEP/FAO 1994b *A suggested national soils policy for Indonesia*. Rome: FAO.

UNEP/FAO 1994c *A suggested national soils policy for Jamaica*. Rome: FAO.

UNFPA (United Nations Fund for Population Activities) 1996 *UNFPA Mission Statement*. New York. Reprinted in *Population and Development Review* 22: 594–600.

USAID (US Aid for International Development) 1994 USAID's policy for stabilizing world population growth and protecting human health. *Population and Development Review* 20: 483–7.

US Department of Agriculture 1989 *The second RCA appraisal: soil, water and related resources on nonfederal land in the United States: analysis of condition and trends*. Washington DC.

Van der Pol, F. 1990 Soil mining as a source of farmers income in southern Mali. In *Fertility of soils: a future for farming in the West African savannah* (ed. C. Pieri, Springer, London): 403–18.

Van Staveren, J. M. and van Dusseldorp, D. B. W. M. 1980 *Framework for regional planning in developing countries.* ILRI Publication 26. Wageningen, The Netherlands: International Institute for Land Reclamation and Improvement (ILRI).

Verboom, W. C. and Brunt, M. 1970 *An ecological survey of Western Province, Zambia, with special reference to the fodder resources.* Land Resource Study 8. Tolworth, UK: Directorate of Overseas Surveys.

Vitousek, P. M. 1994 Beyond global warming: ecology and global change. *Ecology* 75: 1861–76.

Waggoner, P. E. 1994 *How much land can ten billion people spare for nature?* Ames, USA: Council for Agricultural Science and Technology.

Walker, B. and Steffen, W. (eds.) 1996 *Global change and terrestrial ecosystems.* Cambridge University Press.

Warren, A. and Agnew, C. 1988 *An assessment of desertification and land degradation in arid and semi-arid areas.* Drylands Paper 2. London: International Institute for Environment and Development (IIED).

Watkins, K. 1995 *The Oxfam poverty report.* Oxford: Oxfam.

WCMC (World Conservation Monitoring Centre) 1992 *Global biodiversity: status of the earth's living resources.* London: Chapman and Hall.

Westing, A. H. (ed.) 1989 *Environmental security.* Oslo: International Peace Research Institute, for UNEP.

Wilkinson, G. W. 1985 *Drafting land use legislation for developing countries.* FAO Legislative study 31. Rome.

Wischmeier, W. H. and Smith, D. D. 1978 *Predicting rainfall erosion losses - a guide to conservation planning.* Agriculture Handbook 537. Washington DC: US Department of Agriculture.

Wood, A. 1950 *The groundnut affair.* London: Bodley Head.

Woolf, L. 1913 *The village in the jungle.* London: Hogarth.

Woomer, P. L. and Swift, M. J. (eds.) 1994 *The biological management of tropical soil fertility.* Chichester, UK: Wiley.

World Bank 1992 *World development report 1992: development and the environment.* Oxford University Press for World Bank.

World Bank 1993a *The World Bank and the environment: fiscal 1993.* Washington DC.

World Bank 1993b *Poverty reduction handbook.* Washington DC.

World Bank 1994 *The World Bank and participation.* Washington DC.

World Bank 1995a *Mainstreaming the environment: fiscal 1995.* Washington DC.

World Bank 1995b *Social indicators of development.* Baltimore: Johns Hopkins, for World Bank.

World Resources Institute (WRI)/UNEP/UNDP/World Bank, biennial. *World resources: a guide to the global environment.* Oxford University Press.

World's Scientific Academies 1993 *Population summit of the world's scientific academies: a joint statement by 58 academies.* London: Royal Society. Reprinted in Population and Development Review 20: 233–8.

Young, A. 1968 Natural resource surveys for land development in the tropics. *Geography* 53: 229–48.

Young, A. 1973 Cost-benefit analysis where change is irreversible. In *Pollution abatement* (ed. K. M. Clayton, David and Charles, London): 107–8.

Young, A. 1975 Weathered ferrallitic soils: their properties, genesis and management. In *Proceedings of the joint commissions I, IV, V and VI of the International Society of Soil Science Conference on savannah soils of the subhumid and semi-arid regions of Africa and their management* (ed. H. B. Obeng and P. K. Kwakye, Soil Research Institute, Kumasi, Ghana), 171–4.

Young, A. 1976 *Tropical soils and soil survey.* Cambridge University Press.

Young, A. 1978 Recent advances in the survey and evaluation of land resources. *Progress in Physical Geography* 2: 462–79.

Young, A. 1984 Common sense on desertification? *Soil Survey and Land Evaluation* 4: 90–1.

Young, A. 1985 Land evaluation and agroforestry diagnosis and design: towards a reconciliation of procedures. *Soil Survey and Land Evaluation* 5: 61–76.

Young, A. 1987 Methods developed outside the international agricultural research system. In Bunting (1987), q.v.: 43–64.

Young, A 1989 *Agroforestry for soil conservation.* Wallingford, UK: CAB International.

Young, A. 1991 Soil monitoring: a basic task for soil survey organizations. *Soil Use and Management* 7: 126–30.

Young, A. 1993a Agroforestry as a viable alternative for

soil conservation. *Entwicklung und Ländlicher Raum* (DLG, Frankfurt) 27(5): 3–7. Reprinted in *Agriculture + Rural Development* 2 (1995): 45–9.

Young, A. 1993b Land evaluation and forestry management. In *Tropical forestry handbook* (ed. L. Pancel, Springer, Berlin): 811–45.

Young 1994a *Land degradation in South Asia: its severity, causes and effects upon the people.* FAO World Soil Resources Report 78. Rome.

Young, A. 1994b Modelling changes in soil properties. In Greenland and Szabolcs (1994), q.v.: 423–47.

Young, A. 1994c Towards international classification systems for land use and land cover. In UNEP/FAO (1994a), q.v.: Annex V.

Young, A. 1997 *Agroforestry for soil management.* Wallingford, UK: CAB International.

Young, A. and Brown, P. 1962 *The physical environment of northern Nyasaland, with special reference to soils and agriculture.* Zomba, Malawi: Government Printer.

Young, A. and Brown, P. 1964 *The physical environment of central Malawi, with special reference to soils and agriculture.* Zomba, Malawi: Government Printer.

Young, A. and Goldsmith, P. F. 1977 Soil survey and land evaluation in developing countries: a case study in Malawi. *Geographical Journal* 143: 407–38.

Young, A. and Saunders, I. 1986. Rates of surface processes and denudation. In *Hillslope processes* (ed. A. D. Abrahams, Boston: Allen and Unwin): 3–27.

Young, A. and Wright, A. C. S. 1980 Rest period requirements of tropical and subtropical soils under annual crops. In *Report of the second FAO/UNFPA expert consultation on land resources for populations of the future* (FAO, Rome): 197–268.

Yudelman, M. 1994 Demand and supply of foodstuffs up to 2050 with special reference to irrigation. IIMI *Review* 8(1), 4–14.

Zinck, J. A. (ed.) 1994 *Soil survey: perspectives and strategies for the 21st century.* ITC Publication 21. Enschede, The Netherlands: International Institute for Aerospace Survey and Earth Sciences.

Index

INDEX

INDEX